Parity and War

Evaluations and Extensions
of *The War Ledger*

Edited by

Jacek Kugler and Douglas Lemke

Ann Arbor

THE UNIVERSITY OF MICHIGAN PRESS

Copyright © by the University of Michigan 1996
All rights reserved
Published in the United States of America by
The University of Michigan Press
Manufactured in the United States of America
♾ Printed on acid-free paper

2011   2010   2009   2008   5   4   3   2

A CIP catalog record for this book is available from the British Library.

Library of Congress Cataloging-in-Publication Data

Parity and war : evaluations and extensions of the War ledger / edited
    by Jacek Kugler and Douglas Lemke.
        p.   cm.
    Includes index.
    ISBN 0-472-09602-8 (hardcover : alk. paper). — ISBN 0-472-06602-1
    (pbk. : alk. paper)
        1. War.   2. International relations.   3. Politics and war.
    4. Organski, A. F. K., 1923–   War ledger.   I. Kugler, Jacek.
    II. Lemke, Douglas, 1967–   .
    U21.2.P323   1995
    355.02—dc20   95-23699
                                                        CIP

Tables 6.1 and 6.2 reprinted from Henk Houweling and Jan Siccama in *Journal of Conflict Resolution* 31:87–102, copyright © 1988 by Sage Publications; tables 6.3, 6.4, and 6.5 reprinted from Henk Houweling and Jan Siccama in *Journal of Conflict Resolution* 35:642–58, copyright © by Sage Publications. Reprinted by permission of Sage Publications, Inc.

ISBN 978-0-472-09602-2 (hardcover : alk. paper)
ISBN 978-0-472-06602-5 (pbk. : alk. paper)

*Dedicated to A. F. K. Organski,
whose vision this work explores*

Contents

Preface

In the fifteen years since the appearance of *The War Ledger,* a wide range of scholars have added further specification to, and reformulation of, the propositions originally advanced. The purpose of the present volume is to bring these disparate research elements together in an effort to evaluate the impact of *The War Ledger* as an explanation of the causes and consequences of international war.

Most of the selections presented here were initially components of two panels at the 1991 International Studies Association annual meeting in Vancouver, British Columbia. Thanks to the generous support of the Claremont Graduate School, the contributors were able to gather in October 1992 to present revised work and benefit from collective experience. The work before you represents the outcome of these meetings and other exchanges.

This volume is divided into five sections. The first includes the chapters by Lemke and Kugler, by Vasquez, and by Siverson and Miller. These provide an overview of the perspective connecting power parity and war, highlighting the evolution of the link between parity and war from the original formulation of power transition theory by A. F. K. Organski to the present work on parity and war. These chapters explore the inner logic of the various arguments and assess how unresolved issues have been treated in formal and empirical evaluations. In addition, the chapters by Vasquez and by Siverson and Miller point to areas where the power parity perspective may be improved conceptually and empirically.

The second section is composed of selections by Lemke, Kim, Houweling and Siccama, Danilovic, Geller, Wayman, Thompson, and Werner and Kugler. These selections highlight efforts to enlarge the domain of power parity by reformulating the logic of the theory, adding additional components, or combining its propositions with those derived from other structures. Lemke's chapter posits the existence of multiple hierarchies within the international system, suggesting that power parity accounts for symmetric confrontations among minor powers within regional hierarchies. This work indicates that war and peace are not distinct behaviors among major and minor powers. Kim's chapter considers the impact of parity between alliance aggregations and includes a measure of dissatisfaction with the status quo. This extension adds substantively to our understanding of conflict. Houweling and Siccama expand power parity by combining it with Doran's notion of "critical points" in the relative power trajectories of major powers. Their effort adds significant predictive value to the parity condition. Danilovic explores the logical structures by which power parity and status inconsistency arguments complement and interact with one another. Geller's selection explores the connections between dyadic and systemic power attributes, analyzing the relationship between the timing of war, rationality, and the likely identity of the initiator. Wayman's effort considers the relationship between wars and power shifts in general. Thompson's contribution combines power parity, balance of power, and the long-cycle perspective in an effort to account for system-transforming wars from the 1490s through the 1940s. Finally, Werner and Kugler's chapter shows that the combination of power parity and arms buildups where the challenger is winning greatly increases the probability of war. Each of these contributions demonstrates that starting with the power parity perspective as the basic platform provides a consistent and empirically effective perspective on peace and war in world politics.

The third section features the contribution of Arbetman, extending *The War Ledger*'s concern with the long-term effects of wars. The original results showed that advanced European nations experienced a "Phoenix Factor" and recovered within twenty years from the devastation of major war. Arbetman extends this work by assessing the consequences of war on developing countries involved in serious wars, as well as the effects of a major domestic conflict: the American Civil War. The "Phoenix Factor," with small variations, is found in these extensions.

The fourth section focuses on nuclear deterrence within power parity frameworks. To this end, Kugler's chapter contrasts traditional deterrence arguments based on mutual assured destruction (MAD) with those derived from the power parity perspective. He shows how a consistent relationship between nuclear proliferation and deterrence emerges, how stability in the nuclear era can be explained, and why the dramatic changes following the collapse of the Soviet Union result in a peaceful world. Zagare's chapter develops an incomplete information game in which power parity arguments about deterrence are deductively evaluated. The resulting logic is far more consistent with actual behavior than similar evaluations based on assumptions derived from a balance-of-power perspective.

The fifth section introduces formal extensions to the parity perspective. Chapters by Bueno de Mesquita, Kadera, and Morrow develop the inner logic of power parity. Bueno de Mesquita analyzes power parity, using his international interactions game to produce important implications not apparent until preferences and coalitions are included. Kadera's contribution involves the development of a system of differential equations that allows rigorous insight into expectations about dyadic power transitions. Finally, Morrow provides a game-theoretic evaluation of the logic behind the challenger's initiation and the dominant country's resistance, challenging the original logic of power transition theory. After this section a concluding chapter by Organski and Tammen interprets the post–cold war world from the power parity perspective.

In this volume, the original formulation and empirical tests of *The War Ledger* are pushed and pulled in a variety of ways, clarifying ambiguous aspects, highlighting extensions, and resolving ambiguities. Many of the selections are critical of the earlier work, and yet, one cannot escape the conclusion that, overall, the parity perspective survives the assault. Reviewing the results of these efforts, the editors of this volume, somewhat immodestly, suggest that power parity provides a rich account of international relations and is a theoretical structure that has been reinforced by evidence. Indeed, power parity can consistently account for the challenges of the cold war, the changes that have accompanied the demise of the Eastern Bloc, and the current trend toward the creation of potential supranational entities like the European Union (EU) and the North American Free Trade Agreement (NAFTA). This generality and viability strongly suggests that the power parity perspective advanced in *The War Ledger* remains a valuable perspective for the analysis of world politics.

Inevitably in an effort of this magnitude we have amassed a great debt to colleagues, graduate researchers, staffs in our institutions, and editors. We are particularly grateful to Doris Fuchs, who coordinated the academic efforts, to Sandra Seymore, who tirelessly kept this project on line, and to Gwen Williams, who provided superb staff support. Dan Mazmanian provided the financial support and congenial environment of the Center for Politics and Economics that sparked intense and constructive academic exchanges. Our editor, Colin Day, supervised this project, while Malcolm Litchfield was responsible for completing the volume. Lastly, we acknowledge the contribution of John Tryneski at the University of Chicago, who encouraged us from the outset and helped select a superb set of reviewers who improved the book immeasurably. We are very grateful to all.

Part 1
From Power Transition
to Power Parity

CHAPTER 1

The Evolution of the Power Transition Perspective

Douglas Lemke and Jacek Kugler

The War Ledger by A. F. K. Organski and Jacek Kugler explores the controversial proposition that war is associated with power parity and that peace follows from power preponderance. While other perspectives on the relationship between power distribution and war exist, the argument advanced by Organski and Kugler precedes the subsequent concern with hegemony and peace and suggests a vision of world politics consistent with the peaceful decline of the Soviet Bloc (Keohane 1980, 1984; Gilpin 1981; Strange 1982; Russett 1985; Huntington 1988; Kennedy 1988; Choucri and North 1989; Nye 1990). This volume considers subsequent work associated with the idea of power transition and its extensions.

Work on power distributions and war involves several interrelated arguments. Central to the debate is the distinction between power preponderance and balance of power. Does preponderance of power lead to stability, or is a balance of power the key to peace? Discussions have also focused on the question of timing. While there is agreement that major wars will be fought under conditions of parity, will wars be fought before or after transitions? Will the speed of the transition affect the start of war? Who will initiate such a war? Furthermore, the debate raises a number of questions concerning the reliability of alliances. Who can be expected to join the fight when wars are waged? Are alliances stable or highly flexible? Are coalitions determined by overall power distributions? A related argument analyzes patterns of recovery from war, focusing on two main considerations: Can opponents recover from a major confrontation to fight again? Can external aid dramatically affect the pattern of domestic growth in a postwar period? With the balance of terror having replaced conventional balance-of-power arguments, a third controversy addresses the role of nuclear weapons. Has national behavior drastically changed with the introduction of nuclear weapons? Should nuclear proliferation be encouraged or avoided? Finally, following the breakup of the Soviet

3

Union, has the likelihood of major war increased or decreased with the movement from a largely bipolar confrontation to an international system dominated by the United States?

The propositions advanced by power transition theory are a subset of a larger set of dyadic structural theories that concern themselves with power. Like balance of power, hegemonic stability theory, and some cycle theories, power transition theory is concerned with identifying the consequences of particular power configurations on the likelihood of war. The cornerstone of power transition theory is that *parity* is the necessary condition for major war. However, power transition theory does not deny the importance of decision making, nor does it imply that structures determine outcomes. Rather, power transition theory contends that parity sets the stage where decision makers *can, but need not necessarily,* choose major war as a viable alternative. When parity is present, the parties involved can choose to confront one another, with massive consequences for the international system (e.g., World War I). However, even when parity is present and war is chosen, such escalation of hostilities is not inevitable. Power transition theory contends that under parity, long-term friends who share common preferences will fight—if at all—with far less vigor than those opponents who dramatically differ in their original support for the status quo. The point is that in the absence of power parity, choices are restricted to minor wars, having limited long-term effects on the international system.

Power transition theory, like the mainstream balance-of-power theories, places heavy emphasis on power distributions, but is different in that it accommodates Shepsle and Weingast's (1981) notion of structure-induced equilibria. Shepsle and Weingast argue that such equilibria can limit the timing and range of choices available to actors. Paying attention to specific power distributions identifies the periods when intense and potentially system-altering confrontations can be waged. In such rare periods, decision makers can choose either to join or to avoid confrontations to resolve their differences. Structures do not determine outcomes; they shape outcomes. For our purposes, paying heed to structures eliminates the need to consider all dyads across time in search of major confrontations.

The chapters of this volume will, at last, begin the process of clarification and augmentation required for the formulation of better theories of international politics. Of specific interest are such open issues as the role played by the status quo in conflict processes, additional factors in combination with power parity that threaten peace, the comparability of the dynamics operating in the core and in the periphery, and the potential linkage between structural arguments of power transition theory and decision-making models of international relations.

Power Perspectives of War

Balance of Power: Parity Leads to Peace, Preponderance Leads to War

Before we discuss competing theories, let us clarify the use of language in this work. As Kadera has observed, "The literature is inconsistent in labeling, which leads to a good deal of confusion" (1992:1). Following the choice of Organski and Kugler, we will label each argument according to the theory in which it is a condition for peace. Thus, power preponderance arguments correspond with power transition theory, and power parity arguments with balance-of-power theory.

Perhaps the best known contention in international relations is that peace depends on a balance of power (Hume [1752]1990; Wright 1935; Morgenthau 1948; Gulick 1955). Advocates of this perspective claim that equality of power destroys the possibility of a guaranteed and easy victory and therefore no country will risk initiating conflict. Peace is maintained because potential contenders fear the consequences of a large war. When the balance is upset, the opportunity emerges to maximize power and enhance security, and wars are waged. In the anarchical international system, wars can be prevented only when costs are high and equivalent for both parties. Indeed, Waltz argues, wars occur "because there is nothing to prevent them" (1959:232). The actions of powerful decision makers seeking to maximize security and power are constrained by other actors, who ensure that the balance of power is maintained.

Since a power advantage will lead to an attack, each country endeavors continually to ensure that its opponents will not gain preponderance. Consequently, decision makers strive to maintain the balance through flexible alliance alignments. Power maximization is pursued as both a means to an end and an end in itself, where the ultimate purpose is survival. This continual competition for power stabilizes the international system just as competition for profit stabilizes prices in economic theory.

Many variants of balance-of-power theories have been advanced, but they all share the contention that equality leads to peace (e.g., Morgenthau 1948; Kissinger 1957; Liska 1962; Waltz 1979). To buttress this contention, balance-of-power models make a number of assumptions about anarchy, alliances, and the goals nations pursue.

First, a contested assumption of balance-of-power arguments is that decision makers maximize power (Morgenthau 1948). By implication, the desire to accumulate power motivates a drive for territorial expansion. However, it is difficult to maintain the claim that all nations at all times maximize power. For example, the United States chose to avoid conflict after World War I, despite obvious hegemony, and has not taken advantage of the collapse of the Soviet Bloc. Confronting this criticism, neorealist revisions of balance-of-power arguments have centered on the maximization of security (Waltz 1979). Adopting a different angle, recent work assumes that decision makers minimize threats to their security (Walt 1987). These revisions allow some large nations to refrain from incorporating exposed neighbors, even when power preponderance is present. For example, neorealist arguments allow balance-of-power theorists to explain why the United States did not annex Canada at the beginning of World War I or World War II, when Britain could not defend her Commonwealth ally, at least in the short term (for recent recommendations to amend neorealism, see James 1993).

The second key assumption of balance-of-power theory is that the international system is anarchical in structure. No higher authority constrains the interactions of countries. Waltz (1979:chaps. 5, 6) argues that units or countries in international politics have equal status, each being sovereign in its own territory, but that all differ in their power capabilities. Thus, international anarchy creates an environment where nations face a "self-help" system and defend themselves against aggressors. Due to the lack of a formal, legal hierarchy, relations of authority and subordination fail to develop in the international system and there is no formal organization to implement the rule of law. The only constraints on a state's behavior are created by the power capabilities of competing states. For this reason, politics within nations are considered radically different from politics among nations. Domestic systems are constrained by the rule of law. In contrast to international anarchy, domestic governments are explicitly designed with a hierarchy that allows governmental authority to impose political decisions on the population. This hierarchy functions because it creates a division of labor in both the political and the economic domestic arenas. In the world arena, international structures may arise, but they do not affect the interactions of nations who act in relation to others as if they faced an economic market, where purchasing ability is determined by the power of competing countries. Each nation provides its population with the same services and serves the same purposes.

The third assumption made by balance-of-power advocates is that the power of each national unit is relatively fixed in relation to other units. National power is manipulated by the making and breaking of alliances with other countries. Countries interact by creating coalitions for mutual defense against the presumably aggressive intentions of competitors. In order to maintain a power equilibrium, however, nations are forced to change alliances often and swiftly, as the needs of the moment warrant. Thus, under anarchy, alliance agreements have to be fluid and flexible. This need highlights the role of diplomacy in establishing, maintaining, and restoring peace. Effective diplomacy is necessary for the survival of nations who distrust their allies but are even more concerned about the aggressive intentions of potential opponents (Claude 1962; Conybeare 1992).

Alliances can offset changing threats because they are relatively easy to form. The contention is that ideology, religion, ethnicity, and other considerations external to power distributions do not constrain alliance formation. Such external criteria would unduly narrow the set of possible alliance partners, dramatically limiting the ability of diplomats to counter potential threats created by a breakdown in the balance of power. Consequently, alliances can be defined as agreements of convenience permitting countries to adjust to changes in the distribution of power.

Furthermore, a fundamental implication of the assumption that countries can only manipulate the power distribution through alliances is that decision makers are interested in preserving other major international actors. They do not want the pool of potential alliance partners to decrease. As a result, a nation that achieves temporary preponderance following victory in a major war does not utilize its advantage and chance to maximize power by pushing the opposing country over the brink of extinction. This would constrain the possibilities to restore the balance of power after the war. In sum, the logic of balance of power prescribes the maintenance of key actors in the international system (for an illuminating discussion on this point and its implications, see Niou, Ordeshook, and Rose 1989).[1]

Balance-of-power theory enjoys a long history and widespread acceptance among scholars of international relations. It provides a perspective that under a power equilibrium decision makers preserve peace, while a preponderant country or group of countries will choose to engage in war. These claims depend on the assumptions of anarchy, flexibility of alliance politics, and relative constancy of national power described above. The balance-of-power paradigm retains its influence in international politics in the modern era because it provides a comprehensive framework for the preservation of national security in the nuclear era (e.g., Mutual Assured Destruction). We turn now to a counterargument about power distribution and international conflict that challenges the balance of power perspective.

Power Transition: Parity Leads to War, Preponderance Leads to Peace

A. F. K. Organski first formulated power transition theory in 1958, although its intellectual origins can be traced to the work of Thucydides in the fifth century B.C. (1951). Power transition theory postulates a hierarchical international system led by a dominant country. Organski draws attention to the fact that countries grow at different rates of speed. Specifically, he argues that countries that industrialize later grow more rapidly than those who have previously industrialized.[2] Therefore, the relative power of countries can change independently of other countries, as those with faster rates of growth catch up and overtake those that grow more slowly. Organski argues that domestic stages of transitional growth in power result in international power transitions. The two concepts are frequently confused (see Kugler and Organski 1989b).

In connection with the notion of differential rates of growth, Organski reasons that if a dissatisfied country caught up with and overtook the dominant country, then major international war would occur. The dissatisfied challenger would attempt to use its new power to change the status quo, and the dominant country would use its long-held power to resist the attempted changes. The result would be war.[3] However, as long as the dominant country remains preponderant over the other countries in the international system, peace is maintained.

Power transition theory requires a number of assumptions about the structure of the international system, the nature of alliance politics, and the constancy of national power. The first assumption is that the international system is hierarchical rather than anarchical. Organski claims that powerful nations create, impose, and maintain a hierarchy over the weak. At the top of the hierarchy sits the dominant country, which is the most powerful country in the world. The international system is organized to the dominant country's advantage, reflecting its power preponderance. Thus, the status quo in the international system "works" for the dominant power: this country is satisfied and seeks to preserve it. Such efforts are supported by other satisfied powerful countries, which also enjoy benefits from the status quo. However, some countries are not favored by the distribution of benefits within the international system and are the dissatisfied potential challengers to the status quo. These countries may have been very weak when the status quo was established and thus had little say in its creation or operation.

Much like institutionalized groups in domestic politics, the dominant country and the satisfied supporters endeavor to preserve the existing hierarchy as established. As long as the dominant country retains its advantage in terms of the power distribution, this is not difficult. The weak obey the strong with few exceptions. However, as the dominant power's preponderance deteriorates, efforts to preserve the status quo and the international hierarchy become increasingly ineffective.

Following Organski, Keohane (1980) argues that a hierarchical international system exists. According to hegemonic stability theory, the strength and stability of international regimes depend on a preponderant power, which establishes and maintains regular rules of interaction that apply to the other countries in the regime.[4] If one defines peace as a "good," then it is possible to conceive of a security regime that provides or regulates peace. Similar to power transition theory, Keohane's hegemonic stability theory, although designed to explain international economic interactions, implies that preponderance leads to peace, while the absence of a preponderant power may be associated with instability. According to Keohane, a security regime will be strong and function well when there is a preponderant power to support the regime. After changes in relative power relations cause a decline of the preponderant power, the regime will be weakened and war can ensue. The "can" is important here because, as Keohane (1984) later argues, the members of the regime may agree to cooperate in spite of the hegemon's decline.

Gilpin (1981) provides a parallel perspective. According to his theory of hegemonic war and change, shifts in the distribution of power create a disequilibrium between the hierarchy of prestige, the distribution of territory, the rules of the system, and the international division of labor. The resulting disjunctures generate challenges for the dominant power and opportunities for the rising states in the system, who eventually will attempt to change the rules. In reaction, the dominant power will counter this challenge and try to restore the equilibrium in the system through changes in its policies. If the dominant power fails in this attempt, the disequilibrium will be resolved by total war for control of the international system. Gilpin argues that every subsequent international regime has been a consequence of the territorial, economic, and diplomatic realignments that have followed such wars. Victory and defeat reestablish an unambiguous hierarchy of prestige concurrent with the new distribution of power. Peace will follow for as long as the victor from the previous war remains strong enough to preserve this hierarchy.

Organski, Keohane, and Gilpin all describe international systems in which dominant countries sit atop hierarchical arrangements and establish rules of interaction. These notions of international hierarchy contrast sharply with balance-of-power arguments about international anarchy. Furthermore, these hierarchical arguments differ from balance-of-power theory, in that they claim that countries fight wars to achieve net gains through control over the international hierarchy, and not in an effort to maximize power.

A second assumption of theories relating power parity to war is that alliances are fixed rather than flexible. Whereas balance-of-power arguments suggest that alliance politics have to be flexible so that countries can respond to threats as they emerge, power transition arguments hold that alliance politics are rigid. Power transition suggests that with regard to the international status quo countries will be either satisfied or dissatisfied. Those who are satisfied side with the dominant country, and those who are dissatisfied side with a challenger to the dominant country. Countries are unable to shift back and forth between the sides because such shifts would either require a total transformation of preferences or would expose the shifting country as an unreliable and insubstantial ally.

A third assumption is that power transition theory identifies domestic power changes as the source of the greatest disturbances in the international system. Organski (1958) argues persuasively that the rapid growth countries experience during and shortly after industrialization allows them to surpass other states in power, thus inducing them to challenge the international order. Other scholars have since added their voices to his (Choucri and North 1975, 1989; Gilpin 1981).

The power transition argument that preponderance leads to peace while parity leads to war runs directly counter to balance-of-power arguments. Power transition theory also contradicts balance-of-power assumptions of the international system as anarchical, alliance politics as flexible, and the power of each country as relatively fixed. The two theories provide alternate descriptions of the international system at peace and war. Their contradictory propositions can and have been evaluated both logically and empirically. We now turn to discussion of these efforts.

Claims, Counterclaims, and Clarifications of *The War Ledger*

In the years since *The War Ledger* (Organski and Kugler 1980) was published a number of scholars have misunderstood the book's empirical evaluations. The source for these claims is almost certainly ambiguity in *The War Ledger* itself and contradictions between it and Organski's (1958, 1968) earlier work on power transition theory.

The first, and perhaps most serious misunderstanding is that "Organski and Kugler's argument runs from power transitions to the outbreak of war, but their test procedure runs from the outbreak of war to the preceding transition of power" (Houweling and Siccama 1988a:95; but see also Vasquez herein; Siverson and Miller herein). However, Organski and Kugler's argument and test are parallel. In *The War Ledger*, the test procedure focuses on dyads of contenders and identifies whether the power relationship was unequal, equal with no overtaking, or equal and overtaking. The final step determines whether war occurred or not. The test procedure in *The War Ledger* runs from the power relationship within contender dyads to subsequent wars—not the other way around. If the argument and test were not parallel, then the unit of analysis in the tests would be contender wars, not dyads of contenders. That the unit of analysis is dyads of contenders can be seen by noticing the study "*n*" reported in the contingency tables that include both peace and war periods (Organski and Kugler 1980:52).

The second persistent claim is that Organski and Kugler consider inappropriate wars in their evaluation (Houweling and Siccama 1988a; Kim 1989, herein; Thompson herein). It is useful to consider exactly what Organski and Kugler attempt to explain:

> The models we are comparing, it should be recalled, do not claim to establish connections between changes in the international power structure

and the outbreak of wars among small nations, or among large and small nations; nor do the models explain colonial wars. Such conflicts (according to the models) may occur unrelated to fundamental changes in the power structure of the system, and, therefore, power distributions in the period preceding such wars cannot be used to disprove the propositions advanced by any of the models we are considering. The hypotheses in question can be tested fairly only if we locate conflicts whose outcomes will affect the very structure and operation of the international system. (Organski and Kugler 1980:45)

Clearly, such conflicts involve only the most powerful, internationally active countries in the system and further are characterized by all-out efforts to win. Power transition theory at the level of the overall international system really is, as Morrow (herein) accurately claims, a theory about big wars. In an effort to operationalize such wars, Organski and Kugler employ three criteria: (1) major powers on both sides of the war; (2) battle death totals higher than in previous wars; and (3) a mutual threat to core territory or population for the loser of the war. The list of wars that fulfill these criteria for the years 1860 to the present are: the Franco-Prussian War, the Russo-Japanese War, and the two world wars. However, the Russo-Japanese War is dropped from the analysis because at the time it was fought Japan was a peripheral major power and not a contender in the international system. The five big wars analyzed are the Franco-Prussian War, and the eastern and western fronts of World War I and World War II (see Organski and Kugler 1980: bottom contingency table p. 52 and figure 1.2 on p. 59). Thus, the inclusion of the Russo-Japanese War does not, as many argue, affect results in *The War Ledger*. Finally, one might reasonably question the inclusion of the Franco-Prussian War because the Napoleonic Wars had higher casualties, but 1870 was chosen as the starting date of the analysis due to data limitations. Thus, the war is included because Organski and Kugler established their criteria prior to their analysis. The Franco-Prussian War qualifies based on their criteria, and is accordingly included (but see Lemke herein for discussion of a conceptualization of the international system that suggests the Franco-Prussian War may have been fought for control of a local subhierarchy of the international system).

A third repeated claim is that the division of major powers into peripheral and central, and the central major powers into contenders and noncontenders, is arbitrary. Nothing could be further from the truth. Recall that the explanation power transition theory supplies is of conflict behavior related to efforts to establish, recreate, or control the international status quo. Only the very strongest countries can be considered. The designations that Organski and Kugler use to separate major powers into peripheral and central groups are borrowed directly from the careful historical compilations made by experts in the Correlates of War project and reported by Singer and Small (1966b, 1972). Further, the designation of a major power as a contender reflects the fact that only the strongest of states can contest the status quo. Not surprisingly, Organski and Kugler's designation of contenders coincides with the careful identification of global powers presented by Thompson (herein). Organski and Kugler designate as contenders those states with at least 80 percent of the dominant power's capabilities, or in the event that no potential challenger possessed the requisite 80 percent, the strongest three countries in the central system. Organski and Kugler's designation of different types of major powers follows carefully on the logical requirements of power transition theory and makes use of the painstaking efforts of previous scholars.

Related to claims critical of Organski and Kugler's selection of relevant major powers, several scholars have questioned the periods of time Organski and Kugler selected to evaluate power transitions (Geller 1992a,b; Siverson and Miller herein; Thompson herein). Organski and Kugler consider twenty-year intervals, arguing that changes in relative power levels occur slowly, and at average rates between 1870 and 1970, transitions among contenders separated by 10 percent of each other's power took such periods (1980:48). Figure 1.2 from *The War Ledger* (1980:59) shows the amount of time between the actual transition and subsequent wars for the five contender wars in their study. This figure suggests that had Organski and Kugler employed a ten-year interval, four of the five wars would still occur within the requisite amount of time. Indeed, subsequent studies of power transition theory employing ten-year time intervals support the original results (see Geller 1992a, 1992b; Lemke 1993; Werner and Kugler herein).

Other claims arising from ambiguities and contradictions in previous work are discussed below. Claims that power transitions are a necessary *and* sufficient condition for war arise from the fact that Organski and Kugler failed to include an operationalization of evaluations of the status quo in their work. Similarly, contradictions between Organski's (1958) speculation that wars would occur prior to transitions and Organski and Kugler's (1980) empirical observation that wars actually follow transitions, have led to some confusion. There is no consistent theoretical justification to expect war immediately prior to or after transitions. Theoretically, it is parity that is important to war initiation. The closer to parity a dyad is, the greater is the threat of war. Parity, not actual transitions, is of theoretical importance. For this reason it would have been better if Power *Transition* Theory had been named Power *Parity* Theory.

Unanswered Puzzles

Do Dyads Initiate War under Parity or Preponderance?

A variety of empirical studies test the contention that parity provides the necessary condition for war. Over time, empirical support is found at the dyadic level, but less so at the systemic level. In the following pages a number of studies are reviewed to substantiate this claim.

With *The War Ledger,* Organski and Kugler (1980) provide a dyadic empirical test of power transition theory. They conceptualize power as a multiplicative combination of gross national product (GNP) per capita, population, and the political capacity of a country's government.[5] In this test power is operationalized as GNP for the developed world, and as GNP times relative political capacity for developing countries. Power transitions are defined as intersections in the longitudinal power trends between countries over a twenty-year period. The power relations compared are those between great powers for the years 1860–1975. The analysis is divided into separate evaluations of major powers and contenders, reflecting the assumption that only the strongest of major powers can reasonably expect to be able to alter the international status quo.

The results of this test strongly support Organski and Kugler's claims about power parity and war: "If conflicts occur *among contenders,* they do so only if one of the contenders is in the process of passing the other" (1980:51, emphasis in the original). The test clearly gives support to the transition argument, revealing that power overtakings provide the necessary condition for war among the strongest of the great powers, and by extension, the mechanism of change in the international system.

Following on the work of Organski and Kugler, other scholars have investigated the relationship between power parity and war. Analyzing the conflict behavior of the major powers as far back as 1816 using power measures derived from Doran and Parsons (1980), Houweling and Siccama (1988a,b) also observe a strong correlation between power parity and war. Due to this longer time frame, they are able to account for the initiation of major power wars that Organski and Kugler omit from their study. Moreover, Bueno de Mesquita (1990a) reports finding a power transition between Prussia and Austria immediately before the outbreak of the Seven Weeks' War. Maintaining a dyadic perspective, but adding certain systemic considerations, a number of studies have also supported the general contention of parity and war. Using the Correlates of War (COW) data set and his own data collection, Kim (1989, 1992) extends power transition analysis to 1816. By considering war and power parity between members of a dyad but including support the adversaries could reasonably expect from other members of the international system, Kim produces results that support claims that war is associated with power parity.

Moreover, in a study from a related perspective, Rasler and Thompson (1992a) argue that the most dangerous power distribution arises when a deconcentration of power at the global level and a concentration of power among the countries in the international system's most important region occur simultaneously. The greatest probability for war exists when the strongest country at the regional level achieves such a concentration of power that it surpasses the strongest country at the systemic level. According to Rasler and Thompson, power transitions between the regional and systemic levels trigger the cataclysmic wars that lead to a restructuring of the international system.

The contribution of these dyadic studies that include systemic considerations is that they add different triggers for the initiation of war than those anticipated in original power transition propositions. Claims of relationships between parity and war at the dyadic level continue to be reinforced (Gochman 1990; Bremer 1992,1993). Further, Mansfield (1994), who sees some trade configurations as a trigger for war, also finds that parity is a precondition for conflict.

To summarize, a large number of studies linking parity and war at the dyadic level suggest that parity increases the probability of war, and consistently confirm the claim that parity and transitions increase the probability of war.

Does the International System Experience More War under Preponderance?

Numerous tests of power parity as a correlate of war employ a systemic level of analysis. In general, the results of this work have been negative or contradictory. Singer, Bremer, and Stuckey (1972) present a pioneering empirical analysis of propositions about the relationship between the distribution and concentration of power and war at the systemic level. They report an association between preponderance of power and war in the nineteenth century, and a reverse association linking preponderance and peace in the twentieth century (1972:39–40). Statistically, such results should lead to the rejection of both hypotheses. However, the authors speculate that changes in the democratization of foreign policy, industrialization, and urbanization that accompanied the passage of time between the centuries might account for the difference in the observed correlations (1972:48). Their contradictory results generated a great deal of interest in the international relations literature, as many scholars subsequently attempted to determine whether parity or preponderance promotes peace.

Dissatisfied with the contradictory results Singer, Bremer, and Stuckey report, Bueno de Mesquita (1981b) demonstrates that the association between the distribution of power and war depends on the risk propensities of the national actors involved. He argues that individual decision makers find different probabilities of success sufficient to go to war. By constructing nine hypothetical international systems in which power capabilities and risk propensities vary, Bueno de Mesquita argues that a probability of success between the ranges of 0.99 and 0.57 can lead to war. He further suggests that scholars who report support for either the preponderance-leads-to-peace or the parity-leads-to-peace models focus on different types of systems such that selection bias reduces their results to statistical artifacts (1981b:556). Finally, Bueno de Mesquita restructures the original dependent variable of Singer, Bremer, and Stuckey's work as a dichotomy, and reports no relationship between power and war at the systemic level (1981b:567). These results are reinforced in a later study with Lalman (Bueno de Mesquita and Lalman 1988) in which power distributions are shown to have no relationship with the outbreak of war at the systemic level.

Contrary to the results of Singer, Bremer, and Stuckey; Bueno de Mesquita; and Bueno de Mesquita and Lalman, Thompson (1983a) finds a consistently positive correlation between preponderance of power and peace. To achieve this result, he reworks the operationalizations of the Singer, Bremer, and Stuckey study along guidelines established by his and Modelski's long-cycle perspective (Modelski and Thompson 1989). Thompson criticizes previous studies of the relationship between power concentration and war for lacking reference to an overarching theory of international relations.

Dissatisfied with efforts to unravel the relationship between power distributions and war, Siverson and Sullivan (1983) argue that scholars need to be more rigorous in their tests. They suggest that previous efforts must be regarded as inconclusive because scholars focus only on contiguous dyads, or only on major powers, and thus bias their results. Future studies, they demand, must be true to the theory they purport to test, and should take added precautions against introducing selection bias. In spite of their criticisms, Siverson and Sullivan acknowledge that the evidence associating power parity with the incidence of war is building. In a follow-up article, Siverson and Tennefoss (1984) report support for preponderance-leads-to-war arguments, revitalizing balance-of-power claims. However, more recently, Siverson and Miller (herein) revisit the earlier arguments and return to the inconclusive evaluations suggested by Siverson and Sullivan (1983).

Mansfield (1992) provides another installment in this ongoing debate. He evaluates a systemic model of the relationship between power distribution and war, but unlike previous efforts, he does not assume a linear relationship between the two variables. Rather, he casts them in a quadratic-form regression model. He reports an inverted U-shaped relationship between power concentration and war, and interprets this result as support for both parity-leads-to-peace and preponderance-leads-to-peace arguments. However, Mansfield reports the maximum frequency of war at a power concentration level of 0.27, which is much closer to parity than to preponderance of power. Thus, although the relationship he finds is U-shaped, its skewness reveals a higher association of war with low concentrations of power, consistent with preponderance arguments.

Let us now consider the overall results. Dyadic-level analyses combined with some systemic considerations strongly suggest that nations initiate war under conditions of parity and not, as balance-of-power theories argue, under conditions of preponderance. Empirical studies further show that aggregation to the systemic level confounds and diffuses these results. Thus, Organski's early conception of specific countries challenging the dominance of the single leading actor in hope of changing the status quo provides a critically important ingredient that adds strength to propositions about power distributions and war.

What Is the Relationship between Alliances and War?

Recall that the power transition arguments of Organski assume that alliances are relatively fixed and do not affect the hierarchy of power in the international system. Conversely, balance-of-power theories assume that all power changes are accomplished through alliance politics, which are flexible.

One of the less explored, but nevertheless important findings reported in *The War Ledger* is that whereas the dyadic relationship determines the initiation of war, alliances determine the outcome of war (Organski and Kugler 1980:tables 1.8, 1.9). Consider the following, often ignored,[6] conclusion from *The War Ledger:* "When they (noncontender major powers) fought, their intervention was of critical importance in deciding the outcome of the conflict . . ." (1980:56).

Thus, the balance-of-power argument that power is affected by alliances has merit, because the international hierarchy heavily influences participation by allies and determines the outcome. Indeed, Organski and Kugler (1980:fig.1.2) show that contending challengers initiated war against the dominant country in all cases after overtaking the dominant power. However, the coalition that supported the challenger was inferior to that of the dominant country, and this determined the outcome of the subsequent war in favor of the latter (1980:54–60).

Using the original data developed from *The War Ledger,* the implied argument can be made very clear (fig. 1.1). Note that prior to World Wars I and II, the challenger had already achieved superiority over the dominant power. A consideration of the strength of the allied belligerents, however, reverses this picture. In World War I, France, Russia, and Britain towered over Germany and Austria-Hungary, who unexpectedly did not gain support from Italy. The contribution of minor powers muddles this simple picture somewhat, but does not change the overall pattern. A more complex pattern emerges in World War II. The United Kingdom and France are far closer to Germany and Italy in the European theater than had been the case prior to World War I. Much has been made of this relationship, and particularly of the brilliant defense of England during the Battle of Britain[7]. However, the entrance of the Soviet Union into the conflict in 1940 gives the Allies a sizable edge and suggests why they were victorious. The Allies were victorious because they amassed a preponderant coalition.

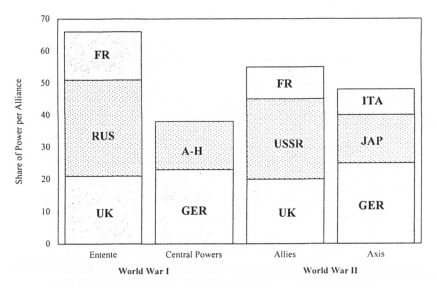

Fig. 1.1. Alliance relations at the onset of major wars. (Original data from
The War Ledger **[Organski and Kugler 1980].)**

The outcome of war can be better explained if one does not look at the
power distributions between alliances at the outset of war but, rather, examines
them during the latter stages (fig. 1.2). Simple comparison between figures 1.1
and 1.2 shows that the dominant country manages to attract support from a vast
coalition of major countries that did not choose to enter the fray at the outset.
This coalescing of forces around the dominant country results from the satisfaction of major powers with the dominant power's regime; countries might
fear the consequences of changes in international arrangements, which would
surely be prompted by the challenger's victory.

Power transition theory argues that countries respond to direct challenges
from actors who overtake their capacity to dominate the international system,
and that the role of allies is to determine the outcome of the conflict. Bueno de
Mesquita (1981a,b), Kim (1989, 1991), and Bueno de Mesquita and Lalman
(1986, 1992), on the other hand, argue each actor is expected to decide on war
based on the calculation of its own strength, as well as on the expectation of
participation by its allies and the allies of its foe. Thus, while there is agreement
that alliances play a critical role in the outcome of conflict, there is disagreement about the role of alliances in conflict initiation. Power transition suggests
that alliances have a limited role while decision theories like those cited above

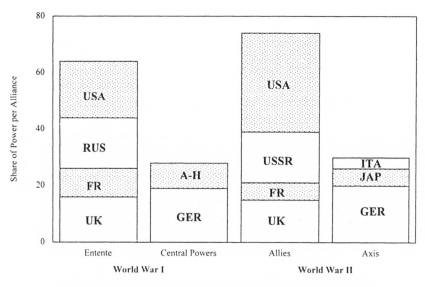

Fig. 1.2. Alliance relations at the end of major wars. (Original data from
The War Ledger **[Organski and Kugler 1980].)**

suggest alliances are the critical actors. Clarity on this point has not been reached. More debate can be found in this very book in chapters 5, 11, 15, and 17.

Given the strength of results on both sides of this question, whether challengers initiate conflict due to their perceptions of the dominant power's capabilities or whether they use a more complex calculus anticipating the participation of other countries remains an open question. It is important to note that power transition implies that fixed, preponderant coalitions ensure stability in the international system. As equality approaches, or as flexibility increases, the probability of war rises. Bandwagoning, where countries join with strong countries rather than against them, follows from power transition logic, but is seemingly inconsistent with balance-of-power presentations that assume international anarchy (contrast with Walt 1987).

In the post–cold war world, power transition arguments imply that Russia and the members of the former Soviet bloc should be interested in joining the West under an umbrella provided by a NATO-like alliance. Thus, rather than raising barriers to the inclusion of the Czech Republic, Hungary, Ukraine, or Russia into NATO, Western allies' interests in maintaining stability should lead them to attempt to attract such countries to join with them (Hwang 1993).

Does Parity Lead to War in All International Environments?

At the time of his introduction of power transition theory in 1958, Organski was unconcerned with the relationship between power distribution and wars below the level of the major countries. Similarly, balance-of-power arguments, such as that of Waltz (1979), focus on the behavior of the major powers, treating minor-power war as a necessary evil to be contained by the major powers.[8]

In spite of this lack of interest in minor-power war behavior, several power parity studies report links at the minor-power level. Weede (1976) tests the preponderance-leads-to-peace model against the conflict behavior of dyads of contiguous Asian countries, most of which are minor powers. He argues that the relationship between preponderance and peace is not a continuous function, but rather a step function with a threshold above which peace is assured. The results of his study strongly support his claims, adding weight to the preponderance-leads-to-peace arguments at the dyadic level. Garnham (1976a,b) investigates the relationship between parity and war at the minor-power dyadic level. He identifies power parity as a necessary condition for war, postulating that a weaker state will prefer to submit to the preponderant rival's demands rather than lose a war and be forced to unconditionally accept the demands. The results of Garnham's studies support the claim of a strong correlation between parity and war, and thus complement Weede's results. Organski and Kugler (1980) report finding links between power transitions and wars among nonmajor powers. Using Relative Political Capacity (RPC) as an additional measure in the analysis of minor power dyads, they observe power transitions prior to the Sino-Indian War in 1962, the three Arab-Israeli wars, and the conflict between North and South Vietnam. In a review of the validity of the GNP times RPC power measure, Kugler and Arbetman (1989a:70) observe a power transition between Iran and Iraq immediately prior to the outbreak of their war in 1980.

Such results may be encouraging to power transition scholars. However, emendation of the conceptualization of the international system provided by power transition theory is a necessary prerequisite for generalization to minor powers. An effort to construct a theoretical framework that includes both major and minor powers is Lemke's contribution to this volume. He extends power transition theory by building multiple regional hierarchies into the hierarchical international system. These multiple hierarchies function as international systems in miniature. They are effectively separated by the decrease of power over distance. Thus, even if Afghans were interested in Australian politics, they would be unable to interfere in any way because they lack the power necessary to be effective so far from their home. Lemke argues that minor-power wars are fought for control of local hierarchical relations, just as major-power wars are fought for control of the dominant international hierarchy. He finds that the power transition argument linking war and power distribution applies to minor countries as well as to major countries. This theoretical scheme allows us to understand the behaviors of different-sized countries and suggests there are no differences in power relations between countries within the periphery and countries within the core. This is a major reformulation of power transition theory.

The extension of power transition theory to minor-power interactions within a regional context allows us to understand how the collapse of the Soviet Empire may be followed by conflicts around important issues in distinct regions such as the Middle East or the republics of the former Soviet Union.

What Is the Role of the Status Quo in the Initiation of War? 4

When referring to the power transition, Organski and Kugler state, "This process is clearly, however, a necessary but not sufficient condition for conflict . . ." (1980:51). A missing, yet theoretically central component is the presence of satisfaction or dissatisfaction with the status quo as a cause of war. Recall that the fundamental distinction between balance-of-power and power transition models is the conception of international anarchy versus international hierarchy. For power transition models, the preservation of the status quo in the international system provides the theoretical reason for the failure of dominant powers to take advantage of their preponderance, as they would be expected to under anarchy. Some scholars have overlooked this distinction and treated power transitions as necessary *and* sufficient conditions. Perhaps the reason for this oversight is that in *The War Ledger* it was simply assumed that the dominant power was satisfied and that the challenger was dissatisfied. It is only in subsequent work by Kim (1989, 1992), Bueno de Mesquita (1990a), and Werner and Kugler (herein), that efforts to measure evaluations of the status quo have been undertaken.

These advances allow power parity proponents to begin addressing constructive criticisms advanced by various scholars. Both Väyrynen (1983) and Choucri and North (1989) have questioned why wars did not occur between the United States and Great Britain in the early twentieth century, and why transitions between NATO allies in the postwar era did not result in wars. A potentially plausible explanation for the lack of conflict between the United States and the United Kingdom in this period is that both are democracies (see, inter alia, Doyle 1983; Schweller 1992; Maoz and Russett 1993). However, the power transition perspective also accounts for peaceful transitions like that between Russia and the United Kingdom in the early twentieth century, where the former was clearly not a democracy.

Organski and Kugler contend that these were transitions between supporters of the status quo; and some have started to test these propositions empirically. Most empirical studies of power transition have only addressed the necessary conditions for war: parity and transition. But the theory suggests that dissatisfaction with the status quo is also of great importance for the initiation of wars. Dissatisfaction with the status quo provides, in Starr's (1978) terms, the willingness to fight a war, while the power transition in which the dissatisfied challenger surpasses the dominant power provides the opportunity to fight. Thus, we might conclude that the United States and Great Britain did not fight each other in the early twentieth century because both were satisfied with the international order. For the same reason, no wars between NATO allies have occurred.

The implications of this fuller specification of the power parity model may also reconcile findings currently at odds. Bueno de Mesquita (1990a) identifies the Seven Weeks' War between Austria and Prussia as a power transition war that changed the international system. He argues that this conflict contradicts power transition theory both because the participants were satisfied with the status quo and because the intensity and severity of the war were limited. Yet, it is exactly the satisfaction which both Austria and Prussia shared that limited their disputes, producing a minor war. In addition, the international consequences were limited. Contrast the gains in prestige Prussia enjoyed following this war with the much more substantial gains and greater severity of the Franco-Prussian War. The latter was a war involving substantial dissatisfaction. Moreover, by taking Lemke's (1993, and herein) conception of multiple hierarchies, we can see the Seven Weeks' War as a regional competition between satisfied countries, but the Franco-Prussian War as a regional competition for dominance over a larger hierarchy.

This reasonable verbal account of the importance of the status quo is now finding its way into operationalizations that treat evaluations of the status quo as a variable. Hussein and Kugler's (1990) and Kugler and Werner's (1993) efforts to conceptualize the status quo characterize the international system as "conditionally anarchic." They formally demonstrate that satisfied countries recognize an order to the international system, whereas dissatisfied countries regard the international system as anarchic, with no constraints on behavior. Attitudes toward the status quo could then, theoretically, be determined from national behavior in the international arena. (For a recent alternative that treats the international status quo as the average issue position in the international system and then deduces some power-transition-like expectations about war relative to evaluations of this type of status quo, see Werner and Lemke 1994.) Kim (1989, 1991) measures satisfaction/dissatisfaction with the status quo, using each country's Tau B alliance score for the dominant country. If the score is positive, that country is satisfied. Conversely, if it is negative, that country is dissatisfied (for a complete discussion of the Tau B measure, see Bueno de Mesquita 1975). Bueno de Mesquita (1990a) operationalizes the level of satisfaction with an intriguing use of money market discount rates. He suggests that a decrease in the value of a nation's currency and the consequent drop in its purchasing power leads to an increasing dissatisfaction among the people. Werner and Kugler (herein) consider the phenomenon of an arms buildup occurring in tandem with power parity. If an arms buildup exists and the challenger is building faster than the dominant power, they find that war is nearly certain to follow (it does in five of six cases). Werner and Kugler suggest that a challenger winning a buildup under these circumstances allows us to infer its status as a dissatisfied country.

It is difficult to disagree with Siverson and Miller's (herein) claim that the prospects for building a more sophisticated model by including evaluations of the status quo are promising.

Timing of War Initiation

A persistent unanswered puzzle that plagues efforts to explain international conflict concerns the timing of war initiation, given power parity. Organski, who originally dismisses the importance of alliances, so central to balance of power, argues that the dissatisfied challenger initiates conflict prior to the transition (1958:333). Otherwise, how could Germany be defeated twice in three decades? However, in *The War Ledger,* Organski and Kugler empirically reported that the dissatisfied challenger initiates war *after* the transition, as demonstrated in their figure 1.2 (1980:59). This finding has been explored and contested.

Thompson claims that: "The logic of the (power transition) model does not permit us to choose between these two rival hypotheses" (1983a:99). However, Thompson reports that his tests of a modified power transition argument demonstrate that the wars occur prior to the transition. Perhaps influenced by Thompson's results, Levy (1987) suggests an alternate plausible explanation of why a stronger country might preempt a rising threatening challenger. He argues, among other reasons, that the dominant country might be risk-acceptant: it does not fight in the present because it does not fear the future situation being worse. In contrast, influenced by results reported in *The War Ledger,* Kugler and Zagare (1990) contend that the dominant country does not preempt the challenger because it is risk-averse. They claim: "The power transition model . . . postulates that the nature of the status quo is negative for the challenger, [and] that the challenger alone is willing to take risks . . . " (1990:263). This suggests a connection between the status quo, risk propensities, and stability. Kugler and Zagare (1990) show that the dominant power values the status quo and does not seek gains by attacking a still-weaker-but-rising challenger. As a result, the dominant power does not launch a preemptive war. Consequently, the challenger continues to grow in power relative to the dominant country, and eventually initiates war in order to change the status quo.

Thus, as long as the dominant country maintains its preponderance, stability prevails and there will be no punitive wars, since the dominant country does not initiate any conflicts. These contradictory but equally plausible arguments can be reconciled by considering the role of alliances. We return to this below. (For an alternate effort to understand the incentives for and against preemption by the dominant country, see Crislip 1994.)

The "democracies-do-not-fight" literature suggests another explanation of the dominant country's unwillingness to preempt the challenger. Schweller (1992) argued that only nondemocracies fight preventive wars, because democracies are constrained by domestic public opinion against launching wars that are difficult to justify. Schweller's claims might well explain the absence of preventive wars prior to a power transition in the past by the democratic political systems of the dominant countries (Great Britain in the nineteenth century and the United States in the twentieth century). As suggested above, this explanation may have merit, but does not have the generality offered by power transition theory.

In this volume, Morrow and Geller specifically address this problem of identifying the initiator and the point of initiation. In a formal treatment of initiation choices, Morrow considers the effects of expected costs of war, risk propensities, and speed of decline on the probability of a war initiation by the challenger or the dominant country. Geller links initiation decisions with the probability of outcomes, and suggests that the challenger will initiate the war after the actual transition. The results of these efforts unpack the underlying logic of the international system as seen through the power transition lens. Scholars continue to disagree about the exact timing of war initiation, however, compelling evidence suggests that wars follow transitions.

What Are the Consequences of Wars?

Chapter 3 of *The War Ledger* specifically addresses the issue of postwar recovery. Organski and Kugler report that countries quickly return to the levels of growth they could be expected to have achieved had war not been waged. They label this unexpected phenomenon the "phoenix factor." Only those countries that are occupied after wars, such as Czechoslovakia under Soviet occupation, are prevented from recovering. Arbetman (herein) extends this argument by comparing political capacity and state-by-state recovery after the American Civil War.

Reporting on the phoenix factor as an empirical observation, Organski and Kugler do not provide a discussion of the reasons behind fast postwar recoveries of countries. Interestingly, Organski and Kugler found no relationship between the amount of foreign aid received and subsequent speed of recovery. In fact, the country that received the most recovery aid per capita after World War

II (Great Britain) actually had the slowest recovery rate, while the country receiving the least aid (Japan) had the fastest recovery rate. In an effort to fill in this theoretical gap, Kugler and Arbetman (1989b) and Chan (1987) test Olson's (1982) argument that wars destroy inefficient political structures that previously prevented market forces from operating. While Chan finds support for Olson's contention, Kugler and Arbetman find no relationship. They suggest relative political capacity as an alternative to Olson's economic reasons for recovery. According to their hypothesis, the more politically capable should recover faster.

Throughout history countries have fought one another, sometimes to a bitter finish, but usually recovered to fight again. During this century Germany was able to initiate massive conflict twice within twenty-five years and has since recovered and become the strongest nation in Europe. Why countries recover from wars is a persistent puzzle for researchers. The phoenix factor points to the uselessness of war as a mechanism of long-term change in the international system. It seems evident now that wars fail to alter the underlying pattern of domestic growth, which is the basis of power. Wars do temporarily rearrange the composition of satisfied and dissatisfied countries. They do not alter the eventual hierarchical leadership of ascendant countries. Wars therefore produce inefficient and temporary distortions of the international system. Whether these distortions will continue to be temporary in the nuclear era is a serious question.

Parity and War in the Nuclear Age

The import of a theory of war lies in what it can tell us of the future. No question can be more important than whether nuclear deterrence is stable. The debate over alternate models' ability to account for the initiation of war is not an empty one, since the policies that emerge from the alternate perspectives vastly differ in their prescriptions.

Building on balance-of-power arguments, traditional deterrence theorists propose that equality of nuclear weapons increases the costs of war to levels that make war initiation unthinkable (cf., Brodie 1946; Kissinger 1957; Intriligator and Brito 1987). This perspective on nuclear deterrence has dominated American strategic thinking in the postwar period.

The power transition perspective squarely challenges classical deterrence. Organski (1968) claims that nuclear weapons do not change the nature of international relations and that power parity in the presence of nuclear arsenals is a recipe for disaster. Even in the presence of weapons of mass destruction, the rise of a dissatisfied challenger heralds war. The only effect of nuclear weapons is to make war much more deadly, not less likely. Since there have been, mercifully, neither major power wars nor major power transitions in the nuclear era, this proposition remains untested.

In spite of the lack of a test of the core of nuclear deterrence, related implications associated with deterrence can be and are tested. In *The War Ledger,* Organski and Kugler consider the escalation of crises before and after the introduction of nuclear weapons, providing chilling support for their argument that nuclear weapons have not changed the nature of international relations (1980:chap. 4). More recently, Kugler (1984) further questions arguments about the ability of nuclear weapons to prevent wars. Huth and Russett (1984, 1990) reach similar conclusions. Huth, Bennett, and Gelpi (1992) suggest nuclear weapons have a deflating effect on the initiation of crises. The record of aggregate testing, as power transition suggests, is that nuclear deterrence is tenuous.

Consideration of nuclear proliferation highlights perhaps the most damaging consequences of traditional theories of deterrence. If, as classical deterrence theorists posit, nuclear weapons have a stabilizing effect that prevents wars, then proliferation of such weapons should enhance peace for the entire world. If nuclear weapons made wars between the United States and the former Soviet Union impossible, they should have the same effect on potential wars between Israel and her neighbors, or between North and South Korea (Rosen 1977; Intriligator and Brito 1981; Waltz 1981; Bueno de Mesquita and Riker 1982; Kugler 1984). Policymakers have not adopted such arguments, perhaps fearing the consequences of nuclear proliferation more than the inconsistencies in their own approach to nuclear weapons.

In this volume, Kugler extends power transition propositions to argue that nuclear proliferation is extremely dangerous because it will not prevent wars, but only increase their severity. Provision of nuclear weapons to additional states will not keep them from enjoying increases in relative power based on their internal development. Likewise, nuclear weapons will not change a dissatisfied country's evaluation of the status quo. The result of proliferation would be the enabling of the country in question to fight a more devastating war once conventional parity is achieved. The decline of the Eastern Bloc and the disintegration of the Soviet Union into four independent countries with intercontinental ballistic missile arsenals gives Kugler's arguments heightened topical relevance. The future holds the potential for disaster.

New Puzzles: Connecting Parity/Transition with Alternate War Perspectives

Macrosystemic Connections

An important consequence of the introduction of power transition theory is that a number of propositions previously seen as disparate can be argued to have commonalities. William R. Thompson (1983a) links power transitions to long cycles of global leadership within the world system. Thompson focuses on those crises that arise when countries are in a transition. These are posited to be especially dangerous, because both rivals may then plausibly expect to achieve success by warring with the other. Thompson further suggests that power transitions themselves might well be "crises of the global political system" (1983a:100). Empirical evidence does not support the evaluation of power transitions within a long-cycle framework. In a related study, Väyrynen (1983) argues that a great many factors converge to create and shape the international system. Väyrynen hypothesizes that periods of ascending growth by a challenger, coupled with an international management regime that is nonrestraining and flexible, produce the most major power war. His empirical evidence supports this multicausal framework.

Taking an independent but related perspective, Houweling and Siccama (1991, and herein) combine power transition theory with Doran and Parson's (1980) notion of "critical points." They argue that power transitions appear as correlates of war only when one member of the dyad is at a critical point. Like Väyrynen, they find empirical support for this combination of approaches.

Another useful insight combines dyadic power transitions with changing levels of power concentration at the systemic level to discover if power transitions are especially dangerous given certain systemic characteristics (Geller 1992a, herein). Specifically, Geller argues that a power transition occurring in a period of power deconcentration in the international system will have a higher potential to result in war. Again, Geller finds empirical support for his hypothesis. Despite some failures, the wide range of studies that successfully connect power parity with other structural and issue-related phenomena illustrates the wide applicability of power transition dynamics and suggests the robustness of the theory.

Several chapters in this volume update or expand previous links between power transitions and systemic phenomena. In addition to Geller's and Thompson's respective contributions considered above, Wayman shows that power transitions are one type of a larger phenomenon of "power shifts." Also, Houweling and Siccama extend their earlier connection between power transitions and critical points and discuss the implications of recent developments in Europe and the former Soviet Union from the perspective of dyadic and systemic power relations. Finally, Danilovic connects power transition theory with work on status inconsistency in an effort to identify especially dangerous instances of parity.

Decision Making and Power Parity

In *The War Ledger* (1980:51), Organski and Kugler reject the deterministic claim that power transitions alone cause wars. If only structural conditions determine the initiation of war, the actions and perceptions of individuals would be meaningless. Work in the last decade shows that decisions are essential in the initiation of war (Bueno de Mesquita 1981a, 1985; Bueno de Mesquita and Lalman 1992; Niou, Ordeshook, and Rose 1989). The combination of power transition and decision-making models promises an opportunity to better understand the process that leads to war and the decision to initiate war.

Kugler and Zagare (1990) make a first attempt by combining power transition theory and the postulates derived from Bueno de Mesquita's decision theory (Bueno de Mesquita 1981a, 1985). They demonstrate the compatibility of the two theories since both focus on the marginal gains anticipated from conflict. Connecting the expected utility concept of probability of success to the power transition concept of relative capability allows Kugler and Zagare (1990:264) to show the logical parallels between the stages of a power transition and the sections of an expected utility plot, anticipating conflict, cooperation, resistance, and capitulation. Using a game-theoretic argument, they derive the conditions under which power parity is stable (no war) and unstable (war). In addition, they are able to deduce that the power transition must precede major war.

Similar results are reported by Bueno de Mesquita and Lalman (1992). They argue that power parity interpreted within their international interactions game, reflecting bargaining dynamics between countries choosing among war, capitulation, acquiescence, negotiation, and the status quo, grants greater power in accounting for the initiation of war. Their efforts show that balance-of-power proposals are inconsistent with the standards of their game. Furthermore, Bueno de Mesquita and Lalman, like Kugler and Zagare, derive that most wars occur after transitions (1992:204–5).

Combinations of the structural argument advanced in power transition with expected utility and game theoretic models bear great promise because their intertwining allows analysts to estimate evaluations of the status quo. Power transition's structural elements indicate which sets or dyads of countries are worth considering, and the decision-making models indicate which of these potential warring groups will go to war. In this volume, Zagare provides an incomplete information game that supplements and extends the boundaries of power transition theory. He shows that while the expected costs of conflict are important, evaluations of the status quo are critical in decisions to go to war. In his contribution, Bueno de Mesquita links power transition theory with the international interactions game, providing greater precision in the derivation of hypothesized relations between power transitions and the risk of wars that lead to systemic transformations. Extensions by Kadera (herein) frame the arguments of balance of power and power transition in a dynamic context, and show the conditions that lead from transitions to different levels of conflictual interaction.

Conclusions: Where We Have Been and Where We Are Going

When introduced, power transition theory arguments ran counter to the then dominant paradigm in international relations: balance of power. Twenty-two years later, Organski and Kugler subjected the theory to its first empirical evaluation. In the years since the appearance of *The War Ledger,* a wide range of scholars have added further specification to, and reformulation of, the propositions originally advanced. This volume brings together these disparate research elements in an effort to evaluate the impact of *The War Ledger* as an explanation of the causes and consequences of international war.

This book wades through and builds on these results and extensions, attempting to determine which aspects of power transition theory are theoretically viable and which should be discarded. Toward that end, the contributors herein have proposed applications and extensions of power transition, enlarging its empirical domain, combining power transitions or parity with other correlates of war, constructing deductive models based on insights that follow from power parity, and considering the current international system from a power perspective.

The work represented in this volume reports efforts over the past thirty-five years to choose between two alternate perspectives of the impact of the distribution of power on war. Balance of power has remained a vital part of our field by generating a number of insightful revisions that allow practitioners to conform theoretical visions with reality. Waltz (1979), for example, restructures Morgenthau's balance of power from multiple actors to bilateral confrontations in response to a world divided between East and West. Brodie (1946) develops notions of Mutual Assured Destruction based on balance of power to assure us of the stability of nuclear deterrence. Neorealists (Keohane 1986) advance more complicated perspectives that further adapt balance-of-power notions. Walt (1987) provides arguments to justify the existence of preponderant alliances while still maintaining claims of balance under anarchy. Closing the circle, Waltz (1993) reconstructs balance of power to account for the peaceful disintegration of the Soviet Union. Many of these changes create profound logical gaps and evaluations of the many-faceted proposals increasingly suggest that the balance-of-power perspective provides inconsistent projections of the reality of the international system. Indeed, empirical work that provides strong support for balance of power is now two decades old (Ferris 1973) and only partial support emerges from alternate tests (Singer, Bremer, and Stuckey 1972; Siverson and Tennefoss 1984).

Concurrently, formal evaluations of the internal components suggest that logical inconsistencies emerge as one pushes balance-of-power theories further. Niou, Ordeshook, and Rose (1989) find that balance of power can account for stability, but only if full information is provided to all parties about the potential outcome of war. For stability, their version of balance-of-power theory also requires that nations grow at equal rates. Both of these critical assumptions contradict reality. More recently, Bueno de Mesquita and Lalman (1992) have demonstrated that balance of power cannot provide stability when one varies risk propensities and preferences. The logical inadequacy of balance-of-power theories can be clearly seen when expectations derived from them are contrasted with those derived using the tools of deductive logic. Figure 1.3 is reproduced from recent work by Bueno de Mesquita and Lalman (1992:204) specifically making such a comparison. In figure 1.3 the bottom line represents the balance-of-power insights of Morgenthau through Waltz, and shows that as an equality of power is approached the probability of war declines. Note that these propositions are not consistent with the logic derived from the international interactions game (nor are they consistent with most of the evidence reported in this review). The power perspective represented by the left-most line is Organski's original version of power transition theory, where the likelihood of war is maximized under parity, but before the transition. The right-most line represents results from the international interactions game and suggests that the probability of war is enhanced by parity, but after the actual transition. These results agree with Kugler and Zagare's (1990) previously developed deductions in the bilateral context of deterrence, were forecasted by the empirical findings of Organski and Kugler (1980), and are effectively supported by far more complex tests developed by Bueno de Mesquita and Lalman (1992).

This extensive exploration of the power transition platform is driven by the combination of empirical support and consistent logical extensions. The claim of this volume is that the bulk of the empirical tests now support the notion that dyadic parity provides the precondition for wars at the global and regional levels. This book shows there is no longer meaningful debate. With the empirical and logical arguments presented here, it becomes clear that power transition theory has evolved into a larger perspective of power parity as a cause of war—thus the title of this book. Parity and war form the core of each chapter.

Over the course of this book, the original formulation and operationalizations of *The War Ledger* are pushed and pulled in a variety of ways. The individual contributions highlight the theory's more ambiguous aspects and suggest resolutions of these ambiguities. Many of the selections are highly critical of earlier work, and yet one cannot escape the conclusion that, overall, power transition survives the assault. Yet, we suggest that power transition should be superseded by a more comprehensive perspective focusing on power parity as a correlate of war, rather than on intersections of power trends. Such a more flexible and general conceptualization provides a richer account of international relations and can consistently account for the changes that have been visited upon the international system.

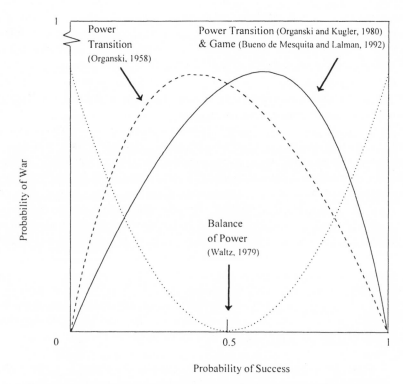

Fig. 1.3. Balance of power theories contrasted with deductive logic. (Reprinted from Bueno de Mesquita and Lalman, 1992:204.)

STATISTICAL APPENDIX

The study of war does not rely on scientifically drawn random samples, but rather draws inferences based on very few cases that approximate the universe of such events, at least for the period under study. Under such circumstances, one is concerned not so much with the statistical as with the substantive significance of results. As Kruskal (1968) cogently argues: "surely such fundamental points as the distinction between statistical and substantive significance must be elementary . . ." in all applications. In the case of nonrandom samples, reporting statistical significance may well be inappropriate. Clearly, in a very large sample even the smallest substantive difference will be statistically significant. For this reason, most critics of significance are concerned with studies that are statistically significant but not substantively so.

The analysis of war suffers from the opposite problem. The number of cases is small, and the universe is approximated. Thus, concern centers on determining if there is sufficient variation to produce significant results when coefficients obtained suggest that substantive results are present. Of course, when a universe is used there is no problem with inference. The power of the coefficients, not the statistical significance of such coefficients, tells the story. Thus, is significance appropriate when inferences are drawn from the approximate universe of major wars in a time series? Morrison and Henkel (1969:134) properly state: "on what basis this assumption is warranted except on the desire of the researcher to apply the statistical inference model is not clear." Faced with data not drawn as a random sample one could easily conclude that significance tests are not useful and that one should rely only on the power of association coefficients disregarding statistical significance. We suggest that such an approach can produce equally problematic results. The study of nuclear deterrence, for example, relies on very, very few cases. Important substantive implications follow since no failures have been experienced (thus far). Should we therefore infer from the existing universe that deterrence is ultrastable? Statistical significance would cast strong doubts on such a strong assertion despite the overwhelming substantive results. Statistical significance in the study of war indicates potential overassertion. We wholeheartedly agree with an anonymous reviewer who suggests that "In the context of historical explanation, the only proper function of significance tests is to cast suspicion on claims. They do not actually provide evidence in support of propositions in any coherent rhetoric of science—their only legitimate function is to cast doubt." Statistical significance tests are reported herein to avoid error and to cast doubt on results; they are not reported to draw undue inferences.

NOTES

1. A number of persistent problems divide balance-of-power theorists and highlight problems with the work that has been accomplished within this perspective. The first is a very basic question of definition. Haas ([1953]1961) has suggested that the term *balance of power* has been used so loosely and by so many scholars that it can be argued to suggest opposite meanings. He criticized Morgenthau for using "balance" as a policy, a description, a specific distribution of power, and any distribution of power, all within the same book (the same criticism has been made by Organski 1958 and Claude 1962). A second area of contention considers the conditions under which balances of power will be more likely to lead to peace. Deutsch and Singer (1964) have suggested that balance of power is more likely to lead to peace under conditions of multipolarity, an international system that has several separate poles or concentrations of power. In contrast, Waltz (1964) has argued that an international system characterized by bipolarity will be most likely to enjoy peace and stability under balances of power.

2. This emphasis on the way that industrialization affects growth rates led Organski

(1958:307) to suggest that power transition theory only applies to the industrial period. For periods after worldwide industrialization and prior to about 1750, he argued that other theories would have to be formulated. Aside from Organski, the connection between industrialization, growth, and development comprises a large theoretical, as well as empirical, literature of its own. For an excellent recent summary, see Jackman 1993.

3. Organski, anticipating the importance of domestic politics and preference structures, suggested we need not be concerned about power transitions between the satisfied dominant and satisfied major countries, such as the United States and the United Kingdom, because such countries have few or no issues to fight over.

4. Hegemonic stability theory, as Keohane acknowledges, depends heavily on the previous work of Kindleberger (1973) on international trade. Keohane defines regimes as alliances around specific issues, embodying implicit rules and norms insofar as they actually guide the behavior of important actors within that given issue area. A regime represents the pattern of international interactions that underlies the provision or regulation of some good.

5. In a large number of studies, scholars have concentrated on developing empirical measures of power. Probably the best-known measurement is the composite capabilities index of the Correlates of War Project (COW). Organski and Kugler (1980), however, suggest the use of Gross National Product (GNP) as a measure of national power. While emphasizing the parsimony of the measure, they also demonstrate that GNPs correlate very highly with COW scores. Later studies by Kugler and Arbetman (1989a), and by Lemke (1991) also report very high correlations.

In addition to measures of national capabilities, scholars have developed measures of the political efficiency of governments. In his early discussions of power, Organski (1958; Organski and Organski 1961) suggests estimating power as national capabilities multiplied by the political efficiency of the government in question. In an effort to operationalize this concept of political efficiency, Organski and Kugler (1980:chap. 2) adapt a measure of tax capacity created at the International Monetary Fund (IMF). This measure of Relative Political Capacity (RPC) compares the resources extracted by the government to the level of resources one might expect the government to extract, given a specific level of development. If the ratio of actual tax extraction to expected extraction is greater than one, that government is characterized as efficient and would score high on the political capacity aspect of power.

In their measure of RPC, Organski and Kugler operationalize only the extraction of resources from the population. There is a second aspect of political capacity they leave untapped. This missing element is political penetration, defined as the ability of the government to be involved in the life of its citizens. Arbetman (1990) provides a first operationalization of this concept. She considers the percentage of the population that is not part of the official work force as an indicator of penetration. High levels of citizen employment in the black market indicate the government does not reach its citizens' lives.

A wide range of research demonstrates that RPC is a useful variable. In independent studies, Organski, et al. (1984), and Rouyer (1987) report strong relationships between RPC and demographic rates, specifically in regard to the ability of governments to control increases in their populations. Furthermore, Kugler and Arbetman (1989b) speculate about links between RPC and the rate at which countries recover from war. Ponder (1992) reports a relationship between RPC and the ability of individual American states to acquire federal grant funds. The wide applicability of this measure strongly supports its validity as an indicator of power.

6. An exception is Ray and Vural (1986), who recognize that subsequent assistance by allies is reported in *The War Ledger* to determine conflict outcomes.

7. A detailed comparison of strength based on measures of power adjusted for political capacity makes this picture much closer to the pattern discovered in World War I (Kugler and Domke 1986).

8. A striking exception to this rule is Haas's 1970 study of the stability of bipolarity and multipolarity between members of subsystems. He is the only scholar we are aware of who has specifically applied balance-of-power reasoning to minor-power interactions.

CHAPTER 2

When Are Power Transitions Dangerous?
An Appraisal and Reformulation of Power
Transition Theory

John A. Vasquez

The power transition explanation of the onset of interstate war has occupied a central place in international relations theory for quite some time. As with any theoretically rich idea, it has undergone various interpretations and changes since its first expression by Organski (1958:chap. 12). In addition, numerous empirical tests, including the major test that was presented in *The War Ledger* by Organski and Kugler (1980), have led to emendations in the theoretical argument. As a result, there is no single power transition explanation, but several, not all of which, of course, are claimed as the legitimate progeny of Organski and Kugler.

This analysis will identify the major versions of the power transition thesis and seek to evaluate their theoretical adequacy and empirical accuracy. It then goes on to reformulate the explanation by identifying the conditions under which power transitions are most dangerous. The first part of the chapter will briefly look at the different versions of the thesis that are in the literature and the role it has played in advancing our theoretical understanding of world politics. The second and third parts of the chapter will review, respectively, the evidence on power transition as a necessary and sufficient condition of war. The fourth section will seek to uncover the role power transitions play in the onset of war by situating the thesis within a broader theoretical explanation. This will be done by identifying what effect power transitions have on interstate relations and under what conditions they increase the probability of war.

The Contribution of the Power Transition Thesis to International Relations Theory

Prior to Organski (1958), the typical way in which power was seen as related to the onset of war was in terms of one or more states disrupting the existing balance of power. The balance of power was said to be a way of maintaining the peace and a shift away from it was seen as dangerous. Organski's main contribution was to show, logically and on the basis of historical example, that a preponderance of power, not a balance of power, was associated with peace. He then went on to argue that war was most likely when a major state supporting the status quo was challenged by another major state that had caught up with and surpassed the capability of the status quo state—hence the label power transition. To this basic model, which emphasized shifts in capability (as did other realist approaches of the time), Organski added the notion that war would ensue only if a challenger was *dissatisfied* with the status quo and the way the dominant nation was running the system.

Various readings of this theoretical argument have led scholars to two different interpretations. One interpretation is that the power transition is a *sufficient* condition of war; that is, once it occurs, war can be expected to follow, ceteris paribus. The other interpretation, and the one Organski and Kugler test and accept in *The War Ledger*, is that a power transition is a necessary condition for war among the strongest states. Although Kugler and Organski (1989b) make it clear in their most recent statement that they accept only the necessary condition explanation, most scholars have tested the sufficient condition explanation (Houweling and Siccama 1988a; Geller 1992a). This approach does not appear entirely illegitimate since the logic of the analysis flows just as smoothly regardless of whether one interprets it as a sufficient or a necessary condition. Of course, part of the problem here is that Organski (1958) did not speak in terms of sufficient or necessary conditions, a terminology that was not prevalent at the time.

Another variation in interpretation has to do with the domain of the analysis. Organski (1958:325–33) referred to the "dominant nation and the great powers" and Organski and Kugler (1980:44–45) to "contenders" (usually the top two or three states).[1] In addition, they limit their analysis to only the strongest states in the central system. Based on the logic of the power transition, some scholars argue that any dyad that interacts frequently should be susceptible to the negative effects of a power transition; that is, all dyads that have a power transition should be more war-prone than dyads that do not have a power transition (Houweling and Siccama 1988a). While Organski and Kugler reject the idea that a power transition would affect all dyads, Kugler (1990) has left open the possibility that regional powers might be susceptible to power transition effects, and Lemke (herein) has provided some evidence to support this for South America from 1865 to 1965.

Lastly, it must be noted that Organski (1958:325–33) saw the power transition as associated with war only if challengers were dissatisfied (see also Lemke, herein). The logical implications of this caveat for the proper specification of the explanation have not been fully explored in the literature. Most scholars who have conducted empirical tests, including Organski and Kugler (1980), have simply ignored measuring dissatisfaction (the major exceptions are Kim 1989, 1991, and Kim and Morrow 1992). Organski and Kugler have used the idea of satisfied states primarily as an ad hoc explanation of why certain dyads have not gone to war.

In order to fully appraise the power transition thesis, the logic and empirical accuracy of each of these interpretations must be assessed. Empirical accuracy will be ascertained by examining the major data-based tests that have been conducted and seeing what conclusions can be derived in light of the adequacy of the research design of the study. The logic of each explanation will be assessed by seeing whether the account provides the kind of explanation of the onset of war we expect in social science; that is, that it provides a generalizable explanation that specifies the variables that bring about war and how they are related to each other so as to cause war.[2] Since Kugler and Organski (1989b) accept the necessary condition as the proper interpretation of the power transition thesis, the evaluation begins with that explanation.

Power Transition as a Necessary Condition of War

The first evidence to support the power transition thesis is found in *The War Ledger*. Organski and Kugler (1980) identify the wars that they think require a power transition and then see whether every war of this type is preceded by a power transition any time twenty years prior to the war (see also Lemke and Kugler, herein). Organski and Kugler seek to demonstrate that war occurs only with equality of power and overtaking. This means that if their hypothesis is correct, major wars will not be preceded by a period where the dominant state and its allies have a preponderance of power.

They show that a power transition affects when the top two or three states (what Organski and Kugler call "contenders") go to war, but it does not influence when other major states go to war (see Organski and Kugler 1980:52, table 1.5). In a study confined to 1860–1939, they find that among contenders, war breaks out only when a power transition has occurred, a finding that produces a moderate .50 (Tau C) (in their table 1.5). They then go on to show (Organski and Kugler 1980:55) that the more rapidly the transition occurs, the greater the danger of war.

Taken at face value, these findings lend support to the thesis. However, before it can be concluded that a power transition is a necessary condition of war between contenders, two cautionary notes about the research design must be sounded. First, the twenty-year time lag that has been used in a number of tests (see Houweling and Siccama 1988a; Doran and Parsons 1980) raises questions. Obviously, the longer the time period in which a transition can occur, the easier it is to find confirmatory evidence. While some time lag is reasonable, a twenty-year lag is a very long time and raises the possibility that any association between a power transition and war could be coincidental. Would a transition be found if only a ten-year test period were used? And if not, why not? Many things can happen in twenty years, and a power transition explanation relying on such a long time lag has to explain why it must take so long for war to break out and, more importantly, why the other things that occur during the long time lag are not the real causes of war (for results produced using ten-year periods, see Werner and Kugler herein).

Second, the sample that is used by Organski and Kugler (1980:45–46) raises some questions. While it is clear that World War I and World War II are very relevant to the explanation, it is not readily apparent why the Franco-Prussian War and the Russo-Japanese War are included when a number of other wars involving major states on each side are dropped (see also Lemke and Kugler herein). Organski and Kugler (1980:45–46) eliminate all interstate wars that have a major state on each side if: (*a*) the war did *not* have a higher level of battle deaths than any previous war and (*b*) the war did not result in a loss of territory and population for the defeated. It is unclear theoretically why the power transition must occur only before wars that are going to have battle deaths higher than any previous wars. It seems more reasonable to include in the sample all wars involving contenders regardless of the subsequent outcome of the war. Alternatively, if one wants to insist that the reason for the battle death threshold is, as Organski and Kugler argue, to insure that the opposing sides make an all-out effort to win the war, then it might make sense to include only definite instances of total war, like the two world wars or Levy's (1983a, 1985) general wars. As it stands, even if one adopts their principle but takes a longer time frame, then the Franco-Prussian and Russo-Japanese Wars should not be included, because they are not more severe than the preceding Napoleonic Wars.[3] These concerns about the sample mean that really only two wars in their sample may be germane and that much greater thought has to go into determining the precise domain of the power transition thesis. It is not enough to say simply that it applies to certain wars; the thesis needs to specify more precisely for what types of wars it is a necessary condition.

Each of these caveats raises questions about the generalizability of the explanation. This is a serious problem, because from the logic of the power transition thesis, it is not obvious why the thesis should not fit all dyads in the system that interact fairly frequently or at least to all major states (Houweling and Siccama 1988a:95; Morrow herein). Nevertheless, Organski and Kugler do not extend the domain of their analysis and their findings do show that the power transition thesis applies only to the very strongest states in the system. Non-contenders—weaker major states, like Italy, and major states in the periphery (i.e., outside Europe prior to 1945), such as Japan, the United States, and China—are unaffected by power transitions. There is no statistically significant relationship between capability (GNP) distributions and when major states in the periphery or other major states in the center fight wars (see Organski and Kugler 1980:52, table 1.7).[4]

It is curious that the power transition effect would turn out to be so limited, given the great theoretical emphasis placed in realist analysis on the role of power and changes in power for shaping behavior. As a result, while Organski and Kugler's findings lend some support to their proposition, the findings also make it evident that the power transition is an inadequate explanation for most of the interstate wars that have been fought in human history. In this sense, the empirical findings can be seen as indicating a deficiency in the explanatory power of the proposition.

The findings raise another question in that they may hold only for the European regional system and not for other regional systems or a more broadly defined global system. Organski and Kugler do not include the United States in the central system until 1945. If they had, of course, there would have been no transition within twenty years prior to either of the two world wars. Given U.S. isolationism, their decision can be justified. However, it is unclear from the more general struggle-for-power perspective underlying their analysis why a country as economically powerful as the United States would not seek to be in the central system. This suggests that power transition concerns may be confined to regions that have been socialized into power politics constructions of reality, and that new states on the periphery, such as Japan and the United States, will not be affected by such transitions.

Lemke's (herein) application of the power transition to South America could undercut this conclusion if further research shows his findings to hold. Presently, they are based on only two wars—the War of the Pacific and the Chaco War. It is essential that earlier wars, for example the Lopez War, and wars in the rest of Latin America, such as the Mexican War, be included. Nevertheless, Lemke's theoretical analysis, particularly the notion of multiple hierarchies and his measure of geographic proximity, is an important contribution. What is of special interest in the Western Hemisphere in terms of differences with Europe is whether the presence of a hegemonic state, such as the United States, can reduce the number of wars by establishing certain rules of the game and restraining minor states from engaging unilaterally in the practices of power politics. Lemke's theoretical analysis provides a framework for addressing this question and thereby explaining why some power transitions may not be followed by war.

All of this means that before the thesis can be accepted, more wars and regions will have to be examined. The most obvious set to look at would be the wars prior to 1815. In an early analysis Thompson (1983a) looked at global wars prior to 1815, and he suggested that attacks are often premature (occurring before a transition) and that some global wars occurred without any transition at all. While Kugler and Organski (1989b:182–84) are prepared to concede that attacks may sometimes be premature, they are (with some reason) not prepared to accept Thompson's (1983a) test as definitive because he measures capability in terms of naval strength rather than in terms of overall economic power, which is their major focus.

The test by Thompson (herein) addresses some of these concerns by expanding the measure of capability to include population and army data along with naval strength. Thompson's main purpose in this chapter is to integrate aspects of the power transition, long cycle, and balance-of-power explanations and subject that new synthesis to a preliminary empirical assessment, rather than provide a formal test of Organski and Kugler's (1980) original proposition. Nevertheless, his findings outline the nature of the evidence both in support of and against the power transition thesis.

Since Thompson traces the long-term capability shifts for major dyads that he identifies as global leader and challenger, his data can be used to assess the original proposition of Organski and Kugler. Thompson looks at five-year intervals, which is very rigorous and permits an assessment of just how directly related transitions are to the onset of war. The results are somewhat mixed and the cases few, despite the long time span of 1490 to 1945. Much depends on what wars are included in the sample. Thompson (herein) focuses on his set of global wars. The first case is between Spain (the rising challenger) and Portugal (the declining leader). A major transition occurs around 1535, too long before 1580, when Portugal is conquered by Spain, to count as a necessary condition for that war.[5] The second case is between France and the Netherlands. Here there is a transition around 1665 and that is followed by the Dutch War of Louis XIV (1672–78) involving France and the Netherlands (and a host of major states), but the global war that fits Thompson's criteria does not occur until 1688 (War of the Grand Alliance), and this is more than twenty years later.[6] The third case is Britain and France, and this nicely fits the power transition proposition, with France passing Britain just before the French Revolution. The fourth case involves the two twentieth-century world wars. If Britain and Germany are taken as leader and challenger, then World War I just misses and World War II holds to the power transition criteria.

What can be concluded from this survey? Two cases do not seem to fit, one clearly does, and the last is split. These results are not overwhelming. Other wars, like the Anglo-Dutch naval wars of the mid-seventeenth century, seem to have the requisite power transition, but do not produce the kind of major war Organski and Kugler predict. How the Thirty Years' War would fit is still an open question. What is clear from Thompson's (herein) data, however, is that because there are so few transitions, and so many interstate wars, it is always possible to find some war near a transition point, or far away from one. In order to test the proposition in a fair manner, it will be necessary to specify unambiguously what *kind* of wars require a power transition as a necessary condition. Obviously, if all wars involving major states on each side are taken as the domain, then the hypothesis will fail,[7] but if it is restricted to a more limited set of global or world wars, then only a careful and rigorous research design will be able to provide a definitive answer.

This review of the evidence and the questions raised by the logic of the analysis suggest that the power transition thesis is probably not going to provide the general explanation of war that it seemed it might when it was first enunciated by Organski (1958). Tests by Organski and Kugler (1980) have limited the domain of the analysis, and questions about the evidence seem to indicate that the domain should be limited even further. Despite these concerns, can it be concluded that a power transition is a necessary condition for the most severe wars among the strongest states? Since the number of cases available for data analysis limits our ability to make a firm answer, one way of addressing the question is to see whether the logic of the power transition is consistent with the specific details of why and how World War I and World War II occurred.

From the perspective of the power transition thesis, one would explain the onset of World War I and World War II by saying that the wars occurred because Germany was surpassing Britain in capability and challenging its leadership of the international system. Does this perspective help us explain the historical events associated with the two wars?

Organski and Kugler (1980:43–45) argue that major wars have their origin in the second-ranked contending state overtaking the first. Indeed, Bremer (1980:69) shows empirically that this is often the case. One would assume then, that the two world wars result from the second-ranked state overtaking and attacking the first. Organski and Kugler (1980:58) maintain that prior to World War I, Germany caught up with Britain (in GNP) by 1905 and by 1913 had surpassed Britain. It then fell behind Britain after 1919 and caught up again in the early twenties, retaining an advantage for the rest of the decade (Organski and Kugler 1980:59). On this basis, the two world wars produced the aggregate statistical findings.

A closer look uncovers some problems. First, Britain is the first-ranked state only because the United States is not included in the contender sample until 1945. To eliminate the United States as a contender because the United States did not view itself as part of the central system or because Britain and the United States did not view themselves as mutually threatening or as members of competing regimes (Organski and Kugler 1980:45; Kugler 1990:209) is too convenient (see also Lemke and Kugler herein). Certainly, the United States viewed itself as a rival to Britain in the Western Hemisphere and as a player in the Pacific, and their relations, prior to 1895, were hostile. During this time, Irish immigrants in the United States also provided a domestic political incentive for "twisting the (British) Lion's tail." All of these factors indicate that the United

States cannot be automatically assumed to be a "satisfied ally," as Organski and Kugler do (see also Kugler 1990:209).

In this regard, the power transition thesis is particularly deficient in explaining why Britain found Germany more of a threat than the United States, given America's economic and naval capabilities. Why, as Britain began to decline, did it choose to resolve its outstanding issues with rising nations, such as the United States (1895) and Japan (1902), and with old rivals, such as France (1904) and Russia (1907), yet remain unable to settle accounts with Germany? There were important acts of cooperation between Britain and Germany, some just before the outbreak of the World War I, as well as significant and longstanding royal ties that formed a foundation for a peaceful settlement. Also, unlike France and Germany, there were no outstanding disputes involving territorial contiguity. Nor were the colonial disputes as serious as those that had been resolved between Britain and France or Britain and Russia.

Even if one accepts the various ad hoc explanations for why the United States and the United Kingdom did not fight or for why Germany and the United States were not global rivals (see Rock 1989), it is hardly established, although often assumed, that World War I started because of Anglo-German rivalry. This view forgets that the actual fighting in World War I did not arise because of German challenges to Britain, but with a dispute between Austria-Hungary and Serbia, which was linked to the French-German dispute over Alsace-Lorraine through alliances with Russia. Instead of seeking to fight Britain in 1914, Germany sought to keep her neutral. This does not sound like a war that is caused by a power transition between the first- and second-ranked states. The dispute did not directly involve Britain. At best, it involved allies of her allies.

World War I does not seem to have evolved the way the findings of *The War Ledger* imply. War did not come about in 1914 with a power transition among the contenders that then dragged in the other major states because of their alliances with one or more of the contenders. Instead, war broke out among weaker allies who then dragged in the contenders. In addition, one of the contenders, namely Germany, sought to keep Britain out of the war.

It is an anomaly for the power transition hypothesis that Germany did not seek to challenge and fight Britain, but Britain's former rivals, France and Russia. Why were the most forceful challenges made toward these states and not the declining Britain or the rising United States? There are some obvious geopolitical and historical answers to this question. The fault of the power transition explanation is that instead of relying on these answers, it points us toward an abstract single factor, the power transition, that seems tangential to the dynamics of what was going on.

Likewise, the explanation for World War II is wanting. The initial events in western Europe in 1939 are consistent with the power transition thesis, since Germany overtakes Britain, but the expansion of the war in 1941 with the attacks on the Soviet Union by Germany and the United States by Japan are not (see also Lemke and Kugler herein). The latter are important because without them, World War II would probably never have become a world war and certainly would not have crossed the battle death threshold of World War I, which is necessary for the war to be included in Organski and Kugler's sample. According to Kugler and Organski's (1989b:181,183) data, Germany never surpasses the USSR in GNP in the interwar period.[8] Therefore, Germany should not have attacked if the power transition were a true necessary condition of war. Similarly, from the logic of the power transition, it seems quite ironic that the most powerful state in the system, the United States, should be drawn into this war by a direct attack from a comparatively weak state, Japan.

In conclusion, if one grants the exclusion of the United States from the analysis for both wars, something that is hard to justify from a systemic power struggle perspective, then the necessary power transition occurs prior to World War I, but it appears tangential to actual factors that permit the war. In World War II, the major events that make it a world war come about in the absence of a power transition. The ways in which the two world wars of the twentieth century occurred and evolved do not seem to conform to the way one would expect war to evolve given the logic of the power transition thesis.

In addition, when the power transition thesis is interpreted as solely a necessary condition of war and not as a sufficient condition of war, it fails to specify what causes war. Delineating necessary conditions of war provides an important tool for predicting how war might be prevented, hardly an unimportant goal. This effort does not, however, discuss what brings about war once the necessary conditions are in place.

The former must also be of concern if war is to be explained, unless, of course, it is assumed that there are so many factors that can bring about war that the mere presence of the necessary condition will insure war. This does not seem to be the case with the power transition thesis, but it is in some power-oriented explanations (like Gilpin's [1981] hegemonic theory) where the struggle for power is seen as so pervasive that once the necessary conditions are present, war occurs. In such an explanation, however, necessary conditions act as necessary and sufficient conditions. Undoubtedly, it was the concern with explaining the actual factors that bring about war that led some to interpret the power transition as a sufficient condition of war.

Power Transition as a Sufficient Condition of War

A number of problems are encountered if the power transition is treated as a sufficient condition for major war. This kind of research design tests for the possibility of there being power transitions among contenders without a war. Houweling and Siccama (1988a, herein) provide such a test using a twenty-year test period. Employing data collected by Doran and Parsons (1980) that measure capability on the basis of economic, demographic, and military indicators,[9] they find a slight relationship between a power transition and war for all major states, and a slightly stronger relationship when the sample is confined to the top three or four major states during the 1816–1975 period. The latter provides further evidence that Organski and Kugler are correct in seeing the power transition as something that affects the top three or four states, rather than all major states, and Houweling and Siccama (1988a:101) conclude by supporting the power transition explanation.

However, a close inspection of Houweling and Siccama's (1988a: tables 7, 8, herein: tables 6.1, 6.2) findings raises questions. It turns out that the association between different types of power distribution and war for all major states is very weak (the highest Tau C equaling .159) (Houweling and Siccama 1988a: table 7, herein: table 6.1), so this general hypothesis should be rejected, even though it is statistically significant and slightly higher than Organski and Kugler's (1980:52) reported correlation of −.03. More importantly, the association for the top three to four states is only .306 (Houweling and Siccama 1988a: table 8, herein: table 6.2). This is quite a reduction from the .50 Tau C that Organski and Kugler (1980:52) found. If one looks at the actual number of cases in which a power transition occurs, then it is clear that, at the most, only about half the cases result in war. For Organski and Kugler five in ten power transition cases are associated with war, and for Houweling and Siccama (1988a: table 8) eight result in war and nine do not.

What this means is that the power transition effect is not very strong. It has no effect on major states in the periphery or on major states that are not among the top two or three states—and, among the very top states that experience a power transition, their chances of going to war are as great as their chances of avoiding war. Clearly, since both the set of eight cases that go to war and the set of nine that do not go to war experience a power transition, there is likely to be some other factor that is causing the onset of war.[10] This raises the possibility, particularly in light of the findings on other major states, that the relationship between power transitions and war is random or spurious.

This suspicion is further fueled by the use of a twenty-year time lag by both Organski and Kugler and Houweling and Siccama. This turns out to be a serious problem, because if a ten-year time span is taken to see whether a war occurs either just before or after a power transition, then in some studies the relationship disappears (see also Lemke and Kugler herein). For instance, in a very systematic test, Peter Wallensteen (1981:80–81) examined all pairs of major states from 1816 to 1976 and found no relationship between war and one major state catching up with and economically surpassing another. Employing a measure based on iron and steel production, Wallensteen found a power transition occurring in eleven of fifty-three pairs of major states, but in these eleven pairs, war took place within ten years only *three* times. Of these three instances, only two are really legitimate examples—Prussia surpassing Austria and the Seven Weeks' War of 1866, Prussia surpassing France and the Franco-Prussian war of 1870. The third case is China surpassing France in 1960 after the Korean War in 1950, when China technically attacked France as part of the United Nations.

Equally important are the instances of transition that do *not* lead to war. Wallensteen points out that Russia surpassed France at the end of the 1890s and instead of war there was an alliance. Likewise, in the classic case, which Organski (1958:323–25) himself recognizes as an exception—the United States surpassing Britain in the 1890s—does not lead to war, but to a collaborative and peaceful transition.

Additional evidence illustrating the limitations of power transition as a sufficient condition is provided by Geller (1992a, herein). In an analysis that demonstrates a great deal of insight, he argues that neither dyadic power transitions nor systemic shifts in capability (as in long-cycle analysis) have much of an impact on the onset of war, but that the right combination of structural and dyadic conditions would have an impact. Specifically, he finds that when the global structure has a deconcentration of power, dyadic power transitions are associated with war, but in the absence of this structural condition there is no association. Geller's (1992a) research design and finding constitute a major conceptual breakthrough in identifying when power transitions are dangerous. At the same time he shows the limits of the power transition as a sufficient condition, since although his findings are statistically significant, the strength of association is not very strong (Kendall's Tau C = .26 [Geller 1992a:279]).

Geller's analysis (herein) extends these findings by comparing dyadic preponderance and parity under varying systemic conditions of increasing and decreasing concentration. Generally, dyadic preponderance in either systemic condition is not associated with war between contenders, but dyadic parity is much more likely to be associated with such wars under conditions of systemic decreasing concentration. Under the latter, seven of twenty-three dyads (about 30 percent) go to war, but under systemic conditions of increasing concentration, only one of nine (about 11 percent) go to war. As with the test of the actual transition, it must be kept in mind that this association is not very strong (phi = .27); indeed it is not even quite statistically significant at the .05 level (*p* = .0631) (Geller, herein: table 8.4b). Much of this weakness, however, is a result of the fact that the relationship between preponderance and war is unaffected by systemic conditions. Thus, this can be taken as evidence that Organski and Kugler are correct that parity is much more likely to be associated with war than is preponderance.

On the whole, Geller's findings seem consistent with Houweling and Siccama's. They both find a statistically significant impact, but a very weak association. These findings are also consistent with what Thompson (herein) finds for a longer period. His analyses are statistically significant, but with associations (Cramer's V) of .26, .52, .30, and .32 (Thompson, herein: tables 10.1, 10.2, 10.3). All of these findings mean that a power transition has some impact, but by itself will probably account for under 10 percent of the variance. Geller advances the research by showing that power transitions are more dangerous when the system is experiencing an overall deconcentration of power. Of course, this in itself may be associated with one or more of the dyadic power transitions, but whether this is always the case must await further research. Geller's (herein) analysis also shows that parity is more likely to be associated with major state war than is preponderance, which rarely is associated with major state war. This analysis raises the question of whether it is parity or the transition that is really associated with the onset of war.

In a series of articles, Woosang Kim (1991, 1992, herein; Kim and Morrow 1992) provides evidence that it is parity, dissatisfaction, and risk propensity that are associated with major state war, rather than a power transition or the rate of growth of the rising state's capabilities. Kim's (1991) contribution to power transition research is to incorporate into the transition framework a measure of the impact of external alliances on capability calculations. He then measures transition and equality in terms of alliance transition and alliance equality. The alliance transition variable, as well as the power transition variable (measured by the Correlates of War Composite Capabilities Index), has no statistically significant impact, but alliance equality and measures of dissatisfaction do. Kim and Morrow (1992) find that measures of risk propensity also have an impact. These findings, for the 1815 to 1975 period, are based on wars among major states and support the notion that parity rather than preponderance increases the probability of war among major states. The findings, however, contradict the notion that a transition in and of itself is important. As Kim and Morrow (1992:917) conclude:

> Our evidence supports Organski's contentions that rough equality of the sides and more dissatisfied rising states increase the chance of war. His other hypotheses do not fare as well. How fast the rising state catches up to the declining state has no statistically discernible effect . . . . Transitions themselves have no effect on the probability of war.

Morrow (herein) elaborates a theoretical rationale from a formal perspective to explain these findings and reformulates and expands a number of the power transition hypotheses. Elaborating on previous work, Kim expands the data analysis to the pre-Napoleonic period, looking at the 1648–1815 period (1992) and the entire 1648–1975 period (herein). Kim's findings are very impressive and consistent with his earlier studies, even though the sample of wars has been changed. Again he finds that alliance transition and growth rate have no statistically significant impact, but that alliance equality and dissatisfaction do. He also finds that splitting the data into pre- and post-Napoleonic periods has no effect; thus the findings are stable.

Kim addresses the question of statistical significance and not strength of association; nevertheless, his findings are important because of the extensive time domain they cover. Kim shows that an overtaking in and of itself may not be theoretically significant. Equality of alliance capability and dissatisfaction have more of an impact. The latter finding is one of the first pieces of systematic evidence that satisfaction is an important variable, even though the measure of it is somewhat indirect. These findings, with Kim and Morrow (1992) on risk propensity, and Morrow's (herein) theoretical analysis, show that power transitions and shifts in capability should not be regarded as some kind of inexorable cycle that brings about war, but as something that has an impact on the choice theoretic operating within the leaderships of states. This impact is worthy of special note since the domestic political context probably makes dissatisfaction and risk propensity correlated with each other (Vasquez 1993:chap. 6).

Last, it is important to note that the dependent variable in Kim's analyses is the presence or absence of war between major states and not war-peace. The findings do not mean preponderance is associated with peace, but only with the absence of wars between major states. It can be inferred from the research design only that equality increases the probability of major-state war and preponderance decreases the probability of such wars.[11] Morrow (herein) explicitly recognizes this issue when he states that power transition is an explanation of big wars.[12] It is highly likely that preponderance is associated with its own type of wars (e.g., imperial wars) and that capability distributions do not so much determine whether war will occur as they determine the *form* war will take once it breaks out (see Vasquez 1986:315; 1993:chap. 2). The precise causal significance of equality and preponderance still needs to be identified. The reduction of major-state wars in periods of preponderance, for example, can simply mean that after major wars where one state emerges as preponderant, there will not be any more wars among major states for a while.[13]

Further evidence on a number of these questions is provided by Wayman (herein). In one of the most systematic analyses of the power transition proposition, Wayman looks at whether power transition, rapid approach, or relative parity is most significant and whether an ongoing rivalry is important. Wayman addresses these questions by examining whether a transition occurs within the same decade or the preceding or subsequent decade as a war. His measure of capability, like Geller's (1992a), employs the Composite Capabilities Index of the Correlates of War project.

Wayman's findings, when compared with the others discussed above, provide some insight regarding the true impact of power transitions. First, there is evidence that equality, in and of itself, is not of causal significance, since Wayman (herein) finds that prolonged parity is not associated with war in the subsequent decade. Parity is associated with subsequent war (a decade later) only when it occurs with a power transition and/or a rapid approach toward parity. However, this picture is muddied somewhat by the finding that prolonged parity is associated with war in the same decade.

Second, Wayman (herein) finds that power transition may be associated with war primarily because capability shifts (whether or not they result in transitions) increase the probability of war. He finds that rapid approaches are more apt to result in war between major states in the following decade than are power transitions (30 percent of the rapid approaches result in war versus 19 percent of the power transitions; Wayman herein:table 9.1). Nevertheless, states having either capability shift are more likely to be involved in a war than states that do not experience such shifts, ceteris paribus. These findings help clarify that the critical factor may be the dynamic of change in capability.

That being said, a close examination of Wayman's findings shows that the association between capability shifts and war is not very strong, which is also consistent with previous findings. Wayman (herein:table 9.1) shows that only five of twenty-six (19 percent) of the power transitions and only seven of twenty-three (30 percent) of the rapid approaches result in war within the subsequent decade. These are not very high rates, since 10 percent of the major-state dyads have wars when there is no kind of capability shift whatsoever. Indeed, Wayman (herein:table 9.3) finds a statistically significant relationship but only a weak gamma of .45 (note Tau B = .16). The magnitude of these findings is consistent with other studies and indicates that capability shifts alone are not a sufficient condition for the onset of war.

Wayman's (herein) own interpretation of his findings steers away from concluding that capability shifts are a sufficient condition; instead he examines what factors increase the probability of war and finds that capability shifts have such an effect. Because these findings involve only weak probabilities and show that many power transitions do not result in war, the question arises: What factors increase the probability that power transitions might be dangerous? Wayman's empirical analysis provides some answers to this question by looking at the role of rivalry, which will be examined in the next section. For now, it should be clear that although a power transition may increase the probability of war, it is not by itself a clear sufficient condition of war; a power transition does not lead to war on any regular basis. Sometimes it does and sometimes it does not.

It seems, then, that the power transition hypothesis has a number of empirical deficiencies that prevent it from being accepted in its classic form as an accurate or dependable account of the onset of war. As a sufficient condition, it is lacking; many wars occur in the absence of a power transition. Half of the time war should occur (because a power transition has occurred), it does not ensue. As a necessary condition, the explanation is also limited, since it does not apply to major states in the periphery, even if such a state is the rising dominant state. Nor when the details of World War I and World War II are examined does the power transition provide much insight; instead it raises a host of unanswerable questions. Finally, if it is taken solely as a necessary condition of war, then it has failed to provide an explanation of the sufficient conditions of war; (i.e., it does not have an explanation of the onset of war).

What the research has found is that power transitions do have a statistically significant impact on wars among major states, but that the strength of this relationship is weak to moderate. Rather than seeing power transition as either a sufficient or necessary condition of war, it is best to see it as one factor that increases the probability of certain types of wars, especially when combined with other factors.

The power transition thesis and *The War Ledger* have also made a number of other crucial advancements in our understanding of war and peace. First, Organski's (1958) and Organski and Kugler's (1980) criticism of the previously dominant balance-of-power explanation of peace is a major contribution. The balance of power does not seem to be associated with a condition of peace. Likewise, their identification of periods with an overwhelming preponderance of power as relatively free of *major* wars seems correct.

In addition, Organski and Kugler's (1980) formulation of the relationship between capability shifts and war is vastly superior to more traditional power explanations; for example Gilpin's hegemonic theory of war, with which it shares a number of propositions. Unlike Gilpin's explanation, which is often very imprecise and ambiguous and often fails to specify what evidence will falsify it, Organski and Kugler clearly specify the conditions under which war can be expected and the tests that can falsify their explanation. These tests seem to indicate that power transitions do have some impact on certain dyads going to war, even though the impact is not so strong that most transitions will result in war. Given the moderate results, such findings require that the propositions that have been tested be reformulated. The next section will undertake this task.

Specifying the Conditions under Which Power Transitions Are Dangerous

One way to begin is to suggest that looking at necessary and sufficient conditions may be too stringent an approach. In fact, there may be no single set of causes of war. Instead, it may be more accurate to say there are various causal paths to various types of war (Vasquez 1993:7,48–49). If this is the case, then what we would want to do is to reformulate the power transition thesis to de-lineate the causal path(s) in which it appears. In doing so we would not say a power transition is a sufficient condition of war, but that in conjunction with other variables it increases the probability of war. In other words, it has a sta-tistically significant impact on the onset of war, although the magnitude of that impact is likely not to be as strong as scholars initially thought.

The main effect that power transitions have, at least among major states, is not to produce war, but to give rise to an increase in militarized disputes. Wallensteen (1981) finds that of the major-state dyads that experience a power transition, 91 percent have at least one militarized confrontation within ten years, but only 27 percent have a war.[14] Wayman (personal communication) also finds that power transitions and militarized disputes are associated. Sixty-eight percent of the power transitions and 44 percent of the rapid approaches in the same decade give rise to more than one militarized dispute between rival states, but only 25 percent of these transitions and 15 percent of these rapid ap-proaches result in war.

Likewise, Geller (herein) finds that while there is no statistically signifi-cant relationship between power transitions and whether the stronger or weaker state initiates *war*, there is a statistically significant relationship between power transition and who initiates militarized disputes. In the latter case, when there is overtaking, the stronger initiates twice as many disputes as the weaker (ten versus five). All of these findings mean that actors who have a dramatic increase in capability can be expected to place new demands on the political system, but that these demands do not necessarily result in war.

One of the key factors that probably explains when these demands result in war is the level of satisfaction of states. According to Organski's (1958) orig-inal formulation, states go to war only if they are dissatisfied, but most existing interpretations and tests of the power transition (including those of Organski and Kugler themselves) give insufficient emphasis to this condition. It has been used primarily as an ad hoc explanation to fend off criticism. This is un-fortunate because focusing on what makes states satisfied or dissatisfied is prob-ably the key to explaining why some power transitions end in war and others do not.

Wayman (herein) provides some evidence to indicate that incorporating a measure of dissatisfaction helps define the conditions under which power tran-sitions are dangerous. Although he does not specifically measure dissatisfac-tion, he examines rivalries in his analysis and rivals by definition must be dis-satisfied with each other. His findings show that being a rival increases the probability of going to war. What is even more significant is that this probabil-ity increases further if the rivalry experiences a power transition or rapid approach. Only 8 percent of the dyads that have neither a rivalry nor a power transition (or rapid approach) go to war in the subsequent decade, while 31 percent of the dyads that are rivals and have a power transition (or rapid approach) go to war (Wayman herein:table 9.3).

Further evidence that dissatisfaction may be a factor is provided by Kim (herein) and Kim and Morrow (1992), who measure dissatisfaction by the similarity of alignments. These studies provide evidence that Bueno de Mesquita (1990a) is correct in his claim that states need only be dissatisfied with each other and not with the way the system is being ruled in order to have major war. This supports the commonsense point of view that some level of dissatisfaction must be necessary for war to occur, but unless dissatisfaction can be defined more precisely it does not provide much explanatory power. What these analyses do not address is whether there are only certain types of issues that are apt to give rise to dissatisfaction so great that it will lead to war.

Too many scholars tend to assume that since international politics is a struggle for power, dissatisfaction is potentially pervasive—that any issue can give rise to dissatisfaction, especially if there is an approaching power transition. To me, not all issues are equally likely to give rise to war. Territorial issues, particularly those involving territorial contiguity, are sources of conflict that have a greater probability of ending in war than other sources of conflict (Vasquez 1993:chap. 4). When not handled correctly such issues poison a relationship. Whether or not states are satisfied, then, depends primarily on whether they have resolved all border questions and territorial claims they have with their neighbors. If they have, the probability of major war significantly decreases; if not, the possibility of war is always lurking in the background, ready to emerge under the right conditions.

Territorial concerns, however, do not make war inevitable. Much depends on how states learn to deal with their proclivities and their need for territory. Attempts to impose borders or acquire new territory through the use of force produce hostility. Imposed borders, especially between relative equals, are much more likely to lead to enduring rivalry than borders established through mutual accommodation and diplomatic agreement. The use of force leads to settlement only when one side attains an overwhelming victory and is able to maintain a preponderance of power over the other side (see Weede 1976; Garnham 1976a). Anything short of that will make territorial issues fester and produce long-term hostile relationships. In this context, power transitions greatly increase the probability of war because they allow a state that has been unable to act upon a festering issue to redress its grievances. This is particularly the case when the global (or regional) culture has institutionalized the ideas of power politics, which legitimize the use of force for dealing with intractable issues.[15]

Whether states deal with territorial issues in ways that increase or decrease the probability of war depends very much on prevailing norms. Territorial conflict can be reduced considerably if norms have evolved for dealing with territorial claims. In other words, political variables (particularly such nonrealist variables as norms, rules of the game, and the political structure of the global system) are very important for determining the probability level other variables will have for encouraging war. Thus, the probability that territorial disputes or power transitions will result in war will vary depending on certain systemic political characteristics.

What determines a peaceful or violent transition is not the power shift, but whether the rising state(s) can be accommodated by the existing political order, including the rules it has for bringing about change (see Kugler 1990:209–11). Because of these complex relationships, power transitions do not always result in war, but sometimes they do, and this is not just a random occurrence. By specifying the conditions under which power transitions are war-prone, we advance our understanding of the onset of war and identify the contribution of power transition theory to the creation of a scientific explanation of the onset of war.

What needs to be done is to place more emphasis on the ability of any given global political system or world order to deal with issues that might give rise to fundamental dissatisfaction, to give more thought to what kinds of issues are prone to war, and, most important, to delineate how *handling* issues one way will increase the probability of war while handling them in another will decrease the probability of war. Elsewhere, I argue that when power politics becomes the principal way of handling war-prone issues, like disputes over territory, they tend to give rise to a series of steps that between equals lead to war (see Vasquez 1993:chap. 5). These steps include making alliances to increase one's capability, building up one's military, and demonstrating resolve through using realpolitik tactics in crises. Power transitions increase the probability of war when they encourage states to take the first of a series of these steps or when they encourage states to move from one step to the next. Sometimes this can occur early on, as when a power transition encourages a state to raise sensitive territorial issues. Sometimes this occurs in the presence of an ongoing rivalry, in which case it may encourage the making of alliances or building up of one's military, or both.

The precise effect of power transitions on the steps to war still needs to be researched, but reformulating and expanding the power transition research program in this direction seems promising (witness the progress made by Werner and Kugler (herein) in dealing with the literature on military buildups and dispute escalation). Identifying the role power transitions play in the steps to war, and mapping the increase in the probability of war they generate, will go a long way in advancing research.

Conclusions

On the basis of the previous analysis, the following conclusions can be made. First, the brief review of the two world wars suggests that the specific details of each war do not conform completely to the logic of the causal explanation offered by the power transition thesis. Much more theoretical analysis must be devoted to how power transitions bring about wars and how they are related to contagion mechanisms, like alliances.

Second, the power transition effect applies only to core actors at the global and regional levels. But it may not apply to asymmetric wars in the periphery. Why the domain of explanation is so confined is not obvious from the theoretical analysis. It would seem from the logic of the explanation that its domain should be larger.

Third, the most serious gap in any explanation of war that looks solely at necessary conditions is that it does not specify the factors that actually bring about war. As a result it does not really tell us why wars occur; nor does it tell us what causes wars.

Fourth, rather than seeing a power transition as either a sufficient or necessary condition, it is better to view it as a factor that may increase the probability of war in the presence of other factors. Power transitions become dangerous in the presence of serious disputes over territory, if the prevailing global political system is unable to handle such issues and, in turn, states unilaterally resort to the practices of power politics to change or maintain the status quo.

These conclusions suggest that a great deal has been learned about parity and war through use of the scientific method, but that a great deal remains to be done. *The War Ledger* and the thought of Organski and Kugler have played a major role in this effort by setting out a scientific research program that has helped us to remove much that is confusing in the power politics and capability approaches and to identify the role power transitions play in conflict and the onset of war. As the studies in this book demonstrate, theorizing associated with the power transition thesis continues to produce new insights, new findings, and new research proposals. Given the still-early stage of our peace science, little more can be asked.

NOTES

My thanks to Marie T. Henehan and to contributors who participated in the conference for comments. The sole responsibility for the analysis is mine alone, however.

1. Organski and Kugler (1980:44–45) operationally define *contenders* as the strongest state in the central system and any other state that has at least 80 percent of the GNP of

the strongest state. If no state meets the 80 percent threshold, then the three strongest states are considered contenders. It should be noted, that because the strongest state in the *central* system is taken as a criterion, the United States is not considered a contender until 1945, even though Organski and Kugler admit that the United States had the highest GNP in the world by the end of the nineteenth century.

2. I find theory construction is facilitated by thinking in terms of causes, but for those who wish to eschew this language, they can think of what I call causes as specifying "how the variables are related to one another so as to make war break out."

3. Kim (herein), for example, drops the Russo-Japanese War because it fails to satisfy the battle death and territory criteria even for the post-Napoleonic period. He keeps the Franco-Prussian War because he looks at the average annual level of battle deaths rather than the total deaths for the entire war (see Kim 1992:160–61fn. 23). Using this average measure, the Franco-Prussian War (because of its short duration) is more severe than the Napoleonic Wars. Still, many might claim, as I do, that total battle deaths in the Napoleonic Wars, which according to Levy (1983a:90) are 1,869,000 compared to 180,000 for the Franco-Prussian War, make the Napoleonic Wars more severe and totalist. Therefore I would not include the Franco-Prussian War under Organski and Kugler's operational criteria.

4. If differences in capability do not affect the other major states, what does? Apparently, they are drawn into wars when there is a tightening alliance structure in the system (Organski and Kugler 1980:55–56).

5. Note that Thompson (herein) rightly discounts the transition around 1570 because it reflects a temporary decline in the challenger and is quickly followed by a rapid decline in Portugal.

6. Because Organski and Kugler (1980) require that a war have a greater number of deaths than the previous war, neither the Dutch War of Louis XIV (1672–78) nor the War of the Grand Alliance (1688–97) would be included under their criteria. In terms of total battle deaths, neither the Dutch War of Louis XIV (342,000) nor the War of the Grand Alliance (War of the League of Augsburg) (680,000) exceeds the Thirty Years' War. If the latter is divided into four wars, as Levy (1983a:67–68) does, then the War of Spanish Succession (1701–13, 1,251,000) exceeds the most severe period of the Thirty Years' War—that is, the Swedish-French War (1635–48, 1,151,000). Presumably this is the war Organski and Kugler's criteria would pick. All data are taken from Levy (1983a:table 4.1).

7. Thompson's (herein) contingency tables show that generally about half of the wars are preceded by power transitions and half are not. For example, in the regional-global power transitions, five wars occur without a transition and four occur with one. Even when the Spain-Portugal dyad is dropped (which raises the Cramer's V from .262 to .522), there are still four wars occurring without a transition and four with one (Thompson herein: table 10.1).

8. Kugler and Organski (1989b:181) list the percentage distribution of GNP for Russia/USSR compared to Germany as: 17.1 versus 16.4 in 1900, 16.9 versus 17.0 in 1913 [note the power transition for World War I], 15.1 versus 12.2 in 1925, and 20.8 versus 15.0 in 1938. According to their note a in table 7.2, they select these years of varying length (including 1938) "to avoid, as much as possible, the direct effect of mobilization for World War I and II . . . ."

Wayman (herein), using the combined capabilities index of the Correlates of War project, shows a slight dip in Soviet capability below Germany in 1925 and then an abrupt increase in Soviet capability by 1930 that Germany does not surpass again while it is a major power (see also Geller, herein). Only if this slight and temporary transition, which is due to World War I, is counted is there a necessary transition.

Organski (1992:table 3) uses a more refined index of national capabilities that discounts GNP in terms of political capability to extract resources. When this is done, then the percent distribution of *national capabilities* for Russia/USSR compared to Germany are: 10 versus 29 in 1900, 20.4 versus 17.7 for 1913, and 19.6 versus 23.1 for 1938. Here Germany does surpass Russia before World War II, but note that with this measure Russia surpasses Germany prior to World War I, which is not consistent with Germany's ultimatum and subsequent declaration of war on Russia in 1914. This measure then "saves" the 1941 attack, but at the price of 1914.

9. This measure, probably because of its emphasis on population and the size of the standing armies, produces certain "power transitions" that Organski and Kugler and most others would not accept as valid, such as China overtaking the United States in 1960 and the USSR overtaking the United States in 1970 (see Houweling and Siccama 1988a:98 for the data).

10. One possible factor would be, as Doran (1989, 1991) argues, a systemic crisis in which many major states experience a shift in their capabilities. Houweling and Siccama (1991, herein) test for this possibility by examining the effect of passing through Doran's critical points and having a power transition. They find that power transition dyads are much more apt to go to war than nontransition dyads if one of the major states in the dyad is also passing through a critical point. This combination, they argue, is a sufficient condition for war. Overall the critical points appear to be a more potent predictor than the transition.

The main problem with the Houweling and Siccama (1991) study is that its findings are based on measures that identify critical points and transitions that lack face validity. For example, three of the twelve cases supporting the claim that transitions interacting with critical points are a sufficient condition of war involve China passing the United States, the USSR, and the United Kingdom after 1945 and the first three states passing through a critical point. If these cases were treated as dyads without critical points or transitions (as many would argue they should be treated), then the critical points would not be a necessary condition. A second problem is that the number of cases for a given war, especially for the two world wars and the Korean War, inflate the number of cases supporting the hypothesis. Finally, the lack of specificity, for example, linking a specific transition and critical point with a specific enemy, make the predictions too easy, especially given the twenty-year time lag. For these reasons, I have not been willing to place too much emphasis on this aspect of Houweling and Siccama's (1991) study, although I do think that Doran's (1989, 1991) general thesis is worth pursuing. In many ways, Geller (1992a) incorporates some of Houweling and Siccama's systemic concerns with fewer measurement problems.

11. Because Kim looks at only certain wars and not all wars, his periods of "no war" would include a host of interstate wars that do not meet Organski and Kugler's criteria. For example, from 1815–1975, Kim (1992, herein) looks at four wars, but there are sixty-two wars that meet Small and Singer's (1982:59) operational criteria of an inter-

state war during this time period. Obviously, periods without one of Organski and Kugler's wars, or even without any major-state wars, could not be construed as a period of peace or "no war."

12. Kim and Morrow (1992:917–18) and Morrow (herein) conclude with some suggestions on how to construct an explanation of big wars. One of the problems of converting Kim and Morrow's analyses into a complete explanation of big wars is that their findings are based on a variety of measures of alliance data. For example, both the indicators of dissatisfaction and risk propensity are based on the manipulation of alliance data. A number of studies have shown that alliances and polarization are correlated with the expansion of war and with world wars (Vasquez 1993:chap. 7). It is likely that the dissatisfaction and/or risk-propensity measures are correlated with these other variables, and if they are, then it will be difficult to determine whether it is these more individual propensities or the practice and structure of alliances that bring about big wars. So, overall, it would be better to have more direct measures of dissatisfaction and risk propensity.

13. Whether there is more to the relationship between preponderance and the absence of war will not be solved until scholars construct an explanation of how power transition, capability, and alliances are related to one another and to bringing about big wars. In another analysis (Vasquez 1993:chap. 7), I try to disentangle the relationship among capability, alliances, and polarization and the onset of world war by presenting evidence to show that there are three necessary conditions of world war: (1) a multipolar distribution of capability in the system, (2) an alliance system that reduces this multipolarity to two hostile blocs, and (3) the creation of two blocs in which one side does not have a clear preponderance of capability over the other.

14. Wallensteen (1981:83) finds that 85 percent of all major-state pairs from 1816–1976 have at least one military confrontation and 55 percent have wars. This means that power transition pairs (as a group) are 6 percent above the "normal" level of militarized disputes and 28 percent below the "normal" level of wars.

15. This raises the possibility that the very idea of power politics and its practices may be derived from an inability to settle territorial questions. Power politics is not the key fact of existence, as the realist paradigm would have us believe, but may simply be an epiphenomenon of territoriality. Realism and the practices of power politics come out of a particular set of struggles and construct a world appropriate to those struggles. Once those struggles end, however, the ideas are less relevant and even counterproductive. The great mistake of realism has been to assume that a struggle for power is a constant verity of history, when in fact it is most characteristic only of [contention on] one kind of issue—territory (see Vasquez 1993:147–48).

CHAPTER 3

The Power Transition: Problems and Prospects

Randolph M. Siverson and Ross A. Miller

This chapter has three parts. First, we survey several promising, indeed, highly attractive features of the power transition theory, including its plausibility and political content. Second, we enumerate and discuss some problems in extant research, focusing mainly on three individual projects, and conclude that, the identified problems notwithstanding, there are some good reasons for remaining interested in the power transition and that the main elements in the theory can perhaps add significantly to other research programs. We close with some suggestions for improving the testing and the theory of the power transition.

Power and War: The Power Transition as a Dynamic Theory

A notable debate in international politics is over the relationship between power and the outbreak of international war (Siverson and Sullivan 1983). One view in this debate is found in that venerable theory generally known as the balance of power, where it is asserted that wars are least likely to occur when power is distributed equally among states. With equal power the outcome of a war is most likely to be in doubt; the high cost of engaging a state with significant capability and the dim prospect of victory provides sufficient disincentives so that peace is preserved. Under such a circumstance rational decision makers will be unwilling to fight. This outcome does not necessarily emerge from the conscious choices of policymakers. While it is true that some early observers of the balance of power saw it as a system that was self-consciously intended to produce peace, more contemporary observers have tended to see the absence of war as the consequences of individual states pursuing their interests. Thus, Claude (1962:18) attributes the peace of the balance of power "to promoting the creation or preservation of equilibrium" by states that act "upon the principle that unbalance of power is dangerous." Gulick (1955:30) is slightly more specific when he says that the basic aim "of the balance of power was to insure the survival of independent states," but continues and asserts that peace is one of the by-products of "equilibrist" policies.

Power transition theory provides a challenge to this point of view that is original and insightful in its statements about the organization of the international system. The position of power transition theory is that if one nation is dominant in power it does not need to resort to war with other nations, since it can get what it wants through asymmetric bargaining. However, if it has a conflict with another state that is about equal in power, its leaders may see no need of deferring to it, and they may reasonably believe that they have a good chance of winning a war. War, in this view, takes place between equals because each side has the expectation that it can win.

Those are the bald essentials of the positions taken by each. *The War Ledger* (Organski and Kugler 1980) provides a summary of each model and a comparison. They are far apart in their explanations, since they make opposite predictions about the effects of power equality. There is an additional feature of the theories that separates them that does not appear in the above account and is apparently barely recognized by Organski and Kugler, but is a major advantage of power transition theory. Put simply, the balance of power is a theory of *statics,* while the power transition is a theory of *dynamics.* Broadly speaking, balance-of-power theory assumes into existence a set of equal powers and then plays out how they will behave given that distribution of power in an anarchic system (Waltz 1979). Given the conditions assumed by the balance of power, a snapshot taken at any instant may produce a single picture more or less in accord with the equilibrium that follows from the model. This is not to say that all theorists describe the balance of power in totally static terms, for some do recognize the possibility of change (Niou, Ordeshook, and Rose 1989, for instance). The change, however, generally takes place within a narrow range or has all nations changing at rates so much alike that dynamic aspects of the system do not emerge. Consider Kaplan's (1957) determined effort to improve the rigor of balance-of-power theory. Chief among his rules is that all essential actors (i.e., major powers) will seek to increase their capabilities. Kaplan, however, does not include in his analysis the possibility that this rate could be differential across actors. This leads him into something of a quandary, for his models are so static that he finds it necessary to construct an ad hoc explanation of how change could ever take place from a balance-of-power system to bipolarity or how the balance-of-power system could become unstable. No part of the explanation he advances derives from his model or recognizes the possibility of differential internal dynamics of growth (Kaplan 1957: 27).[1]

If the balance of power approximates a snapshot of international politics, the power transition is a motion picture. Like several other theories of the international system (Choucri and North 1975; Doran 1983b; Thompson 1988) the power transition sees the international system as unceasingly and, more important, unevenly dynamic. Thus, across states the properties that combine to produce power are seen as changing at an uneven rate, and, in fact, some may not be growing at all in either absolute or relative terms. This, of course, is not a novel approach. It can certainly be traced back to Hobson [1902] (1965), his follower Lenin [1917] (1975), and perhaps to Adam Smith [1776](1981). Whatever its origins, there is no denying that this description of international relations is far more faithful to the unfolding of history than is the balance of power. To be sure, policymakers may react to any one situation of conflict as if it were static, but the process that brought them there was dynamic and took place over time. Consider, for example, the account of the coming of World War I given by Choucri and North (1975) in which the war is the outcome of a forty- or fifty-year process abetted by innumerable uncoordinated decisions.

A second advantage of the power transition over the balance of power is less tangible. International politics is often described as anarchical, and inasmuch as it generally lacks institutions able to enforce binding agreements this is correct. But if it is anarchy it is decidedly not chaos, since its centripetal tendencies are tempered by a healthy dose of order. Neorealists agree that order is present. Waltz, for example, says that in international politics:

> Patterns emerge and endure without anyone arranging the parts to form patterns or striving to maintain them. . . . Order may prevail without an orderer; adjustments may be made without an adjuster; tasks may be allocated without an allocator. (1979:76)

Power transition, too, believes there is order, but it exists because the most powerful state orders, adjusts, and allocates. To Organski, the international order is not a passive state of existence that follows from a seemingly hidden hand, as in some versions of the balance of power or neorealism, but rather it is the manifestation of the effective political regime of the system. This order is directed by the most powerful state in the system, perhaps with the help of a few major-power allies, who themselves also share in the benefits. Organski is not overly explicit about what these benefits are, but from his discussion of the international order (1968:351–54), it is not too unreasonable to infer that they would include security, access to resources, trade advantages, and even something so simple as deference from other states.

Not all states share in the distribution of these benefits, and, more important, not even all the powerful states share in them since not all states will necessarily share the interests of the most powerful state. Among the *most* powerful, some who do not share in the distribution of the benefits constitute the dissatisfied states. As such they often try to increase their benefits from international politics by making distributional demands upon the satisfied states. If the relative power of the dissatisfied state(s) increases, the demands will eventually lead the most powerful state and its coalition partners toward making a choice between accommodation and confrontation. If adopted, the latter policies have obvious consequences for increasing the probability of war.

This is not the place to detail why states make the strategic choices they do.[2] However, it is worthwhile to observe that this picture of international politics suggests that leaders are motivated toward policy choices that go beyond the lexical preference for security present in virtually every realpolitik or balance-of-power model of international politics. Leaders, wishing to maintain themselves in power, are motivated to seek benefits for their state that very often go beyond security maximization. Policy failures can thus result in governmental crises and the removal of the leader from power (Bueno de Mesquita and Siverson 1993). In this respect it is instructive to reconsider some of the "diplomatic" history of Europe. Taylor's (1954) justly esteemed *Struggle for Mastery in Europe* can thus be read not as a search for security by the powerful European states, but rather as a story of ever-shifting coalitions seeking benefits and advantages of various kinds over other states. The results of these benefits were then used to sustain claims on further tenure in power. International politics is about *politics* and not just the search for some balanced state of affairs.[3]

The Power Transition: An Evaluation of Theory and Evidence

In this section we evaluate the theoretical status of the power transition and the adequacy of the evidence used to evaluate the theory. Our attention is directed at three particular studies: (1) the original statement by Organski (1958) with its scion (Organski and Kugler 1980); (2) a subsequent, more broadly based follow-up by Houweling and Siccama (1988a); and (3) a recent article by Kim and Morrow (1992).[4] We evaluate each of these efforts in terms of its theoretical coherence and empirical relevance. We believe this analysis will show that whatever its advantages, the power transition is not without problems, some that are fairly substantial.

The Early Work

We begin by evaluating some critical aspects of the theoretical status of the power transition. Theories may be evaluated in several ways, but chief among these are their domain and their degree of specification (Reynolds 1969). By the domain of the theory we mean the breadth of the class of activity it purports to cover. Power transition theory does not have as its domain all wars between states. Rather, it has generally been limited to explaining the big, system-transforming wars that are asserted to establish the central political status of the state that possesses sufficient power not just to defeat others, but to take on the task of organizing the system, even loosely, and allocating values and benefits. Lesser wars and conflicts are not of relevance to the theory of the power transition.[5] Although it is clear that this class of system-reorganizing wars is substantively important, a theory useful for explaining only them will not be, other things equal, as valuable as more general theories. For example, several other theories that include central hypotheses about the power cycles of major power states within the international system have some common features with power transition theory, but have more general domains inasmuch as the type and number of wars they explain are broader and more numerous (Doran and Parsons 1980; Modelski and Thompson 1989). Still more generally, the domain of all these theories of systemic war is significantly more limited than expected utility, which offers an explanation not just for major-power wars but for the necessary conditions for wars of all types (Bueno de Mesquita 1981a). Thus, when seen only as a theory of war onset, the power transition's scope is more limited than its plausible alternatives.

Is the dynamic responsible for the onset of the war portrayed in such a way that the model provides a specification sufficient to identify data consistent with the model? In the first version of the power transition, Organski gives a description for the motivation of war onset that is driven by rising powers who are dissatisfied with the organization of the international system because they do not share in the distribution of benefits (1968:366–67). Although in the later version of the power transition Organski and Kugler state that the power transition "provides no general rule to explain and predict the circumstances in which elites move toward war," they continue on to state that it is "a general dissatisfaction with its position in the system, and a desire to redraft the rules by which relations among nations work, that move a country to begin a major war" (1980:22–23). Determining ex ante which states are dissatisfied is a matter not touched upon by either Organski or Organski and Kugler (Bueno de Mesquita 1990b). This is unfortunate since, given the content of the theory, the relative potencies of changing power and dissatisfaction require specification. As the gap between the rising nation and the main contender grows narrower, does dissatisfaction become greater? The need for better specification of dissatisfaction in the theory is clearer when it is noted that the transition of power from Great Britain to the United States takes place without a war, and this is described purely in the ex post facto terms that Britain did not object to the change. We return to this point below.

We now turn to the correspondence between the theory and data. Since Organski's (1958) original statement of power transition did not contain a test, here we confine ourselves to discussing the testing of the theory as reported by Organski and Kugler (1980:28–63). Their test is restricted, in accord with their theoretical position, to the states at the top of the power pyramid. These states constitute the major powers and also contain a subclass of the most powerful states, termed the contenders. The entire class of major powers is identified by measuring power, quite reasonably, through an aggregation of national productivity. Using the presence or absence of alliance involvement as an indicator, they also draw a distinction between states at the center and the periphery of the international system. This distinction removes the United States, Japan, and China from significant parts of the analysis, although their productive capacities qualified them otherwise. Not all wars are included, nor are all wars included that involve the major powers. For a war to be included it must be a serious war, one that involves (1) major powers on both sides, (2) the possible loss of territory, and (3) battle deaths larger than in any previous war. The application of these criteria produces a set of wars that consists of the Franco-Prussian War, the Russo-Japanese War (see also Lemke and Kugler herein), World War I, and World War II.

To circumvent the problem of testing the theory against only these wars, Organski and Kugler decomposed them into all the pairs of nations on their list of central system major powers for the appropriate period.[6] The power relationship for the dyad was then traced for the twenty-year period prior to the war to discover whether the state that trailed at the beginning of the period had closed the gap. The measure of power was the ratio of the national capability of the less powerful state at the beginning of the period to that of the more powerful state. A ratio that was greater than 80 percent was taken as an indicator of equality and ratios less than that indicated inequality. Further, a state was deemed to have passed another when the state that was "less powerful at the beginning grew more powerful than the other member before the end of the period" (1980:49). Following these procedures produces an analysis that fits the power transition only when the data for the contenders are separated from other states and the direction and "velocity" of power relationships, or, more specifically, transitions, is taken into account (1980:52, table 1.7). As Organski and Kugler summarize: "if conflicts occur *among contenders*, they do so only if one of the contenders is in the process of passing the other" (1980:51; emphasis in original).

The fit between the theory and the data, however, is less impressive if we note the fact that the five cases providing the evidence in support of the theory are all dyads involving Germany and its opponents in the two world wars and the Franco-Prussian War (1980:59, fig. 1.2). This is a fairly limited segment of international politics, although, as Organski and Kugler might fairly protest, it is a centrally important part. That notwithstanding, we should expect a *theory* of international war to have considerably more explanatory power than appears in their tables, even when recognition is given to the admittedly limited domain of the theory's coverage.

In addition, there are two other problems. The first relates to two aspects of ex post facto research design, while the second concerns possible unrecognized influences of diffusion. Organski and Kugler's tests have two problems in this respect. The selection of the test periods is made on the basis of the fact that a war was known to have occurred. From the date of the war onset the capability data are traced backward to examine the ratios across time. Selection bias is present in the data, since across time not all major-power or contending pairs have an equal chance of being in the data. Put differently, the major powers are known because history is known, or, alternatively, the states in the population are selected because what happened later is known. Thus the decline of France from its position of a contender can be recognized on the basis of knowledge that appeared later. The same is true of Great Britain as well as of the rise of the United States, China, and Japan. The ex ante procedure that should be followed is to use *all* states in, say, 1815, not just those who become contenders and major powers, and track each pair looking for transitions and the attendant opportunities for war. As Houweling and Siccama describe it, "Organski and Kugler's argument runs from power transitions to the outbreak of war, but their test procedure runs from the outbreak of war to the preceding transition of power" (1988a:95) [7] (see also Lemke and Kugler herein).

Second, we note that some of the contending dyads that are reported as fighting are not exactly independent events. There is a connection between the Germany-United Kingdom dyad in 1939 and the Germany-France dyad in that same year. The connection is also close between these same pairs in 1914. We shall return to this point more than once.

The Houweling-Siccama Follow-up

In their effort to improve on the work of Organski and Kugler, Houweling and Siccama state that all power transitions should be identified and it should then be determined "whether power passings coincide in time with war outbreaks" (1988a:95). In their own research they attempt to set things right by following this prescription, but perhaps without total success.[8] At the outset we need to point out that they do make three important improvements: they (1) broaden the indicators of power, (2) drop the distinction between major powers and peripheral states, and (3) take power measurements every five years between 1816 and 1975. All of these sharpen the precision of their test. They establish their test periods as twenty-year spans beginning with 1816–35, then 1836–55, and so on. Identifying any war in which at least one major power was on each side, they record all warring dyads, which they may then compare with respect to power transitions to all possible major-power dyads in the test periods. Using three levels of power differences across the dyads (5, 10, and 20 percent), under circumstances of unequal power or equal power with overtaking and no overtaking, they produce quite robust results in favor of the power transition.

Despite these robust results, the work of Houweling and Siccama is not without its problems. To be sure, one advantage is that they work forward in time with respect to the relationship between power transitions and war outbreak. However, the major powers are again introduced on an ex post facto basis. That is, the major powers are selected on the basis of what we now know. For example, Spain is not included in the data set, presumably because we know that state had declined in power and would not rise again. By the same token, other states are included. This is not without its consequences, for when Japan appears in the power data (with slightly more than 5 percent of the power in 1895) and is shortly followed by the United States (with 22 percent in 1900!) the data on the other nations are wrenched about fairly sharply; 27 percent of the data are reallocated over a five-year period simply because two nations "enter" the major-power group. How many transitions involving these two nations were missed because they were not included over the entire period? The United States, because it enters the system with more power than any other state, single-handedly produces what appears to be seven power transitions, two of which ultimately lead to war.[9]

The unrecognized problem of war diffusion also creeps into the Houweling and Siccama data, and they are not aware of the theoretical and methodological problems this causes. Diffusion enters the research design in two ways. First, to generate their war data set, Houweling and Siccama identify among their 119 dyads 33 who actually fought each other. Many of these cases of warring dyads are not strictly independent of each other, and, consequently, war diffusion is included in the data (Siverson and Starr 1991). For example, the fact that Great Britain is fighting Germany in 1939 is not unrelated to the fact that France is at war with Germany at the same time. We do not suggest that war diffusion is necessarily the only process operating here, although it might be. Minimally, we do suggest that the research design could be recast quite easily to estimate the extent to which power transitions add to the probability of becoming involved in a war both at its initial outbreak or after. No listing of the wars is given, so the exact magnitude of the problem is difficult to gauge.

A second problem is related to the first. Houweling and Siccama use only those wars in which their identified major powers participated on different sides. Again, this is based upon ex post facto knowledge. There were wars during all these periods in which one major power fought with a minor power and no other major power became involved on behalf of the minor power. Put simply, the power transition theory also makes statements about when wars will not take place and major powers refusing to help minor powers under attack can indeed be evidence used to evaluate the theory. Wars that do *not* diffuse are significant. We return to this point in the next section.

We also note that the test periods are twenty years. As everyone will admit, twenty years is an arbitrary selection. There is nothing obviously wrong with it, and it does furnish a lengthy enough period for the effects being investigated to manifest themselves. At the same time, there is nothing to prevent investigators from retesting the same relationships on periods of, say, fifteen, twenty-five, or thirty years to explore the robustness of the results. We noted above that Houweling and Siccama followed this strategy with respect to varying the amount of power difference across the dyads with positive results that strengthen their analysis (see also Lemke and Kugler herein).

Although we think the problems identified in Houweling and Siccama's work indicate that the research should be revisited, it does represent a significant improvement over the original analysis. The model is better specified, the data are considerably broader, and the results are more robust.

The Kim-Morrow Innovations

Finally, we come to the most recent effort to use the theory of the power transition. To improve research, Kim and Morrow (1992) (1) use a rational choice framework to propose a far more rigorous theory than has previously been used, and (2) undertake a more general test than those reported earlier.

One major step forward in this work is the degree of theoretical specification they present. While Organski and Kugler may have been less than exact about some aspects of their theory, and Houweling and Siccama made some refinements that constrained the predictions of the theory, Kim and Morrow endow their work with a degree of theoretical specification that is a huge step beyond the earlier work. The theory is so well developed, and the concepts so well related to one another, that Kim and Morrow present a model in which they hypothesize the *relative magnitudes* of several variables in a logit model in which war/no war is the dependent variable. We will not go through the derivation of the hypotheses Kim and Morrow propose, but we will say that their conclusions are highly plausible and the method of their generation appears to be appropriate and carefully done. In essence, they propose a set of relationships in which the chances of war are increased by:

1. the existence of risk-acceptant rising states and risk-averse declining states,
2. the level of dissatisfaction with the status quo by the rising state,
3. low costs of war, and
4. rough parity in resources of the two states.

In order to test this model Kim and Morrow use all the conventionally identified major powers. Beginning with 1820, they construct a series of test periods of twenty years each and look at all major-power dyads over those periods; this effort produces 156 dyads to examine for possible wars.[10] They choose the wars of relevance on the basis of a major power being on each side of the war and the seriousness of the fighting as indicated either by battle deaths or by the loss of territory. This results in the selection of the following wars: Crimean, Italian Unification, Seven Weeks', Franco-Prussian, Russo-Japanese,

World War I, and World War II. Across these wars, Kim and Morrow claim that 31 of the 115 dyad periods (almost 27 percent) ended in war.[11]

Despite the considerable care taken with the theoretical development of this work, we are troubled, once again, by ex post facto aspects of the research design and a modest indifference to effects of diffusion; in some ways these problems are connected.[12] We begin with a consideration of the ex post facto problem.

As we have argued above, one of the difficulties of doing research on international politics with historical data is the simple fact that history has taken place. This means that the investigator is aware of what happened. (It also means that there may have been unsurprising disruptions in the data.) This can lead to innocent, well-intentioned decisions that have subtle effects on the results obtained from the data. It is possible that these factors influenced the results of Organski and Kugler, Houweling and Siccama, and Kim and Morrow. We suggest that using a list of major powers that is derived ex post facto may have an impact on the results, as may the particular twenty-year test periods that are selected with no theoretical justification whatever. That is, we can accept the necessity of looking at power dynamics over twenty years, but why these twenty-year periods? Why do they start in 1820? Would the results obtained look different if the time periods had begun in 1816? We do not know the answer to this question, but we do know that at least two of the time periods would not have ended with the outbreak of the two world wars. The selection of the major powers is similarly determined by ex post facto considerations. Kim and Morrow, like most others, including one of the writers of this chapter, draw upon the list of major powers produced by the Correlates of War (COW) project. In essence, this list represents the consensus of historians about who was important in European politics (Small and Singer 1982:45). Thus, the United States fought Great Britain to a tie in the War of 1812, sent a squadron to the Mediterranean in 1801 to suppress the Barbary pirates (who were protected by the Dey of Algiers[13]), forced Japan to open foreign trade in 1855, forestalled foreign intervention in the Civil War, ordered the French from Mexico, and, since roughly 1880, had the world's largest industrial output, but did not become a major power until roughly 1900. To be sure, the United States did not actively enter into the cut and thrust of European international politics, but it had no necessity of doing so as long as it could preserve its autonomy while simultaneously getting what it wanted in the way of security and resources.

What is the justification given for the entry of the United States into the class of major powers in 1900? The reason usually given is the defeat of Spain in the Spanish-American War, just as the entry of Japan is explained in terms of its defeat of Russia in the Russo-Japanese War. Using this method, however, involves a modest sort of circularity with respect to the identification of a state as a major power. That is, states are identified as major powers at $t0$ because they act like major powers at $t + 10$, 15, or 20, not because of their capabilities and involvement at $t0$. We suspect that the United States became a major power in 1895 or 1900, depending upon the data set, not because it defeated Spain in 1898 (hardly a major power), but because the United States entered World War I in 1917! This is teleological reasoning unwittingly taken in as causal reasoning.[14] The consequence of such sudden assignments as Japan and the United States to the major powers in 1900 is that, as we have seen above, this is likely to produce drastic changes in the allocation of power among the states, including their rates of change. It might be argued that the United States and Japan fought their way into the major power class, thus affecting the perceptions of the other major powers, but the theoretical specification given does not adequately take such sudden changes into account. None of the theories specifies how a state enters the major power group.[15]

A larger problem is that the analysis rests on a data set that excludes all the wars that did not happen. This statement will certainly strike most readers as odd. Recall that Kim and Morrow include a war only if there is a major power on both sides. If a war takes place between a major power and a minor power with no major power on the side of the minor power, the war is not included. Relying on the selection of wars that involve major powers on both sides is, once again, an ex post facto decision. When one major power begins fighting with anyone, the opportunity exists for another major power to join in. Do they do so? If not, did the conditions of the model suggest that they should or should not have come into the war? Recall that World War II began as the German-Polish War. The British and French decided to join, thus causing the case to be selected. Not including all wars in which a major power participated discards information that could be used for a more rigorous test of the model. Consider, for example, the changes in the relationship between France and Austria-Hungary between 1848 and 1859. In 1848 Austria-Hungary by itself defeated Sardinia, Tuscany, and Modena. In 1849 France joined Austria-Hungary to defeat the Papal States, but by 1859 France joined Sardinia in defeating Austria-Hungary. What does the theory of Kim and Morrow tell us about these gyrations on the part of France with respect to its standing against Austria-Hungary?

There is another aspect of this issue. Data on international disputes are now widely available. Presumably rising dissatisfied states can raise issues that can become disputes. Kim and Morrow recognize this in their "Extensive Form of Challenge" (1992:fig.1). Their game represents a situation in which a rising state makes a demand that can be met by the declining state or resisted, with the former choice being a concession and the latter being a demand. However, historically speaking, other outcomes are possible. Perhaps the declining state resists and the rising state backs down. No such outcome is possible in their game. Hence, disputes are not of interest. In our view a fuller test of the model would require a data set including disputes that escalate to war and that do not.

We now turn again to the question of contagion. Whatever difficulties we may have had reconstructing the data set used by Kim and Morrow, approximately twenty-four of the thirty-one warring dyads (77.4 percent) they record were the major-power participants in World Wars I and II. Kim and Morrow recognize that contagion took place; that is, the onset of a war between two major powers increased the chances that another major-power warring dyad would emerge. This recognition significantly informs their conclusions, but its impact on the reported analysis is unclear.[16] The problem, as pointed out by Siverson and Starr (1991), is that such procedures conflate the onset of war and the spread of war once fighting begins. This goes beyond the problem of abstract statistical dependence noted by Kim and Morrow as a process in which warring dyads activate other warring dyads, thus triggering wars that otherwise might not have happened. Under such a condition it is likely that the "extra" cases could have worked against the Kim and Morrow model. Consider, for example, the fact that both Houweling and Siccama and Kim and Morrow include the case of the war between Japan and the United States that began in 1941. What does this have to do with any power transition? In the Japanese view they were fighting the United States over the issue of Japanese supremacy in East Asia, but their power was so much less than that of the United States (Japan had about 41 percent of the power of the United States according to COW capability data, and it was not much better when adjustments are made for distance) that there was no transition in prospect. The case is still included but, given Japan's ongoing war in China, should it be? Interestingly, Kim and Morrow are keenly aware of this issue in their conclusions, where they devote considerable attention to the question of war growth and the possible effects of alliances on the growth of war. But having done that they do not go back and reanalyze their data using just the initial participants in a war. Such a procedure may have shown something different.

An Assessment

Despite our reservations about some aspects of the hypothesis testing, we remain firm in our conviction that the theoretical development in the power transition theory made by Kim and Morrow is a major step forward. Previous inconsistencies are removed, the political substance of the theory is enhanced, the domain of the theory is enlarged (now going beyond the contenders to all the major powers), and the specification of the theory's content is considerably tightened. These are no small achievements.

Thus far the theory of the power transition has fared unevenly. While we think the theory makes a good deal of sense (i.e., has high face validity) and provides a political framework, its overall content and empirical relevance leave something to be desired. Although the balance sheet is a bit mixed, there are two observations we wish to note. First, the theory of the power transition can still be useful if there is no better theory. Second, even if there is a better theory, it may be possible to use the insights of the power transition to amplify a superior theory so that its domain is larger without a substantial reduction in parsimony.

Is there a better theory? The two most plausible rivals are balance of power and expected utility.[17] It is not clear that the logic of the balance of power is superior to the power transition. In fact, the theoretical apparatus needed for the balance of power is considerably more complex than the power transition. Nor is the evidence in support of the balance of power stunning. In fact, one review of the evidence on balance of power and power transition was inconclusive with respect to both (Siverson and Sullivan 1983).

In the last ten years expected utility theory has quickly emerged as the most comprehensive, rigorously formulated, and elaborately tested theory in the field of international politics. Whatever its advantages, it has been a more or less static theory, not making much allowance for the dynamic structure of the relationships of the international system. That is, at any one point it could identify friends and foes and make predictions about the outcomes of conflict situations, but it was not able to make any statement about what might happen in the next year. What is known is well dealt with, but the theory is agnostic with respect to the direction of political tendency in the world. Using the concepts and methods advanced by Kim and Morrow appears to move in the direction of overcoming those shortcomings of expected utility theory. For example, it is possible that the approach outlined by Kim and Morrow could be used to construct (with acknowledgment to Richard Nixon for the concept) an enemies list; that is, a list of which nations are most likely to emerge as enemies of one another in the near future. Further, it moves power transition theory away from being a special case of expected utility theory and thereby makes it more useful.

A Future for the Power Transition?

In this final section we discuss two considerations that we believe would improve the testing of the power transition and put forward one suggestion for future research. No matter how appealing a theory may be, it must stand the test of empirical relevance, of being confronted with data that are not biased in some way by their selection. We have commented above on this aspect of past tests, and it is incumbent upon us to propose an alternative method. We think a test of the power transition theory should be quite straightforward. Here we give the simplest version of what such a test might look like. First, use a set of all nations that might reasonably be considered to be major powers at any time in the period between 1816 and, say, 1985. Then, beginning in 1816, track each state's share of the international power distribution and use these percentages, together with the alliance configurations, to measure attitudes toward risk. As power transitions, or critical points, occur in the data and if risk attitudes are appropriate, it is relatively easy to determine if the nations in the transition fought each other within ten years of either side of the transition.

Second, we think the idea of dissatisfied states needs much more work. Earlier we commented on the fact that the idea of dissatisfied states was critical to Organski's theory of war onset. Organski and Kugler do not specify a meaning for dissatisfaction, nor do Houweling and Siccama. Kim and Morrow do. Their theory appropriately finds a key place for the satisfaction or dissatisfaction dimension, one that is very close to that originally articulated by Organski. We are, however, uncertain as to the validity of their measure. Essentially, they measure satisfaction or dissatisfaction as the utility between any given state and the policies of the dominant state. One state's utility for another is measured by the similarity of their alliance portfolios (Bueno de Mesquita 1981a). Although such a procedure is often very useful, particularly given the paucity of consistent international relations data over long time periods, in this instance it makes the United States into a dissatisfied state between roughly 1900 and 1949, when it entered NATO, since between those two dates the United States had no alliances with any of the European major powers. Because Japan does not have any alliances now, this measure would also make Japan a dissatisfied state today.[18]

As an alternative to this we propose that the data on disputes be used to measure satisfaction. Disputes are raised because a state wishes to change some aspect of its situation. Tracking a state's dispute record over some time horizon (which we do not specify) in relation to whether it is the initiator or target of the dispute and whether the dispute is with a nation that is stronger or weaker ought to give us some estimate of whether or not the state is satisfied.

For reasons too lengthy to pursue in detail here, we believe that dissatisfaction is far more important than the place accorded it in most of the empirical tests. Briefly, we assume that political leaders are interested, above all else, in staying in power. This may be for personal or political reasons or both. For whatever reason, they will act to maximize their opportunities for staying in power. Broadly speaking, all political leaders are subject to removal by their opponents inside or outside their states or both. Leaders are aware that they may be removed and within the constraints of their systems seek to remain in power; in fact, under some circumstances they may even seek to change the constraints of the systems in order to remain in power. Although their ability to remain in power will be the result of many factors, we believe that a major component is the ability of leaders to use foreign policy to produce benefits for the members of the political system. These benefits may be material or they may be related to the extent to which the state is secure against its external enemies. In essence, the necessity of producing benefits joins external and internal politics, since the leaders must pursue security in order to avoid exposing their societies, and thus their own political positions, to risk. This recognition, that rearranging some aspect of international politics to benefit the state, and thus the political leader, creates incentives for states to be dissatisfied (for a formal elaboration of this, see Bueno de Mesquita and Siverson 1993).

Finally, we urge that the ideas of the power transition be used to attempt to cast light on the future of large-scale international conflict. Over the last five years the world has changed in ways that could not have been imagined earlier. What does the future hold? Will there be an era of another "long peace" among the major powers? Alternatively, will some state become dissatisfied and stronger at the same time and make unacceptable demands on other states? One may be tempted to find comfort in the fact that all the major-power states, save China and what remains of the Soviet Union, are democratic, and even the latter is moving in the democratic direction, albeit perhaps at a slow and uncertain pace. One of the most interesting findings to emerge in recent research is the robustness of the absence of war between and among democratic political systems. Such a fact suggests that, in the future, we need to worry most about China and the nature of its political system. Unfortunately, no matter how much we wish all this to be true, it is not automatic that any or all political systems will retain their current, highly desirable characteristics in perpetuity. Consider the unpleasant facts that France has a rather mixed history with democracy and Japan was democratic for a period between the two world wars until ultranationalists transformed the political system and destroyed democracy. Germany, too, was democratic until domestic economic and political crises led them to a tragic choice. The experience of Russia with democracy is, to say the least, severely limited. Among major powers, only Britain and the United States have long, uninterrupted histories of democratic institutions. It is possible to hypothesize that regimes have life patterns much like those ascribed to political leaders by Bienen and van de Walle (1991). Specifically, the longer the institutions, arrangements, and rules stay in place, the longer they are likely to survive. We do not know the answers to the questions we have posed, but we think they are well worth asking in the context of the power transition.

NOTES

The authors gratefully acknowledge the support of the Institute for Global Conflict and Cooperation of the University of California.

1. At one place, Kaplan recognizes that the capabilities of a coalition could increase, but that if other states know of this they will compensate (1957:26).

2. See the emerging literature on grand strategy and the conditions that shape national choices (Rosecrance and Stein 1993).

3. It is not clear whether or not it is simply the level of abstraction that is achieved, but some accounts of the international system seem almost apolitical. Waltz's (1979) impressive account of how international relations works, for example, is so systemically determined that the political content of what is being sought by states often gets lost from view. Of course, the view put forward here (and in Bueno de Mesquita and Siverson 1993) is that domestic political considerations drive security policy; Waltz flatly denies this is a useful approach to theory, but in so doing he may drive out some of the politics.

4. We discuss these studies because at the time this chapter was originally written they were the best of the attempts to evaluate the power transition against data. For some earlier research that bears generally on the relationship between power distributions and the onset of war or conflict, see Weede 1976 and Garnham 1976a, 1976b. For a review of these, see Siverson and Sullivan 1983.

5. Some of the chapters in this volume move toward covering a broader set of wars, but the original scope of the theory was limited to what might be called system-transforming wars.

6. It should be noted that the list of contenders given by Organski and Kugler (1980:45) contains a typographical error in that Germany is listed as entering this class in 1890 when the actual date should be given as 1860.

7. For an excellent discussion of this problem in a general context, but with great relevance here, see Bueno de Mesquita 1990b.

8. The Houweling and Siccama paper (1988a) has a number of eccentricities, including gratuitously infelicitous language, tables that require some effort to comprehend, and an attack on the use of tests of significance with the kind of data employed just before tests of significance are used.

9. In a later follow-up on the power transition, Kugler and Organski (1989b) lay out the relative power in ten-year segments of the major powers between 1870 and 1970, including both the United States and Japan for the entire time period. This is certainly a step in the right direction. Unfortunately, neither Austria-Hungary nor China are included in the data. Although they do not attempt to use the data for theory-testing purposes, it draws our attention once again to the possibility that empirical results might be influenced by the set of nations chosen, the period covered, or both.

10. For unexplained reasons, the period immediately following the end of World War II is abbreviated to ten years (i.e., 1946–55).

11. There may be a problem with these data, since we are unable to reproduce thirty-one warring dyads from the list of wars and major powers given by Kim and Morrow; the best we can do is twenty-nine, unless we include "wars" between Japan and Austria in World War I and France and Japan in World War II. It is not clear that either of these cases should qualify for inclusion.

12. We also have a modest quibble with the reporting of the results. Logit does not readily lend itself to something like amount of variance explained. However, it is possible to compute the proportionate reduction in error from a cross tabulation of predicted and observed results. Instead Kim and Morrow report the percentage of cases classified correctly. This ranges between 73 and 75 percent. The same results could be obtained by placing all cases in the no war category, since the 84 dyads of no war in the 115 cases

(or 73 percent) would be correctly predicted. We hasten to point out that all the models fit the data at quite reasonable levels of significance.

13. In fact, the Dey declared war on the United States. Although the "war" does not make it into the Small and Singer list, it went on until 1805, when the United States forced the Dey to accede to its demands (Blainey 1988).

14. It matters less what historians and political scientists say after World War I about the power status of the United States before then than what was said at the time. Do contemporaneous accounts of United States policy describe the ascendancy of a new major power prior to, say, 1910?

15. In fairness it must be admitted that the failure to offer a theory of what constitutes a major power is not limited to the research discussed here.

16. However, in their footnote 3 they do comment on diffusion as making tests of significance problematic.

17. There are numerous theories of the onset of war. We compare these not because of any faults in the others, but because their domain is limited and their statements are of a different character. Thus, for example, Modelski's (1987) excellent work on long cycles makes statements about a broad span of time, but has relatively little to do with human choice. On the other hand, some of the interesting work on deterrence is not directly tied to the domain of war as opposed to the success/failure of deterrence (Huth and Russett 1984).

18. Japan has a security agreement with the United States, but because this agreement does not carry reciprocal arrangements it functions more as a protectorate and is not included in the COW set of alliance data.

Part 2
Empirical Tests and Extensions of the Power Transition

CHAPTER 4

Small States and War:
An Expansion of Power Transition Theory

Douglas Lemke

My goal is to reformulate power transition theory in such a way that both major- and minor-power wars can be considered. I argue that for both major and minor powers, when parity between countries is observed the necessary condition for war prevails; consequently, when there is an imbalance, or preponderance, of power the necessary condition for peace prevails. In essence, I argue that the difference between major- and minor-power war is one of degree, not of kind.

The frequent selection bias against minor-power war requires one to assume that such war is different in kind from major-power war (cf. Levy 1983a; Kugler 1990). The reason for doing so is that many claim that minor-power war is unimportant due to its smaller scale (cf. Organski and Kugler 1980:45; Levy 1983a:4). I explore this assumption in hope of challenging this claim. When one considers that roughly twenty-two million people have been killed in wars since World War II (Sivard 1989) and that all of these were minor-power wars, an exclusive focus on the conflict behavior of the major powers must be questioned. Minor-power wars are significant, and efforts must be made to construct theoretical frameworks that include both minor- and major-power wars.

A previous effort to account for major as well as minor-power wars is presented by Bueno de Mesquita, whose expected utility theory was designed to explain all wars (1981a:x). His framework does not differentiate wars by the size of the countries involved, and his tests include both major- and minor-power wars. However, the main drawback of Bueno de Mesquita's framework is that it cuts the international system in temporally static cross sections. This prevents consideration of a time framework of developing relations between countries. As a result, Bueno de Mesquita is unable to determine which dyads of countries deserve consideration and is forced to consider all regional dyads and all dyads containing major powers. Due to this broad view his expected utility model includes dyads that had no opportunity for war between 1820 and 1975. What is needed is a structural theory of war initiation that incorporates both major- and minor-power wars, but provides the researcher with clues as to where to look for wars to occur. This chapter is intended as an effort at the construction and test of such a theory.

I argue that power transitions (Organski 1958; Organski and Kugler 1980; Gilpin 1981; Houweling and Siccama 1988a), which alter power relationships and lead to parity between countries, precede wars among minor-power countries just as they precede wars among the major powers. I contend, unlike previous parity efforts, that this is so because the minor-power countries are members of isolated local hierarchies of the international system, just as the major powers are important members of the dominant hierarchy that characterizes international relations. In making this claim a multiple hierarchy perspective of power transition theory is developed.

Power Transition Theory

Organski's (1958) power transition theory, further developed and tested in Organski and Kugler 1980, has generated a considerable amount of interest over the years and fares quite well in empirical evaluations (Garnham 1976a,b; Weede 1976; Organski and Kugler 1980; Siverson and Sullivan 1983; Houweling and Siccama 1988a,b, 1991; Moul 1988; Kim 1989, 1991; Gochman 1990; for negative evaluations see Ferris 1973; Siverson and Tennefoss 1984). In spite of the increasing support for power transition arguments, there is a persistent problem with the theory.

This problem is that the theory considers wars fought for control of the international system, and thus only addresses the war behavior of countries at the very top of the international pyramid. This is so because the conflicts involve the very order of the international system, and only the very powerful can hope to affect that. In its present form, power transition theory is incapable of explaining the war participation of lesser countries, and is symptomatic of the selection bias complained of in the introductory section of this chapter.

One tantalizing result of empirical efforts to evaluate power transition theory has been the discovery of power overtakings prior to non-system-changing wars (Garnham 1976a,b; Organski and Kugler 1980:chap. 2; Houweling and Siccama 1988a; Kim 1989; Kugler and Arbetman 1989a; Bueno de Mesquita 1990a). Power transition proponents may point to these results as support for their theory, but if they do so they are not being true to the original argument. This is so because studies of overtakings in such dyads have not considered competition in a hierarchy over the distribution of benefits, which is argued to motivate international wars. My reformulation of power transition theory to include the multiple hierarchy perspective is intended to reconcile the theoretical demands of the theory with its application to minor wars.

The multiple hierarchy perspective, as the name suggests, posits that the international system is composed of a series of hierarchies, differentiated by power and distance. Within these hierarchies countries interact, sometimes peaceably and sometimes violently. Hierarchy means that there is differentiation between countries in the international system. Specifically, countries are differentiated by the amount of power they possess. The more powerful are located at the top of the hierarchy and are able to make demands and set rules that are heeded by countries with less power, located at the bottom of the hierarchy. Within each hierarchy we can test the claim that war is associated with power parity, for when one country is much more powerful than another, resistance by the weaker is futile. Conversely, when there is rough equality between the two countries—when the hierarchy has broken down—it is reasonable for the demanded country to resist the demands to which it had previously acquiesced. In this chapter the multiple hierarchy perspective is tested against power transition propositions alone; it is not tested against balance-of-power propositions, because the multiple hierarchies contradict the fundamental assumption of international anarchy.

Following Organski and Kugler, I posit that within the multiple hierarchies a given country will be interested in the relations and interactions of the hierarchy it is a member of, at least more so than it will be interested in relations between members of other hierarchies. It is within a specific hierarchy that a country is able to follow foreign policies and interact with others on a meaningful level, including the waging of war.

The Multiple Hierarchy Perspective

It may be easier to picture the international system that the multiple hierarchy perspective posits by considering figure 4.1, which depicts the international system as a three-dimensional cone. The entire cone represents the international system, with countries arranged throughout its area.

The international power cone comes to a point at the top. This point is the summit of power in the international system. It is the position of the most powerful country in the world, the dominant power. Movement downward from this point indicates decreasing power. The length across the cone, or left-right diameter, increases as one moves to lower levels, to represent the fact that as one moves lower in the cone there are more countries. The powerful are few, the weak are many.

The entire international power cone represents a hierarchy, specifically the dominant international hierarchy. This power relationship includes all countries and has the dominant power at the summit. In addition to the dominant international hierarchy, the multiple hierarchy perspective posits that there are "local hierarchies" within the overall international hierarchy that operate as international systems in miniature. These are conceived as segments of the surface

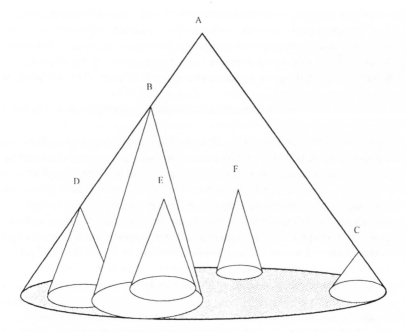

Fig. 4.1. The international power cone.

area of the international power cone (an example is the area beneath B in fig. 4.1).

These local hierarchies are diminutive parallels of the dominant international hierarchy in that there is a relationship of power dominance and subordination within each one. Each local hierarchy has its own dominant power that has established relations within the local hierarchy to its liking. When a dissatisfied member of a local hierarchy achieves parity with the local dominant power it has the opportunity to go to war to alter the local status quo. Thus, the multiple hierarchy perspective declares that minor powers fight wars to change the status quo in their specific local hierarchy and that power transitions make such wars possible by providing power parity, the necessary condition for war. Minor-power wars are thus parallel to major-power wars.

The dominant power of a local hierarchy is necessarily weaker than the overall dominant power that sits at the very top of the cone. As a result it can exert influence over a smaller section of the international system. This is represented by the third dimension of the cone: the increasing circumference as one moves to lower levels. Countries at lower, that is weaker, levels are less able to "throw their weight around." The increasing circumference represents the fact that weaker countries, unable to overcome distance, cannot exercise influence around the globe and restrict their politics to their local hierarchy. For example, while the United States is able to exert its influence around the globe, Brazil is restricted to South American interactions.

The local hierarchies consist of geographically proximate states. Geographic proximity is extremely important because power must be transported across borders (or at least to them) if it is to be of use for its possessor against another. The further apart these borders are, the more difficult it is to transport power, and the more of the same power that must be used for transport rather than influence (see the discussion of the loss-of-strength gradient in Boulding 1962). The point is that the local hierarchies really are local, comprising states that are able to challenge only each other, but unable to exercise resources in areas farther away from their borders.

There may be some interaction by countries in different local hierarchies such that a member country of B's local hierarchy (from fig. 4.1) might also be able to interact with a country from a different nearby local hierarchy. Graphically this is depicted by an overlap between B's local hierarchy and one of the neighboring areas. This complicates the picture, because it creates the issue of many subhierarchies.

I have resolved this issue operationally by recourse to the original theoretical basis driving this analysis. Recall that the motivation for transition wars is competition over control of a hierarchy. Minor powers, or small states, which are so weak as not to be able to influence the countries above them can only be victims, not initiators, of aggression. However, small is a very relative term. Compared to the United States, Argentina is a minor power. Compared to Brazil, Argentina is a contender. Likewise, compared to Argentina, Paraguay is a minor power, yet compared to Bolivia it is a contender. Therefore, I construct hierarchies from all potential interactions among countries that have the ability to reach each other's capital. Countries in the region unable to reach the capital of *any* other state are disregarded. However, if two actors who are excluded from such interactions themselves can reach each other, they create an even smaller local hierarchy. Thus, during the 1800s, Argentina, Brazil, and Uruguay composed a hierarchy excluding Paraguay. After recovering from the Lopez War, Paraguay grew to the point where it had the ability to threaten Bolivia, as did Bolivia against Paraguay after the War of the Pacific. Neither Bolivia nor Paraguay could challenge the countries in their respective and separate hierarchies, headed by Chile and Peru in Bolivia's case and by Argentina and Brazil in Paraguay's. Since Bolivia and Paraguay could only challenge each other, a local hierarchy, beneath those of the powerful countries on the coasts, is identified. This nested hierarchy structure allows full exploration of competition among major and among minor powers. Clearly, the smallest of the hierarchies is affected by all others above it, yet the lower hierarchy members cannot affect those above. The difference between this conceptualization and that of the original power transition theory is that the empirical domain is no longer re-stricted to the most powerful members of the international system.

Another type of overlap or interaction is also possible. The major powers are located at the upper regions of the international power cone. Many (if not all) of them are powerful enough to exert influence over the entire globe. As a result, they have the opportunity to interfere in local hierarchies. If this inter-ference is pervasive enough to determine, in some sense, the power relations of the members of the local hierarchy, then it is meaningless to talk of these mi-nor powers as fighting for control of their local hierarchies. There is no point in Vietnam challenging Cambodia for local control, for example, if the Chinese dictate international relations in Southeast Asia. In general one might argue that such major-power interference is likely to be inconstant, or to be cancelled out by the interference of other major powers (Milstein 1972), but it is of concern. The direction of interference is from the top to the bottom. Thus, leading mem-bers in the top of the nested hierarchies can affect members below and not the other way around. South America, used in this analysis, has the unusual char-acteristic of being a region only very marginally affected by the overall hierar-chy considered by Organski and Kugler or by local hierarchies in other regions. The results reported below attest to the validity of the power transition per-spective on conflict initiation. My claim to generality is based on the contention that leaders of minor powers within their local hierarchies act just as leaders of great powers in the overall international hierarchy do. In future work I will sub-stantiate this claim of generality and resolve this question of major-power in-terference.

In summary, the multiple hierarchy perspective on power transition the-ory posits that the international system is composed of a series of hierarchies, differentiated by power and distance. Within them countries interact, some-times peaceably, sometimes violently. Overall and local hierarchies are sug-gested, functionally very similar, yet largely independent of action. The simi-larity between the two types of hierarchies is that, in addition to having nearly identical structures, wars within them are fought when the same conditions pre-vail and for parallel reasons. Regardless of which type of hierarchy one is con-sidering, the necessary condition for war is the same. The local hierarchies are the result of the effect that distance has on the exercise of power—preventing many countries from having substantial relations with one another and certainly preventing some countries from fighting one another. Distance cuts off inter-action opportunities between many countries and prescribes who will signifi-cantly interact with whom. The conception of multiple hierarchies is useful for applications to international conflict but is also a general scheme that could prove useful in other areas, such as international political economy. It remains now to evaluate this theoretical reformulation empirically. The first step is to operationalize each of the variables.

Operationalizations

Local Hierarchies

It is a considerable task to define a subset of countries that affect one another in a way that one can speak of a status quo between them. For the purposes of power transition theory, recall that Organski posits that when countries fight, they fight for control of something. For the countries at the very top of his international hierarchy it is control of the international system. I suggest that for the minor countries the impetus for fighting is attempts to control the local hierarchical system, to establish relations within the local hierarchy to the liking of the challenger, or to preserve the local status quo for the local dominant power.

A great deal of work in international relations and comparative politics has attempted to divide the international system into meaningful subunits given one of the following names: spheres of influence, regions, subordinate state systems, subordinate international systems, international subsystems, geographic zones, regimes, regional subsystems, enduring rivalries, politically relevant neighborhoods, and clusters of nations. The criteria by which the various subunits have been constructed have included cultural similarities, trade patterns, common membership in international organizations, alliance patterns, demographic similarities, and, most often, geographic proximity. (For a discussion of why geographic proximity matters see Starr and Most 1976; Vasquez 1993.)

Another common method for cordoning off a subset of the international system has been to focus on a single area without addressing why it might be distinct. In this way Europe has been studied (Choucri and North 1975; Bueno de Mesquita 1990a), Asia has received independent attention (Brecher 1963; Hellman 1969; Weede 1976), the Middle East has been considered (Binder 1958; Lebovic 1986), and Africa has been set apart (Bowman 1968; Sigler 1969). Here geographic proximity is the key distinguishing variable. Regime analysis provides a conceptually similar but independent approach to slicing up the international system (Krasner 1983), in which functionally separate sets of interacting countries that share norms, rules, decision-making procedures, and expectations are analyzed separately from the rest of the international system.

From a methodological perspective, different techniques have been used in attempts to create subsystems. Russett (1967) uses factor analysis to categorize countries along a variety of variables. Wallace (1975) relies on hierarchical cluster analysis to delineate sets of nations. Most studies, however, have relied on claims of cultural, linguistic, or developmental similarities between proximate countries as the criteria for establishment of a subunit (see Thompson 1973 for a conceptual review).

For subunits to be meaningful in a test of power transition theory they must reflect the preconditions required by that theory. The reliance on demographic, cultural, or international organization membership does not provide the preconditions. The use of factor or cluster techniques is also not useful in this context because they aggregate along similar characteristics but do not differentiate along the dimension of influence. For this reason, the conceptualization of hierarchies I present is an attempt to delineate an analytical tool that is useful for international analysis. Like a nation, which is an analytical tool not defined by size or culture but rather by the notion of sovereignty, the notion of a hierarchy as a unit within which countries interact, is an artificial construct created to simplify analysis.

These theoretical considerations lead me to operationalize local hierarchies using Boulding's (1962) principle of the loss-of-strength gradient, by which the further afield a country attempts to exert its influence the weaker it is. I suggest that the "politically relevant neighborhoods" (Gochman 1991), or the group of countries a given country can influence, of minor powers are confined to their geographically and topologically reachable neighbors. Thus, the local hierarchies I focus on are close-knit geographic sections of the South American continent that are separated by the ability of the member countries to influence one another. Similar hierarchies are to be found in the Middle East, Asia, Africa, the Balkans, and so on. They await definition in future work. The model is globally generalizable. The focus on South America reported here is dictated because I study a group of minor-power countries that have existed as independent countries for a reasonably long period.

Bueno de Mesquita's (1981a:103–8) effort is the most promising operationalization of the loss-of-strength gradient.[1] He conceptualizes the loss-of-strength gradient as involving three considerations: a country's power must decline monotonically with distance, the rate at which that decline occurs must be greater the weaker the nation is at home, and the rate of decline must itself decline with major advances in technology. From discussions with military officers and selected historical readings, he determines the transport range one could cover in a day was 250 miles for the years 1816–1918, 375 miles for the years 1919–45, and 500 miles after 1945. He then treats the loss-of-strength gradient as an exponent on the power score of a country that penalizes power over distance. The actual result is as follows:

$$\text{Power} = \text{Capabilities}^{\log[(\text{miles/miles per day}) + (10 - e)]}$$

In order to delineate local hierarchies I adapt the loss-of-strength measure that Bueno de Mesquita developed. The major difference in my treatment is that, given the underdeveloped state of transportation in South America until recently, I find his claims about possible travel miles to be overstated.[2] I solve this problem empirically. To address the "miles per day" question I gathered data on naval, rail, and highway transportation possibilities in South America for the years 1865 to 1965. Data on ships were drawn from *Conway's All the Worlds Fighting Ships* (various years). Modelski and Thompson's (1988) ships-of-the-line criterion indicates that South American countries did not possess navies to any degree until after World War II. Focusing on numbers and types of ships that were possessed suggests that, for most of the period studied, Argentina and Brazil had adequate naval capacities to move on their rivers and defend their coasts, but nothing else. Data on railroads is gleaned from a variety of sources (Duncan 1932; Heath 1972; Wright 1974; Worldmark 1988). This data suggests that until recent decades, the major states have had moderate rail development internally, but international links have been few, and differences in gauges made such transportation difficult. There were few highways that would facilitate transportation for the period I consider (Delper 1974).

I discovered that until the 1920s, and even then only for the stronger South American states such as Argentina and Brazil, constant estimates are inappropriate for South America. Naval capacities, railroads, and highways were distributed in such a way that great variation in mobility existed.

The task then faced is to determine how many miles per day was realistic within South American local hierarchies. I differentiate type of movement by the nature of the topography to be crossed, such that movement over open ground is easier than movement across mountains, which is in turn easier than cutting a path through the Amazon rain forest. In order to determine how many miles per day can be covered over different types of terrain, I consulted the records of explorers and missionaries who actually crossed these territories, distinguishing between rain forests, mountains, and open territory (Mulhall 1881; Ray 1914; Roosevelt 1914; Siegfried 1933; Huxley 1962). These sources suggest transport ranges of 8 miles per day through the rain forests, 10 miles per day through mountain regions of high altitude, 20 miles per day through mountainous regions of lesser altitudes, and 30 miles per day over open territory. When a road, railroad, or navigable river with sufficient boats was available, I considered the number of miles per day to be 250, 375, and 500, as Bueno de Mesquita identifies them. My adaptation of the loss-of-strength measure accurately reflects the daunting distances and conditions that nature has placed between countries.

One final difference between Bueno de Mesquita's formula and mine is that he determines distance as being from the "locus of power of the potential attacker and the closest point of its intended victim" (1981a:104), but I define distance as being from capital city to capital city. My operational choice makes distances between countries equal and does not require any data about where the "locus of power" might be (see also Russett 1967:chap. 10).

I then use the amended loss-of-strength equation to determine how much of a country's power would be spent in moving forces into other countries. I assume that if 50 percent or more of a country's power is "spent" in transit to another country, then the aggressor country will not undertake operations within that potential target and is effectively deterred by distance. When a group of proximate countries are symmetrically reachable they constitute a local hierarchy. I construct local hierarchies by identifying a list of countries a potential attacker can reach while maintaining at least half of its power. This list constitutes the politically and militarily relevant neighborhood for that country. When a group of countries are members of each other's politically relevant neighborhoods, they constitute a local hierarchy.

For South America four consistent groupings of countries satisfying the local hierarchy criteria correspond to the Atlantic Coast (Argentina, Brazil, and Uruguay), the Pacific Coast (Chile and Peru), the Northern Rim (Colombia, Ecuador, and Venezuela), and the central states (Bolivia and Paraguay). Each of these groupings is a local hierarchy by the definition outlined above and can be produced by applying my adaptation of the loss-of-strength gradient to each country's power score (for more detail, see Lemke 1993).

Let me elaborate on the structure of these local hierarchies. There is no overlap (until recently) among the Atlantic Coast, Pacific Coast, and Northern Rim local hierarchies. However, in the Central State local hierarchy, Paraguay could be influenced by Argentina and Brazil, members of the Atlantic Coast local hierarchy, while the reverse is not true. A similar situation exists for Bolivia with Chile and Peru, and for Ecuador with Peru. This structure allows for hierarchical competition within relatively independent hierarchies that can be influenced by larger local hierarchies, as well as by the overall international hierarchy considered in all previous analyses of power transition theory.[3]

Power

In a recent evaluation, Merritt and Zinnes (1989) identify at least eleven different alternatives used to measure power over the last several decades. A widely used measure of power is the Composite Capabilities Index created by the Correlates of War Project (COW). This measure is chosen here because Organski and Kugler (1980:38) and Kugler and Arbetman (1989a) report high correlations between it and Gross National Product, another commonly used measure of power. The two measures are excellent substitutes.

I adjust national capabilities for political capacity. The measure of political capacity proposed by Organski and Kugler, a measure of political capacity based on a government's ability to penetrate its society and extract resources, is unavailable. A close approximation was obtained by relying upon data of government revenues, collected by Banks (1971). I select government revenue as my measure of political capacity because what a government can do in the international sphere is largely constrained by the resources it has directly available.

Thus, my measure of power is:

$$\text{Power} = \text{COW} \times [(\text{Government Revenue})/(\text{Population})]$$

Organski and Kugler also standardized political capacity, but in a different way. Although my version is quite simple, it allows me accurately to approach their measure; efficient states will have high capacity while inefficient states will have low capacity.

Relevant Wars

In the period 1865–1965 two wars are identified, the War of the Pacific and the Chaco War (Richardson 1960; Wright 1965; Singer and Small 1972; Small and Singer 1982).[4] If power transition theory is correct, the relative absence of war suggests that unusually few transitions must have occurred in the South American local hierarchies from 1865 to 1965. This study suggests that the predominantly peaceful interstate history enjoyed by South Americans is the result of relative stability in the local hierarchies of the continent. Anticipating the results reported below, I suggest the absence of war is not attributed to some learning process that South Americans have undergone. South American peace has been preserved because there have been relatively few periods of parity, that is, periods in which the initiation of war was possible. The multiple hierarchy version of power transition theory successfully explains the long periods of peace in South America. This explanation simultaneously accounts for why peace has occasionally failed.

Results

The model I evaluate divides South America into local hierarchies and suggests that their war and peace behavior can be understood by paying attention to the power transitions between member countries. Specifically, when there is an unequal or preponderant distribution of power, peace is expected, but when there is a power transition that changes relative power levels and leads to parity, war is expected.

I trace the power levels of each country within each local hierarchy and correlate power distributions with the incidence of war. Specifically, I compare the power of the local dominant country with the power of each member of the local hierarchy. The direction of my inquiry runs from the distribution of power to wars. When I discover transitions I then look for subsequent wars (see discussions in Vasquez herein; Siverson and Miller herein). According to the theory, no wars should occur within local hierarchies unless they are preceded by power transitions. It is not essential that all transitions be followed by wars, as power transitions are a necessary and not a sufficient condition for war.

The sample period runs from 1865 to 1965. I break it down into twenty-year periods and consider the power of each local hierarchy member for each interval. The theoretical reasoning behind the twenty-year interval is that the changes in national capabilities that lead to transitions are slow and require a long time frame to be observed. Also, it is the time scale Organski and Kugler selected (1980:48), and I maintain it for the sake of comparability.

The distribution of power within each local hierarchy in each interval is considered one observation. I maintain continuity with previous studies because my interest is in generalizing the theory to minor powers and comparing with results already obtained for major powers. The cases are coded for power distributions as "unequal," "equal, no overtaking," and "equal and overtaking." An observation is coded unequal if the strongest country from the beginning of the interval is still the strongest country at the end of the interval and if the weaker countries have less than 80 percent of the strongest country's power. An observation is coded equal, no overtaking if a weaker country has 80 percent or more of the stronger country's power or if the relative power levels fluctuate back and forth in any given period. An observation is coded to be equal and overtaking if a weaker country at the beginning of the interval is the stronger by the end of the interval. All of this is consistent with Organski and Kugler (1980).

There are five twenty-year periods for each local hierarchy, with one exception: the Central States are not considered in the 1865 to 1885 period because they do not qualify as a local hierarchy. Thus, the sample size for this study includes nineteen observations. The results are as shown in table 4.1.[5] As table 4.1 indicates, the only instances of war occur when the countries in a local hierarchy are equal and the weaker is overtaking the stronger. When they are equal but the hierarchy is maintained there is no war, and when they are unequal, so that the hierarchy is as rigid as possible, there is no war. Due to the

TABLE 4.1. **Power Distributions within Each Local Hierarchy**

		Unequal	Equal, No Overtaking	Equal and Overtaking
WAR	No	12	4	1
	Yes	0	0	2

$N = 19$, Tau C = 0.66, $p < 0.01$

overall stability of power hierarchies within the four South American local hierarchies, power transition predicts overall interstate peace. The one observation in the upper right-hand cell may represent a satisfied transition, a situation not addressed in the present work, and there are no observations in the lower left-hand and lower center cells. The hypothesis that power transitions are associated with wars in local hierarchies is strongly supported.

This result is consistent with previous results for major powers (Organski and Kugler 1980; Houweling and Siccama 1988a). The important finding of this study is that results previously found for a single international hierarchy are generalized by similarity to results for these regional local hierarchies. Like Kim (herein), who successfully extends the temporal domain, this effort shows the power transition logic extends to local hierarchies in the international system. The evidence can be considered by comparison between this study and earlier results. The multiple hierarchy perspective generalizes and extends power transition theory. Combining the results reported in table 4.1 with those for major-power contenders from table 1.7 of *The War Ledger* (1980:52) produces table 4.2. Table 4.2 demonstrates that this combination of power transition studies at the major- and minor-power levels improves upon both. This improvement demonstrates that the generalization is consistent with theoretical expectations. The results reported here, if confirmed outside the South American context from which they are drawn, would address consistent criticism of lack of generality of power transition theory (see Vasquez herein; Siverson and Miller herein). To achieve generalization, the multiple hierarchy perspective should be adopted.

Conclusions and Implications

The multiple hierarchy perspective is used to generalize power transition theory. The restructuring improves upon the original theory by allowing simultaneous consideration of major- *and* minor-power wars within a unified theoret-

TABLE 4.2. Power Distributions for Major and Minor Countries

		Unequal	Equal, No Overtaking	Equal and Overtaking
WAR	No	16	10	6
	Yes	0	0	7

$N = 39$, Tau C = 0.74, Significance = 0.01

ical framework. Rejoicing may be premature, since the test of the multiple hierarchy perspective is limited to South America. In future work I will attempt to extend these promising insights to other local hierarchies. However, the combination of minor-power results with those of Organski and Kugler demonstrates immediate generalizability. The evidence marshaled here supports the hypothesis that power distributions and war follow the same pattern regardless of the size of a country. The successful extension of power transition theory to the minor-power level supports Bueno de Mesquita's claims that the big war/little war debate is overblown and refutes claims that major-power war must be studied separately (cf. Midlarsky 1990b; Thompson 1990).

A number of other implications arise. The first is that peace, whether between minor or major powers, can be explained by the same theoretical structures used to explain war and does not require some new input or learning process, as has been suggested by others (cf. Mueller 1989; Kegley 1991). A second implication is that power transition applies to nonindustrialized as well as industrialized countries (see also Kim herein). Organski (1958:307) suggested that power transition theory applies only to the industrial period. Critics (cf. Thompson 1983a; Levy 1989) have focused on this as an especially negative and limiting aspect of the theory. This chapter refutes this claim.

Extensions of multiple hierarchy applications should be able to account for other problems left unresolved by power transition theory's previously exclusive focus on the overall international hierarchy. Bueno de Mesquita's (1989) specific argument that his expected utility framework is superior because it can include both large and small wars is challenged. In addition, the multiple hierarchy perspective allows for dynamic analysis not possible in the similarly general expected utility model. Bueno de Mesquita's (1990a) discussion of the Seven Weeks' War is now logically incorporated within the multiple hierarchy perspective as a war between members of a Central European local hierarchy, for example. Moreover, Choucri and North's (1989) criticisms of power transition theory as unable to account for why the United States and Germany did not fight each other from the start of World War II is addressable by identifying Germany as a member of a European local hierarchy and the United States as a member of a Western Hemisphere local hierarchy. Finally, differences between Houweling and Siccama (1988a), Kim (1989), and Organski and Kugler (1980) can be refocused through the multiple hierarchy perspective, analytically identifying the appropriate wars for each competitive hierarchical arrangement.

The multiple hierarchy model of power transition theory offers promise of greater generalizability to the study of war and unifies currently disparate segments of this branch of international relations research.

NOTES

1. I am aware of Wohlstetter's (1968) criticism of the loss-of-strength gradient itself and of Diehl's (1985) criticism of Bueno de Mesquita's effort, but find neither critique compelling enough to reject using Bueno de Mesquita's formula.

2. My adaptation of Bueno de Mesquita's operationalization of the loss-of-strength gradient is thus very similar in purpose to Moul's (1988). However, I believe that my adaptation is superior to Moul's because it is based on more tightly defined criteria.

3. In a fascinating connection to Wright's *A Study of War,* this differentiation of South America corresponds almost exactly with his "regional classification of primitive peoples" map (1965:545). It seems the indigenous peoples of South America were constrained by the same geographical and topological barriers that affect present-day countries.

4. A third war, the Lopez War pitting Argentina, Brazil, and Uruguay against Paraguay, was already underway when my study period begins. Since power transition theory focuses on changes in power relationships prior to wars, it is impossible to include this war in my empirical evaluation.

5. Significance is reported for the benefit of those who prefer such details and due to common practice. The author is aware that it is perhaps unnecessary and potentially misleading when a population of cases or other nonrandom sample is being studied.

CHAPTER 5

Power Parity, Alliance, and War from 1648 to 1975

Woosang Kim

Introduction

The idea of balance of power is one of the oldest in the literature on international politics and has been widely accepted, not only by political scientists and historians, but also by journalists and foreign policy decision makers. Many scholars contend the concept of balance of power is one of the most important means of maintaining the status quo in the system (Gulick 1955; Kaplan 1957; Liska 1962; Morgenthau 1973; Waltz 1979). Historians have also described the stability of the nineteenth-century European political system as the result of the balance of power (Taylor 1954; Albrecht-Carrié 1973). Decision makers have often justified their policy responses to crises and demands in the international system by suggesting that balance of power promotes peace and stability in the international system (e.g., Kissinger 1979).

In fact, the majority of balance-of-power theorists argue that the presence of an equality of power among nations tends to discourage war. Wright (1965:755) asserts that for balance-of-power theorists "stability will increase and the probability of war will decrease . . . as the parity in the power of states increases." Claude (1962) and Levy (1989) also find that most balance-of-power theorists advocate the proposition that an equal distribution of power promotes peace.[1]

However, Organski (1958), in his book, *World Politics,* challenged the balance-of-power perspective and argued that the balance of power, defined as the approximate equal distribution of power among great powers, is conducive to war rather than peace or stability in the international system. Organski and Kugler (1980), in their book, *The War Ledger,* further developed the power parity thesis and have provided some empirical evidence to support it. A growing body of literature has provided similar theoretical arguments. Industrialization, population growth, and increase in the state's political capacity to extract resources from its population lead to increases in a nation's capabilities. The financial burdens of foreign commitments and a large military establishment, the need to

93

service substantial debts from prior wars, and the failure to dominate new leading economic sectors slow the growth of a nation's capabilities. Uneven rates of growth of capabilities drive the rise and fall of great powers and such shifts in capabilities lead to power transitions among them. Power transitions, characterized by approximately equal distribution of power, lead to cataclysmic wars (Organski 1958; Organski and Kugler 1980; Gilpin 1981; Modelski 1983; Kennedy 1988; Thompson 1988; Doran 1989; Kugler and Organski 1989b; Modelski and Thompson 1989).

Several studies covering the nineteenth and twentieth centuries have supported power transition theorists' view of the relationship between the distribution of power and the outbreak of war (Weede 1976; Garnham 1976a,b; Doran and Parsons 1980; Organski and Kugler 1980; Bueno de Mesquita and Lalman 1988; Houweling and Siccama 1988a; Kim 1991; Geller 1992a; Kim and Morrow 1992).[2] Little, if any, rigorous empirical analysis of the power transition theory has been done using the empirical record from before the Napoleonic Wars.[3] Much of what we know about the causes of great power war is based on evidence after the Napoleonic Wars. Among others, Singer, Bremer and Stuckey's (1972) national capability data, Singer and Small's (1966a) and Bueno de Mesquita's (1981a) alliance data, and Gochman's (1975) interstate war data cover the period from after 1815, and have encouraged many empirical studies about the distribution of power, alliance, and war. However, the lack of availability of the national capability and alliance data prior to the Napoleonic Wars has discouraged rigorous empirical analyses about the causes of war prior to 1816.

Using similar evidence I have developed elsewhere (Kim 1992) for the period 1648 to 1975, I will show that the key hypotheses of the power transition theory supported in studies of the post-Napoleonic era are also supported in the period from the Peace of Westphalia to the present.

Organski (1958:345–46) has argued that the power transition model can only be applied to the "period of industrial revolution" (i.e., nineteenth and twentieth centuries), when "differential industrialization is the key to understanding the shifts in power." I believe that power transition theorists' exclusive focus on internal means of augmenting national power and dismissal of such external means as alliance formation confine the theory's direct applicability to only the industrial era. In this study, therefore, I will also demonstrate the importance of alliance relationships to power transition theory.

Revised Hypotheses of the Power Transition Theory

Organski's restriction that the power transition theory applies only to the period after the industrial revolution stems from an emphasis on the role industrialization plays in stimulating differential growth rates. Internal development through industrialization is often considered to be the principal means by which relative power changes. Alliance formation, by contrast, is generally over-looked by power transition theorists as an important means of augmenting na-tional power (Organski 1958; Organski and Kugler 1980; Kugler and Organski 1989b).

I question power transition theorists' assumption of power augmentation, particularly in the context of the years leading up to the Industrial Revolution. According to Organski, prior to the period of the Industrial Revolution, all na-tion-states were preindustrial. Only during and after the Industrial Revolution have some nation-states been preindustrial, some industrializing, and some in-dustrialized, thereby creating a context of uneven growth and transition of power.[4] Conversely, others assert that the structures of the global political sys-tem were well in place prior to the Industrial Revolution. There seems to be no compelling reason to assume that differential growth rates among nation-states require some degree of industrialization as a prerequisite. The processes of dy-namic changes in power can be observed in an earlier agrarian and trade-ori-ented phase of the development of the modern world system (Thucydides 1972; Gilpin 1981, 1988; Levy 1981, 1983a, 1985; Modelski and Thompson 1988, 1989; Thompson 1988).[5]

I reconstruct the power transition model by relaxing the assumption that internal growth is the only method of augmenting power and suggest that a na-tion's power can be augmented not only by such internal means as industrial-ization, but also by such external means as alliance formation (Altfeld 1984; Most and Starr 1984; Levy 1987; Iusi-Scarborough and Bueno de Mesquita 1988; Kim 1991,1992; Morrow 1991). The revised power transition theory can be applied to the study of great-power wars during the preindustrial era. Here, I offer a brief summary of the revised power transition arguments.[6] Differences in growth rates and the formation of alliances and counteralliances produce a redistribution of power in the international system. During the redistribution of power, some catch up with their rivals. The newly strengthened great power challenges the dominant power. The great power's willingness to challenge its rival increases as its relative capabilities increase. That is, as the challenger in-creases its capabilities through internal development and alliance formation and approximates the dominant power, it is more likely to attack. At the same time, demands for new arrangements and changes in the existing international order are a threat to the dominant nation. The dominant nation, enjoying the benefits and privileges from both collective and private goods in the existing interna-tional community with its allies, has a large stake in preserving the status quo. So, the dominant nation, with support from its allies, tries to intercept the chal-lenger's progress. During such periods great-power war is more likely.

The challenger's degree of satisfaction is also an important factor influ-encing the likelihood of major-power conflict (Organski 1958; Organski and

Kugler 1980; Howard 1983; Kennedy 1988; Kugler and Organski 1989b; Bueno de Mesquita 1990a; Bueno de Mesquita and Lalman 1992). Organski (1958:366) writes that in the international system some great powers are "satisfied with the present international order and its working rules, for they feel that the present order offers them the best chance of obtaining the goals they have in mind. The dominant nation is necessarily more satisfied with the existing international order than any other since it is to a large extent *its* international order." However, others are not satisfied with the status quo because they "have grown to full power after the existing international order was fully established and the benefits already allocated" (1958:366). During the power transition period, if the challenging power is dissatisfied with the existing international order and refuses to abide by the rules, then major-power conflict is hypothesized to be most likely. On the other hand, if the challenger is satisfied with the existing international order and wishes merely to take over its leadership, then major war is less likely.[7]

Power transition theorists (Organski 1958; Organski and Kugler 1980; Kugler and Organski 1989b) also argue that the overtaking of the dominant power by the challenger leads to major war and the speed with which the challenging power overtakes is an important factor in understanding the likelihood of conflict. If the challenging power's internal and external capabilities are increasing slowly, the problems arising from an overtaking may have a greater chance of being resolved without conflict. However, if the challenger's capabilities increase rapidly, then the dominant power is unprepared for the resulting shift in the international power order, and war is expected to be more likely.

Based on the above arguments four hypotheses can be stated.

H1.   Major war is most likely when the two great states' internal and external capabilities are equally distributed.

H2.   During the power transition period, if the challenging power is dissatisfied with the status quo, then major war is more likely. If, however, the challenging power is satisfied with the existing international order, war is less likely.

H3.   Major war is more likely when a great power's internal and external capabilities are overtaken by those of another great power.

H4.   Major war is more likely the faster the challenger increases its internal and external capabilities relative to the dominant nation.

Research Design

The hypotheses of the revised power transition theory pertaining to all power relationships between pairs of great powers during the period from 1648 to 1975 are examined. The definition of the great powers used in this analysis for the seventeenth and eighteenth centuries is from Modelski and Thompson 1988 and that for the nineteenth and twentieth centuries is from Small and Singer 1982.[8]

To create the set of test cases, I use the same procedure as Organski and Kugler (1980) and Houweling and Siccama (1988a). First, the whole period is divided into twenty-year periods.[9] Then, in each twenty-year period each great power is paired with each other great power. This procedure creates 189 dyad-periods.[10]

The objective is to separate the dyad-periods that include a war from those that do not. The observations in this analysis include all possible pairs of great powers and not just those that eventually fought. Hence, the dependent variable in this analysis is a dichotomy of "war" or "no-war."

The selection of great-power wars in this analysis is based on Organski and Kugler's war criteria: (1) whether one or more great powers participated on each opposing side; (2) whether or not the two opposing sides made all-out efforts to win the war, judged by the severity of battle deaths of each war, and (3) whether the war resulted in the loss of territory for the losing side. Six wars satisfy these criteria for the pre-Napoleonic era: the Dutch War of Louis XIV (1672–78), the War of the League of Augsburg (1688–97), the War of the Spanish Succession (1701–13), the Seven Years' War (1756–63), and the French Revolutionary and the Napoleonic Wars (1792–1802; 1803–15). Four wars satisfy Organski and Kugler's criteria for the post-Napoleonic era: the Seven Weeks' War (1866), the Franco-Prussian War (1870–71), World War I (1914–18), and World War II (1939–45). Unlike the previous studies on the great-power wars since the post-Napoleonic era (Organski and Kugler 1980; Houweling and Siccama 1988a; Kim 1989, 1991; Kim and Morrow 1992), in this analysis the Crimean War (1853–56) and the War of Italian Unification (1859) are not included. Although both wars satisfy the first and the third criteria, they do not satisfy the second criterion. That is, they were not as severe as the previous war, the Napoleonic Wars. The Russo-Japanese War is not included in this war list because it does not satisfy both the second and the third criteria.

To estimate each nation's power during the nineteenth and twentieth centuries the following procedure is used. First, a nation's internal capabilities are measured as the Composite Capability Index developed by the Correlates of War project. Second, the support that nation expects from other great powers is added to the internal capabilities. The amount of support a particular third party contributes to a great power depends on its own capabilities and the closeness of relations between the two. Countries with greater internal capabilities can contribute more, and those with better relations contribute a greater fraction of those capabilities.

The following equations represent great power i's and great power j's adjusted national capabilities, which consist of both internal and external capabilities of each great power i or j:

$$NC_i = IC_i + EC_i \quad \text{and} \quad NC_j = IC_j + EC_j$$

where

$$EC_i = \Sigma_{k \neq i,j} IC_k \times \frac{(U_{ki} - U_{kj})}{2} \quad \text{if } (U_{ki} - U_{kj}) \geq 0$$

$$EC_j = \Sigma_{k \neq i,j} IC_k \times \frac{(U_{kj} - U_{ki})}{2} \quad \text{if } (U_{kj} - U_{ki}) \geq 0$$

where i (or j) is great power i (or j) in each dyad and k is a third-party great power; NC_i (or NC_j) is i's (or j's) adjusted national capabilities; IC_i (or IC_j or IC_k) is i's (or j's or k's) internal capabilities; EC_i (or EC_j) is i's (or j's) external capabilities; U_{ki} (or U_{kj}) is the utility of k attached to i (or j) calculated as the Tau B coefficient for similarity in alliance portfolios between k and i (or j).[11]

The adjusted national capabilities of great power i in a dyad, denoted NC_i in the equation, add the proportion of third-party resources great power i can draw on to augment its internal capabilities. That proportion is based on those third parties believed by i to prefer i to j. It is estimated as the sum of the products of the capabilities of each such third party multiplied by great power i's approximation of each third party's intensity of preference for an outcome favoring i over j. The adjusted national capabilities, then, measure a great power's internal capabilities, augmented by the assistance it expects from other great powers (for more detail, see Kim 1991, 1992).

To estimate each nation's power during the seventeenth and eighteenth centuries, I developed the national capability data and the alliance data for that period elsewhere (Kim 1992). Three indicators—army size, population, and sea power—are equally weighted and are transformed into a quinquennial relative great power capability index from 1648 to 1815. I have also developed the alliance data for the same period, which are utilized in calculating the Tau B coefficient of alliance portfolios (i.e., the utility index) and in measuring one of the independent variables, the dissatisfaction variable.

An independent variable, "alliance equality," which measures the power relations between the two great powers in each dyad after taking into account alliance effects, is defined as the mean adjusted national capability of the weaker great power divided by the mean adjusted capability score of the stronger great power during each twenty-year period.

Another independent variable, "alliance transition," which dichotomizes whether there is a power transition, is defined as the overtaking of one great power's adjusted capability by the opposing great power's adjusted capability. When a great power that was less powerful at the beginning of the twenty-year interval grew more powerful than the opposing great power before the period ended, transition is recorded as occurring.

The "alliance growth rate" variable is defined as the difference between the growth rates of the two great powers after taking into account their allies' support. The growth rate of each great power during the period from 1660 to 1679, after considering alliance effects, for example, is calculated by ($NC_2 -$ NC_1)/NC_1, where NC_1 is each great power's mean adjusted capability for the period from 1655 to 1664, and NC_2 is each great power's mean adjusted capability for the period from 1675 to 1684. The difference is measured as the smaller growth rate of one great power subtracted from the bigger growth rate. So, if the growth rate of one great power is fast and that of the other is slow, then the difference will be great, whereas if the growth rates of both great powers are slow or fast, then the difference will be small.

Organski argues that the dominant power is the most satisfied nation with the existing international order and that if a nation is in favor of the status quo in the international order, it is "satisfied" too. On the other hand, if a nation that does not like the status quo seeks to upset the existing international order and establish a new order in its place whenever it has power to do so, it must be a "dissatisfied" nation. Based on this argument, I argue the dominant nation is always satisfied. The dissatisfied challenger is the nation that has little or no common interest with the dominant nation. The more dissatisfied a challenger is, the less common interest it shares with the dominant nation.

To measure the degree of satisfaction, I first determine which nation is dominant in each twenty-year period. The dominant nation is, of course, the nation whose national capability score is the highest in that period. Then each nation on the list of great powers is paired with the dominant nation and an estimate of the utility score between the two is calculated. If the mean utility score (U_{ij}) over a twenty-year period between the dominant nation, i, and a great power, j, is positive, then the great power, j, is a "satisfied" nation. If the mean utility score (U_{ij}) for a twenty-year period is negative, then the great power, j, is a "dissatisfied" nation for that period. With this measurement of the degree of satisfaction of each great power, the "dissatisfaction" variable is operationalized as the challenger's level of satisfaction in each dyad. That is, the degree of satisfaction of the weaker power in each dyad is used as the measurement for the dissatisfaction variable.[12]

Model Specification

To examine how much influence each independent variable has on the likelihood of war, these four variables are included individually. The first model, then, is specified as follows:

Model I

$$\text{War} = \beta_1 + \beta_2 \times \text{Alliance Transition} + \beta_3 \times \text{Alliance Growth}$$

$$\text{Rate} + \beta_4 \times \text{Alliance Equality} + \beta_5 \times \text{Dissatisfaction} + U$$

where β_i's are parameter estimates and U is the stochastic error term.

Some studies indicate that the equality of power has a significant individual effect on the likelihood of major war, that neither the power transition nor the rate of growth has a significant individual effect, but might have interactive effects with the equality of power on the incidence of war, and that when the challenging power is dissatisfied and catches the dominant power, major war is more likely (Organski and Kugler 1980; Kim 1989, 1991; Kugler and Organski 1989b; Kim and Morrow 1992). To examine the possible interactive effects of the independent variables, the interactive term of the four independent variables is included in the second model. I also include the interactive variable for alliance equality and dissatisfaction because the empirical findings in the previous studies suggest that the level of dissatisfaction and the equality of power between the two great powers have a significant interactive effect on the likelihood of major war (Kim 1991). The second model, then, includes the alliance equality variable, the interactive term for alliance transition, alliance growth rate, alliance equality, and dissatisfaction, and the interactive term for alliance equality and dissatisfaction.

Model II

$$\text{War} = \beta_1 + \beta_2 \times \text{Alliance Equality} + \beta_3 \times \text{Interactive I} + \beta_4$$

$$\times \text{Interactive II} + U$$

where Interactive I = Alliance Transition $\times$ Alliance Growth Rate $\times$ Alliance Equality $\times$ Dissatisfaction; Interactive II = Alliance Equality $\times$ Dissatisfaction; β_i's are parameter estimates and U is the stochastic error term.

Results

Because the dependent variable, war, is dichotomous, logit analyses are employed. Logit analysis estimates the effect of each independent variable on the log of the odds ratio of the dependent variable, using a maximum likelihood procedure. Results of the logit analyses for models I and II are summarized in table 5.1. Significance levels are reported based on one-tailed tests, as I have expectations about the direction of predicted effects as well as their magnitude.

The results in table 5.1 demonstrate an important finding. The value of the maximum likelihood estimate of the alliance equality variable in model I is 2.501 and in model II is 2.616. This means that in model I the probability of war increases about 0.36 for a one-unit change in the alliance equality variable, holding other variables constant, and in model II it increases about 0.37 for the same amount of change in the alliance variable, holding other variables constant.[13] The findings from both models strongly support the main argument of power transition theorists that the rough equality of power among nations leads

TABLE 5.1.   Logit Analysis Results

Independent Variables	Model I	Model II	Model Ia	Model IIa
Intercept	−3.189	−2.805	−3.108	−2.659
Alliance				
transition	0.526		0.488	
(s.e.)	(0.451)		(0.455)	
(prob)	(0.122)		(0.142)	
Alliance				
growth rate	0.403		0.434	
(s.e.)	(0.369)		(0.372)	
(prob)	(0.138)		(0.121)	
Alliance				
equality	2.501	2.616	2.613	2.701
(s.e.)	(1.011)	(0.851)	(1.028)	(0.855)
(prob)	(0.007)*	(0.001)*	(0.006)*	(0.001)*
Dissatisfaction	1.950		1.799	
(s.e.)	(0.837)		(0.860)	
(prob)	(0.010)*		(0.018)*	
Interactive I[a]		0.024		0.287
(s.e.)		(1.787)		(1.838)
(prob)		(0.495)		(0.438)
Interactive II[b]		2.138		1.815
(s.e.)		(1.205)		(1.235)
(prob)		(0.038)*		(0.071)
Period			−0.237	−0.304
(s.e.)			(0.383)	(0.374)
(prob)			(0.269)	(0.208)
−2 × LLR[c]	21.730	16.520	22.110	17.180
(prob)	(0.000)*	(0.001)*	(0.000)*	(0.001)*

Note: Significance levels are reported based on one-tailed tests and * indicates statistical significance either at .05 or .01 level.

[a]Interactive I = Alliance Transition × Alliance Growth Rate × Alliance Equality × Dissatisfaction
[b]Interactive II = Alliance Equality × Dissatisfaction
[c]−2 × LLR = −2 × Log Likelihood Ratio (model chi-square, χ^2, value)

to war. The findings also suggest that alliances are important external means of augmenting power. I argue that this means we should take into account not only such internal means as development, but also such external means as alliance formation, when measuring a nation's power.

Second, one of the arguments of the power transition theory that is overlooked by both Organski and Kugler's (1980) and Houweling and Siccama's (1988a) empirical analyses is the hypothesis about the level of dissatisfaction of the challenger. Organski (1958) argues that when the challenger approximates the dominant power, if it is satisfied the power transition can be resolved without conflict. However, if the challenger is not satisfied, war is more likely. This argument is strongly supported in the analyses. In model I the value of the parameter estimate for the dissatisfaction variable is 1.95. Results for model II also indicate that the alliance variable and the dissatisfaction variable have an interactive effect on the dependent variable, demonstrated by the value of the parameter estimate for the interactive variable, 2.138. The interaction between alliances and dissatisfaction matters.

Further Empirical Considerations

Note in table 5.1 that I have two sets of results for each of the models described above. The results of models I and II are based on the pooled data for the two different periods, (i.e., the seventeenth- and eighteenth-century data and the nineteenth- and twentieth-century data). The sources for the national capability data and the alliance data for the seventeenth and eighteenth centuries are different from those for the nineteenth and twentieth centuries. Pooling two different data sets into one could be problematic. To examine the possible different influence of the two data sets for the two different periods, I have included a period dummy variable in models Ia and IIa. The findings of models Ia and IIa parallel those of models I and II and the parameter estimates of the dummy variables in both models Ia and IIa are not significant at the 0.05 level.[14] To assess the potential danger of pooling two data sets into one more carefully, I have also examined the change in the χ^2 value associated with model Ia as compared with model I. The inclusion of the dummy variable in model I increases the χ^2 value only 0.38 with one degree of freedom. This difference is not significant at the 0.05 level. For model II, the inclusion of the dummy variable increases the χ^2 value 0.66 with one degree of freedom. The difference in the χ^2 values between model II and model IIa is not significant at the 0.05 level either.

I have also run the logit analyses by including not only the period dummy variable but also the interactive variables for each independent variable and the period dummy variable. For example, in model I, I have added the period dummy variable, dummy $\times$ alliance transition, dummy $\times$ alliance growth rate, dummy $\times$ alliance equality, dummy $\times$ dissatisfaction. In model II, I have added the period dummy variable, dummy $\times$ alliance equality, dummy $\times$ interactive

I, and dummy $\times$ interactive II. To assess the impact of the variables I added in each model, I have examined the change in the χ^2 value associated with model I including these additional variables as compared with model I without these additional variables. The inclusion of these variables in model I increases the χ^2 value only by 3.67 with five degrees of freedom. This difference is not significant at the 0.05 level. For model II, the inclusion of the interactive variables between each independent variable and the dummy variable increases the χ^2 value to 1.13 with four degrees of freedom. The difference in the χ^2 values between the original model II and model II with additional interactive variables is not significant at the 0.05 level, either.[15] These findings suggest that there is no serious problem in pooling the seventeenth- and eighteenth-century data and nineteenth- and twentieth-century data into one in this study.

Conclusion

Many studies of power transition theory have evaluated the explanations advanced for the recurrence of great-power wars. They have argued that the growth of challenging states increases their ability to carry out demands for change in the existing international order. Such demands accumulate over time until the challenging state, dissatisfied with the status quo, approximates the capabilities of the dominant power defending the status quo. The accumulated weight of the dissatisfied challenger's demands for change then triggers a cataclysmic war. These explanations, however, have not been subjected to rigorous empirical tests with the historical evidence since the emergence of the modern nation-state system. In this chapter, I have attempted to investigate the power transition hypotheses with the empirical record from the Peace of Westphalia to the present.

Unlike some of the previous empirical studies on power transition theory that overlook the importance of alliance in increasing a nation's power and the effect of the challenger's level of dissatisfaction on the likelihood of war, I have demonstrated how important alliance relationships are to power transition theorists' thesis about power equality and how much the challenging power's level of dissatisfaction influences the onset of major war. The evidence in this study strongly supports the main argument of the power transition theory: the rough equality of power among great powers and more dissatisfied challengers increase the probability of major war.

NOTES

1. The ambiguity of the concept of the balance of power among balance-of-power theorists makes it difficult to delineate central balance-of-power propositions (Haas 1953; Claude 1962; Hartmann 1978). There are only a few empirical studies for the balance-

of-power arguments (Singer, Bremer, and Stuckey 1972; Ferris 1973; Siverson and Tennefoss 1984). Recently there have been efforts to develop formal models of the balance of power (Zinnes 1967; Wagner 1986; Niou, Ordeshook, and Rose 1989).

2. Blainey (1973) also suggests that major war is most likely to occur when the distribution of power between the two opposing sides is approximately equal. His survey of the major wars of the period from 1700 to 1815 suggests that "the traditional theory which equates an even balance of power with peace should be reversed. Instead a clear preponderance of power tended to promote peace" (Blainey 1973:113).

3. Some have tested the relationships between sea power and a global war cycle and between alliance formation and war using evidence prior to the Napoleonic Wars (Levy 1981, 1983a; Modelski and Thompson 1988, 1989; Thompson 1988). However, Thompson (1983a) has been the only one to examine the power transition thesis with empirical data prior to the Napoleonic era. Although his analysis does not employ rigorous statistical techniques, he examines the trend of relative naval capabilities among great powers to test a key power transition hypothesis.

4. Organski (1958) also argues that once all states are industrially advanced, the circumstances presumably will once again preclude the possibility of transition struggles since "great and sudden shifts in national power" will be less likely. So, he suggests that we need a "new" theory for the future.

5. Toynbee (1954), Levy (1981, 1983a), Wallerstein (1984), and Modelski and Thompson (1988) consider the modern nation-state system to have emerged around 1500. Russett and Starr (1981), on the other hand, note that many scholars date the modern nation-state system from 1648 (after the Treaty of Westphalia).

6. These arguments might also be applied to major-war cases in general where the rapidly growing great power challenges the declining stronger rival (in those cases, the declining stronger rival is not the dominant power but a great power). The hegemonic war arguments seem to be special cases of power transitions in general. For the original arguments of the power transition, power parity, and the rate of growth, see Organski 1958 and Organski and Kugler 1980.

7. The previous theoretical and empirical studies by Kim (1991, 1992) and Kim and Morrow (1992) have supported only these two arguments (the power equality and the dissatisfaction hypotheses).

8. The great powers during that period are as follows: France (1648–1940/1945–75); England/Great Britain (1648–1975); the Netherlands (1648–1810); Spain (1648–1808); Russia/the USSR (1714–1917/1922–75); Prussia/Germany (1816–1918/1925–45); Austria-Hungary (1816–1918); Italy (1860–1943); the United States (1899–1975); Japan (1895–1945); China (1956–75). For more discussion of the definition of the great powers during the pre-Napoleonic era, see Small and Singer 1982, Levy 1983a, Modelski and Thompson 1988, and Kim 1992.

9. Organski and Kugler (1980) believe that a long period of time is required to produce sufficient changes in the power distributions between possible adversaries for war to break out. They claim that approximately a twenty-year period is reasonable.

10. This procedure decomposes disputes between the two groups of nations into all possible dyads. There are advantages and disadvantages in doing this. It increases the number of cases and avoids aggregating nations whose actions may be independent. On the other hand, it also isolates dependent decisions and increases the effect of random

error in the measurement of the independent variables. For example, there are 21 dyads in the period from 1919 to 1939. The behavior of some nations in those dyads was not independent. Cases that are not statistically independent can be problematic in statistical tests. The significance tests used here should be viewed as a heuristic device to suggest the relative strength of associations. Decomposing multilateral disputes is not novel. Statistical analyses of arms race disputes often decompose multilateral disputes (e.g., Wallace 1979; Morrow 1989). Other studies have also used the dyad-year data that might have the problem of statistical dependence (e.g., Bueno de Mesquita and Lalman 1992; Maoz and Russett 1993).

11. This "utility index" comes from Bueno de Mesquita's expected utility research program. The utility index is a measure of the degree to which the policies pursued by two nations are congruent. The utility index is measured as the Tau B coefficient of alliance portfolios. For details about the utility index, see Bueno de Mesquita (1981a).

12. For more details about the definition and measurement of the dissatisfaction variable, see Organski 1968, or Kim 1991, 1992.

13. The minimum value of the relative power ratio in the data is about 0.1 and the maximum value is 1.0. So, we can change the value of this independent variable from 0.1 to 1.0 and obtain about 0.36 increase in the probability of war for model I, holding other variables constant ($p = 1/[1 + e^{-1.0*2.501}] - 1/[1 + e^{-0.1*2.501}] = 0.9242 - 0.5622 = 0.362$) and about 0.37 increase for Model II ($p = 1/[1 + e^{-1.0*2.616}] - 1/[1 + e^{-0.1*2.616}] = 0.9319 - 0.565 = 0.3669$).

14. I do not have any specific expectation about the direction of a predicted effect of the dummy variable. So, a two-tailed test might be suitable. With the two-tailed test, of course, the parameter estimates of the dummy variables in both models Ia and IIa are not significant at the 0.05 level.

15. As anyone might expect, however, there is a multicollinearity problem when I add interactive dummy variables to either model I or model II. For example, in model I the correlation coefficient between alliance transition and dummy × alliance transition is 0.72, between alliance growth rate and dummy × alliance growth rate is 0.79, between dissatisfaction and dummy × dissatisfaction is 0.80, and interactive I and dummy × interactive I is 0.85. The significance test may not be interesting when we have multicollinearity problems. Therefore, I do not present the results of these analyses, although the results are very similar to those of Models Ia and IIa in table 5.1.

CHAPTER 6

A Two-Level Explanation of World War

Henk W. Houweling and Jan G. Siccama

Between 1495 and 1975, 119 wars were fought with the participation of at least one great power. Of these, 64 wars were fought with at least one major power on each side (Levy 1983a). Within this category, only a very small number of military struggles escalated to world wars.

This study addresses the question of why some great-power wars escalate to the global level, while others do not. There is a widespread belief in a significant portion of the contemporary literature on international relations that the category of "world wars" is a separate class of warfare within the wider set of major-power war (Wallerstein 1980; Gilpin 1981; Modelski 1987). Though scholars differ in their views of which processes lead to world wars, they share the functional and historicist form of explanation in which transitions in the international order are considered to be the result of "world wars" involving "structural causes." These causes summarize all those forces that bring down a hegemon and let other states rise relative to the leader, ushering in a period of capability deconcentration in the interstate system, followed by a period of global war or world war.

The research reported in this chapter consists of a further inquiry into the onset of world wars. Instead of a functional form of explanation at the level of the international system, we introduce a two-level answer to the question of when and why world wars break out. Our approach is multilevel in the sense that we focus on the effects of *dyadic* power transitions at the *systemic* level, it is data-based, it is causal in nature, and it is devoid of speculations on the functional meaning behind the bloody affair of warfare. We believe that we will not understand the phenomenon unless we are willing to consider causes at multiple levels of analysis. We turn now to discussion of the first level used in our approach.

The First Level: The Power Transition Hypothesis

The War Ledger

According to Organski (1968), it is the loss of power of the status quo power(s) relative to the state(s) that is prepared to use force to change the status quo that determines whether conflicts escalate. In *The War Ledger,* Organski and Kugler phrase the power hypothesis as follows:

> The dominant nation and the challenger are very likely to war on one another whenever the challenger overtakes in power the dominant nation. (1980:206)

Their test procedure consisted of three steps. First, total output was selected as the sole indicator of power resources. This means that other indicators are rejected as measures of capabilities. Second, all major powers were subdivided into central and peripheral major powers. A central major power has alliance ties with one or more major powers, while a peripheral major power is not allied with any other major power. Consequently, Japan enters the center only in 1900, the United States in 1940, and China in 1950. Central major powers were further divided into contenders and noncontenders. Contenders have at least 80 percent of the capabilities of the strongest power in the international system. When no nation meets this criterion in a given period, the three strongest states are defined as contenders.

In the third part of the test, the dependent variable, major-power war, was defined by three criteria: (a) on each side at least one major power participates; (b) the death toll is higher than in previous wars; and (c) the struggle results in the loss of territory or population for the loser.

Assuming that power transitions between contenders will trigger war at some point, Organski and Kugler narrow down the research question as follows: Is the outbreak of the major-power war preceded during the previous twenty years by a power transition?

It should be noted that Organski and Kugler put very severe limitations on the test of power transitions as a cause of war. By applying their operational criteria, they reduced the number of wars available for the test to four: the Franco-Prussian War, the Russo-Japanese War (see also Lemke and Kugler herein), and the two world wars of the twentieth century. Excluding the years of the actual fighting from their test periods, they settled on researching dyads of major powers in six periods: 1860–80, 1880–1900, 1900–1913, 1920–39, 1945–55, and 1955–75.

Table 1.6 in *The War Ledger* shows the test results for dyads of all major powers. For the entire set of major-power countries, both equal and unequal shares of power resources between adversaries are associated with war (1980:50). In addition, there is no case of military conflict among the most powerful nations of the world when power is shared equally by both members of each pair and one member is not in the process of overtaking the other. At the level of the great powers, wars occur if the balance of power is equal if and only if one member of the pair is overtaking the other in power. This fact, signifying that equality may be accompanied by war, contradicts balance-of-power thinking. Table 1.7 from Organski and Kugler (1980:52) shows the results of the test for each subgroup of major powers. The power transition hypothesis cannot be rejected for the subclass of contenders. Leapfrogging among the top three is a necessary, though not a sufficient condition for conflict. Within the other two subgroups of major powers, transitions are not related to the outbreak of war.

Evaluation of "The War Ledger"

First, it is not clear whether Organski and Kugler believe that the fact of power overtaking as such creates the perception of threat, or, alternatively, that the notion of contention already implies the perception of threat. If power transitions are believed to be the source of the perception of threat, the hypothesis should be tested against all cases of transitions among major powers, whether they are contenders or not.

Second, the authors make their test procedure confusing by identifying only cases of war among contending states and searching only for power transitions in the twenty-year periods preceding the outbreak of these wars. In our view, the power transition hypothesis implies that all power transitions should be identified. It should be established subsequently whether power transitions coincide with outbreaks of war. Organski and Kugler's argument passes from power transitions to war, but their test passes from war to preceding power transitions (see also Lemke and Kugler herein). Third, the power transition hypothesis points to war if a contender surpasses the dominant power, while their test procedure is concerned with all dyads in the subclass of contenders (for a potential resolution to this issue, see Lemke herein).

Extensions of Power Transition Research

Since we consider differences in growth rates of capabilities as one of the most important factors in explaining the outbreak of war, we have reproduced Organski and Kugler's analysis, while trying to achieve a more satisfactory modus operandi in a number of ways (Houweling and Siccama 1988a). First, we broadened the measurement of national power to demographic and military variables. We used the relative power indicators constructed by Doran and Parsons (1980:953). A disadvantage of Doran and Parsons's shares as indicators of power is that they are sensitive to the entry or exit of major powers. However, since we are using the shares only to analyze the power distribution on a dyadic basis, this disadvantage is not relevant to our analysis.

The second adaption we made is that in defining the category of major powers we adopted the definition proposed by Doran and Parsons. We did not use the distinction between peripheral and central powers.

Third, our dependent variable, major-power war, includes all military struggles in which at least one major power participated on each of the opposing sides. Using the war data collected by the Correlates of War (COW) group (Small and Singer 1982:82–95), this amounts to ten wars and thirty-three warring dyads.

Finally, in an effort to avoid any bias in the connection between outbreaks of war and test periods, we defined test periods simply as lasting roughly twenty years. Beginning in 1816 this yields a total of eight test periods.

To replicate Organski and Kugler's analysis we have adopted the following procedure. First, in each test period the number of major powers determines the number of dyads in that time frame. Major powers entering the subsystem during a test period were included in the analysis for that period. Additionally, using the relative power shares, we divided the major-power dyads in each subperiod into three categories: (a) An *unequal* power relationship. The power distribution in a dyad is considered unequal when, in any year for which scores are available during the test period, the relative capabilities of the nations differ more than x percent. We used three values for x: 20 percent (the criterion selected by Organski and Kugler), and in order to consider the sensitivity of this criterion, also 10 and 15 percent. (b) An *equal* power relationship without overtaking. The power distribution is defined as equal if the relative capabilities differ less than x percent. In this category, cases where one major power is overtaking the other are excluded. (c) *Overtaking,* defined as the passing of one major power by the other during a test period. A major power entering the subsystem during a test period with a higher relative power score than other nations (e.g., the United States in 1900, China in 1950) is considered to have overtaken these other nations.

Finally, we have made two computations: one for all major powers, yielding a total of 119 dyadic relationships for all subperiods, and another for only the three or four strongest major powers.

For all major powers, this results in the relationship between power distributions and the successive outbreak of war represented in table 6.1.

The relationship varies from 0.12 to 0.16 (Tau C), contrasted with Organski and Kugler's (1980:50) 0.05 from their table 1.6. In comparison with the re-

TABLE 6.1.   **Power Distributions and the Incidence of War, All Major Powers, 1816–1975**

	Unequal			Equal, No Overtaking			
	20%	10%	5%	20%	10%	5%	Overtaking
No war	58	63	70	14	9	2	14
War	17	20	21	4	1	0	12
Total	75	83	91	18	10	2	26

Source: Houweling and Siccama 1988:100.

Note:		Kendall's Tau C	Significance
	20%	.15931	.0255
	10%	.12796	.0477
	5%	.14914	.0167

sults in *The War Ledger,* the relationship we found for all major powers is much *stronger.* Since we are analyzing the population of wars itself, significance tests are reported solely for the purpose of comparing our results with those of Organski and Kugler. Finding such results here can be interpreted as supporting our objections to the definition of the contender class.

For the three strongest major powers in each twenty-year period, the results are shown in table 6.2. Organski and Kugler's (1980:table 1.7 "Contenders") Tau C is 0.5. Ours vary from 0.27 to 0.3. Thus, our results are weaker than theirs and, as is the case with their results, the result for the contenders is stronger than that for the complete class of major powers. Contenders act in ways more consistent with the expectations of power transition theory than does the overall set of major powers.

We explore the nature and strength of this relationship further, presenting table 6.3 as our object of analysis, abstracted from table 6.2.

The DRF-index of association between power transition in a dyad and warfare in that dyad is one if and only if: (1) all dyadic transitions are followed by outbreaks of war in that dyad, and (2) no dyad fights without a preceding transition. Consequently, the index, being the difference between two ratios, equals $8/17 - 3/19 = 0.32$. Dyadic capability relations are clearly important to the outbreak of war. But, since they provide neither necessary nor sufficient conditions, we must augment them with other factors. This is where the second level of our approach comes into play.

The Second Level: Critical Point Theory and Power Transition

The National Level

Charles Doran (1989) hypothesized that critical points (peaks, troughs, and inflection points) in the relative capability trajectory of nation-states cause the in-

TABLE 6.2. Power Distributions and the Incidence of War, Three of Four Strongest Powers, 1816–1975

	Unequal			Equal, No Overtaking			
	20%	10%	5%	20%	10%	5%	Overtaking
No war	10	12	15	6	4	1	9
War	2	3	3	1	0	0	11
Total	12	15	18	7	4	1	36

Source: Houweling and Siccama 1988:101.

Note:

	Kendall's Tau C	Significance
20%	.30556	.0327
10%	.27469	.0459
5%	.30247	.0276

volvement of that state in war. The results of his research are summarized in table 6.4.

The presence of a critical point on a nation's capability trajectory, signifying changes in modernization or decline, is a necessary condition for war. As table 6.4 shows, each of the eleven war participations is preceded by a critical point. There are, however, nine critical-point nations that do not fight during the relevant period. The condition at the national level, thus, is not sufficient.

In comparing the effects of power transitions and critical points, it may be of some interest to recall that the dependent variables differ. Power transitions relate to dyadic warfare, while critical points explain national war involvement against any other major power. Our restructuring of power transition theory allows us to connect it to critical point theory, because we have expended the set of wars that apply. This connection was not possible given Organski and Kugler's original formulation. In table 6.5 we study the performance of the transition hypothesis in explaining war involvements of dyad members against *any* major power (Houweling and Siccama 1991).

The two cases in the bottom left cell, the United Kingdom in the second period (1836–55) and the United States in the sixth period (1920–39), fighting in wars without power transitions, appear to contradict power transition theory. However, the nations against which they fought (Russia during the Crimean War and Germany in World War II) did experience a transition in these periods. If these two cases were to be added to the top left cell, only transition states would fight wars. Accordingly, being involved in a power transition would be a necessary condition for war (in this case the DRF jumps to 0.73 and Q to 1.0). The six cases in the top right cell prevent involvement in a power transition from being a sufficient condition as well. Of these six cases, two are allied in the Holy Alliance and three states are involved in nuclear deterrence relationships after World War II. This leaves only one state, Russia, that "should" in some sense, have fought against another major power in the period from 1856

TABLE 6.3. Relationship between Power Transitions and War in All Dyads

		War	
		Yes	No
Transition	Yes	8	9
	No	3	16

Source: Houweling and Siccama 1991:650.
DRF = 0.32 Q = 0.65

to 1875—if being involved in a transition were also a sufficient condition for war. However, Russia had just ended a war and became subsequently involved in a period of domestic unrest. The transition hypothesis clearly provides a very powerful explanation of war participation.

In the structural approach of leadership, system-level forces directly determine lower-level behavior. Consequently, systemic properties affect affairs at the dyadic and national levels. The alternative approach we prefer is multiple determination. This approach is the subject matter of contextual analysis (Hummell 1972). In contextual analysis the values of dependent variables at the individual level are predicted from the combined impact of individual-level predictors and group-level predictors.

With regard to the matter under discussion here, dyadic capability transitions between two major powers substantially increase the probability of warfare in the transition dyads; involvement in a transition predicts involvement in war. In this respect, world wars are not different from other major-power wars. What is different is the systemic context in which each type of war breaks out. We suggest that dyadic wars spread into world wars when capabilities become less concentrated and when the concentration index reaches an all-time low. Indeed, both world wars of this century have originated in dyadic transitions. In addition, in 1914 and in 1939, the British share of the Doran-Parsons capabilities set is lower than at any point in time since 1816. On the eve of World War I, naval assets, as measured by Modelski and Thompson (1988), are more evenly spread among the major powers than in any year since 1815. When the distribution of capabilities becomes ever more deconcentrated, a barrier against the spread of dyadic warfare is removed. It may be that world wars evolve from dyadic fights simply because there is no firefighter around to stop the blaze. This explanation is illustrated by British involvement in the "War in Sight" crisis of 1875 and lack of immediate involvement in the crises of July 1914. Similarly, on the eve of World War II, Britain and the United States followed a wait-and-

TABLE 6.4. **Relationship between Critical Points and War in Dyads of Three or Four Strongest Major Powers**

		War	
		Yes	No
Critical Point	Yes	11	9
	No	0	16

Source: Houweling and Siccama 1991:647.
DRF = 0.52 $Q = 1.0$

see policy. However, the change in German foreign policy from Bismark's status quo to the conquest and mass murders in high-tech death factories of the Third Reich is not implied by the deconcentration of capabilities at the system level. Systemwide deconcentration only opens the door. The things that get through that door are created at lower levels of analysis.

Two Caveats

The conclusion from the previous sections is that power transitions in major-power dyads correlate with outbreaks of war in these dyads. Correlations reveal a pattern, but they do not explain why things go together. Attention should be paid to the nature of the linkage between the variables investigated. First, we point out the problem of time frames. We have divided our study into twenty-year periods to enhance comparability with previous works. Choices about the time frame have to be made in one way or another, and no general theory is

TABLE 6.5.   Involvement in Power Transitions and War
against Any Top-Ranking Power

		Involvement in a War	
		Yes	No
Involvement in a Power Transition	Yes	1836–1855 France 1836–1855 Russia 1856–1875 France 1856–1875 Germany 1896–1914 UK 1896–1914 USA 1896–1914 Germany 1896–1914 USSR 1920–1939 Germany 1920–1939 UK 1920–1939 USSR 1946–1965 China 1946–1965 USA 1946–1965 UK	1816–1835 Russia 1816–1835 France 1856–1875 Russia 1966–1975 USA 1966–1975 USSR 1966–1975 China
	No	1836–1855 UK 1920–1939 USA	1816–1835 UK 1856–1875 UK 1876–1895 UK 1876–1895 USSR 1876–1895 Germany 1946–1965 USSR

Source: Houweling and Siccama 1991: 654.
DRF = 0.45 Q = 0.75

available to help. However, choices in research design can conceal rather important assumptions about the time nations and dyads need to respond. If we assume that the war response is "immediate" upon "the moment" of power passing, we have to define "immediate" and "the moment." In this study, immediate is within twenty years, and the moment is a five-year period. In respect to the choice of the time frame, we should emphasize the rather primitive nature of the findings reported above. Similar concerns about the spatial distribution of war activities (i.e., regions versus the whole system) may be in order too (Barraclough 1961).

Our second caveat concerns motivation. What about inserting the will to fight as an intervening variable? Constructs such as Hobbes's preference order, "propensity to plunder" (Skocpol 1981; Bauer and Matis 1989), or "lateral pressure" (Choucri and North 1975) could be useful. Intervening factors may be important if they appear to have some relationship with the actors' expressions of subjective experiences. What people are fighting for is of the essence here, yet we often only know that once fighting has begun. For this reason, Holsti's (1991) effort to introduce actor's purposes as explanations of war initiation is unfortunate. Purposes are a part of the behavior to be explained, not an explanatory tool on their own. Intervening variables are very likely critical to the linkage between capability transitions and war. Do transitions put nations in a state of readiness to fight, with other events required to trigger the actual combat? Ultimately, causes of war are dependent upon a state of readiness to fight, and transitions provide only the opportunity.

In order to consider our arguments more fully, we now assess the implications of recent world events, taking into account the two levels of our approach. Although we are optimistic about the future, we must still tread carefully.

Policy Relevance

For many decades, the political relevance of the power transition hypothesis has been sought in the demise of United States hegemony (Russett 1985; Rood and Siccama 1989). The Soviet Union and China were competing to succeed United States leadership of the world. In the light of recent events, how should we evaluate the peaceful abdication of the Soviet Union as the main contender for world leadership? What are the consequences of the reunification of Germany?

Our two-level explanation of war suggests increased uncertainty and insecurity as a consequence of the demise of the Soviet Union, the decline of the United States in Europe, and the increased power potential of Germany. A hierarchical structuring of power reduces uncertainty about the behavioral reactions of all parties. Shifts in the power distribution create uncertainty and elicit attempts to reduce this uncertainty by purely national initiatives. This change is most clearly visible in the tendency to "renationalize" defense in Central and Eastern Europe and, to an extent, in NATO. With the United States less prominent, France and Britain could potentially encourage nuclear proliferation and a nationalistic German foreign policy if they attempt to remain superior to Germany through their nuclear status and veto power in the UN Security Council. In this respect, renationalization still competes with integration. Assigning a special status to Germany in these issues, however, boils down to confusing German tanks with German Marks.

CHAPTER 7

Modeling Power Transition: An Extended Version

Vesna Danilovic

Theoretical Framework

For several decades, the international relations literature was dominated by the classical realist approach, which states that a balance of power creates peaceful structural conditions. This approach was later challenged by power transition theory on both logical and empirical grounds. Related theories of hegemonic stability and global cycles also attempt to demonstrate that the history and logic of major power relations run contrary to the classical realist model.

These alternative approaches argue that systemic instability and the likelihood of major wars do not decrease, but rather increase during periods of power parity when there are power shifts between major contenders in the system. Power preponderance, therefore, creates a peaceful structural condition since a preponderant nation is satisfied with its advantaged position in the system, while other great powers are not capable of challenging the order. According to power transition theory, the Pax Britannica and Pax Americana offer historical evidence of this argument. By contrast, the condition of parity between two or more great powers creates a much less stable condition. Under these circumstances, more than one great power is capable of assuming the leading role in the international system. The period of power parity, therefore, when an overtaking occurs between a rising challenger and a dominant status quo power, is a period of competition for leadership and may be critical for systemic stability. Two world wars in the twentieth century are examples of this unstable condition.

The arguments of both schools have been tested in formal models and systematic empirical analyses (see Lemke and Kugler herein; Siverson and Miller herein; Vasquez herein). Though these tests tend to give more credence to power transition theory, they also demonstrate that the theory needs some refinements and an extension with microlevel variables. In particular, power transition is advanced as a necessary, but not a sufficient, condition for major wars. As a result, several studies have looked for additional factors as sufficient conditions for major wars under power parity.

In search of further refinements of power transition theory, the present analysis attempts to amend the original argument with a more precise specification of the challenger's dissatisfaction and the implications for war. It also extends the original model with actors' credibilities as an additional factor that may help to account for exceptional cases of peaceful transitions or unstable preponderance conditions, or both. A simple formal model that combines game theory and expected utility theory is developed to structure the assumptions in a logically consistent way. The analysis concludes with a discussion of the implications of the formal model for the debate between classical realism and power transition theories.

The model contains three predictor variables that define the necessary and sufficient conditions for the outbreak of major-power war. The emergence of a dissatisfied major power—measured by an underrecognized diplomatic status—signifies the presence of a potential challenger to the international order. Critical values of two other factors must coincide with the emergence of a dissatisfied power for a major war to occur. One is the relative power potential of a challenger as compared to the principal defender of the status quo, which is either equal or marginally different (depending on the stage in the power transition process). The other critical variable is the credibility of both powers, defined in terms of the stakes that they have in areas where an attack might occur. This theoretical framework, then, is formulated in the tradition of the power transition theory as it explores the destabilizing effect of power parity, but supplements it with a credibility variable and uses the concept of status inconsistency to assess dissatisfaction.

The set of predictor variables comprising capabilities, credibility, and dissatisfaction, reflected by attitudes toward the status quo, was proposed by Kugler and Zagare (1990), and extended by Zagare and Kilgour (1991, 1993) in their game-theoretic analysis of the power transition model. The present analysis suggests an alternate conceptualization of the interrelationship of these variables.

Power Capabilities

Power transition theory (Organski 1958; Organski and Kugler 1980) challenges the prevailing balance-of-power paradigm (Morgenthau 1948; Waltz 1979). Three of its contending assumptions are particularly important for this study (the differences between these two models are also summarized in Kugler and Organski 1989b).

First, the classical realist assumption about the anarchical nature of the international system is challenged by the alternative assumption that the international order is hierarchically structured. In a hierarchical system, some states are presumed to be satisfied with their place in the system, while others are not. Satisfaction with the existing hierarchy is one of the critical factors in the power transition model. Satisfied states do not, in general, initiate conflicts. Negative attitude toward the status quo distinguishes a potential challenger from the defender of the order. In the classical realist model, however, where all states are presumed to be dissatisfied, preponderance in power capabilities is sufficient to provide an incentive to initiate war (for an extension of this argument, see Kugler and Werner 1993).

The second difference between the two theories concerns the goals states are presumed to pursue. In the realist framework, where anarchy reigns, states are seen as power maximizers intent on preserving and enhancing their security. An advantage in capabilities facilitates the pursuit of this goal, making power advantage a critical requirement for initiating a conflict. Conversely, parity constrains conflict initiation, creating a more stable environment.

By contrast, assumptions about the ordered nature of the international system bring power transition theory to opposite conclusions about more stable power distributions. Similar to the hegemonic stability and long-cycle theories (Gilpin 1981; Modelski and Thompson 1989), the power transition model maintains that the large power gap dividing a dominant nation from the next layer of major powers facilitates the maintenance of the international order. Instability arises when the gap between the dominant nation and a challenger narrows. It is the competition for leadership in the international system between a status quo power and a dissatisfied challenger under parity that creates the critical condition for the occurrence of major wars.

The third difference is an underlying issue in their disagreement over more destabilizing power distributions. The argument about power-maximizing goals of states is consistent with actors' calculation of their potential absolute gains and losses from conflict. An alternative understanding of the goals of national elites (i.e., the competition for leadership) brings forth the importance of relative costs and benefits in calculating utilities from war and peace (for a discussion of this issue, see Kugler and Zagare 1990). Empirical evidence for either power framework has been inconclusive (see Lemke and Kugler herein; Siverson and Miller herein; Vasquez herein). Since analysis of power variables cannot go beyond the necessary conditions for major war (Organski and Kugler 1980:51), this analysis attempts to strengthen the argument by looking at dissatisfaction and credibility to approximate sufficient conditions for major wars.

Dissatisfaction: Attitude toward the Status Quo

Though Kugler and Organski (1989b:173) contend that "[d]egrees of satisfaction as well as power are critical determinants of peace and conflict," power transition theory does not specify sources of discontent with the international order, except for the decrease in the power gap between a dominant nation and a challenger (for alternative ideas about dissatisfaction, see Kim herein; Werner and Kugler herein). While not rejecting possible correlations between changes in power distributions and variations in support of the international order, it would be more appropriate to create a distinct measure for dissatisfaction.

Midlarsky (1975) notes the existence of analytical compatibility between the concept of status inconsistency and dissatisfaction toward the status quo as formulated in power transition theory. If status inconsistency is defined as a discrepancy between the achieved status of a major power (its position in capabilities) and its ascribed status (its prestige within the existing hierarchy), then underrecognized status could cause dissatisfaction. The uneven differential rate of change in the achieved and ascribed status of a major power may be regarded, then, as a principal source of dissatisfaction (Wallace 1972). Still, while dissatisfaction may be rooted in underrecognition, a challenger's relative power potential is critical in determining whether the challenger will initiate conflict.

Most previous research on status inconsistency is set at the systemic level: the aggregate amount of status inconsistency in the system is correlated with the aggregate amount of war (East 1972; Wallace 1972; Midlarsky 1975). A major weakness of this approach is that it does not reveal whether a war was initiated by an underrecognized state or whether some of those states were even actual participants in wars. To avoid the above-mentioned problems, status inconsistency is conceptualized here rather as a property of dyadic relationships. This consideration is similar to that of Gochman (1980), who conceptualizes and operationalizes status inconsistency at the national level. However, a consistent dyadic focus is appropriate for work in the power transition tradition.

Credibility

Deterrence theory provides another potential clue to understanding conflict initiation, that is, credibility as closely related to the consideration of power capabilities as a determinant of conflict. A similar debate divides theorists over whether credibility stems from calculating costs and benefits of retaliation in absolute or relative terms. Despite the differences in answers, credibility is regarded as a subjective category in both. Still, defining credibility as a result of a subjective evaluation involves empirical indeterminacy as to whether the would-be attacker was actually threatened or not, making empirical evaluation problematic.

For these reasons, credibility is defined here as an objective category that reflects an actor's willingness to defend against attack. Furthermore, it is defined separately from an actor's capabilities, since both credible and capable threats are necessary for deterrence to work. Put differently, though the defender might be capable of retaliation, it does not mean that it is willing to do so.

One possibility is to connect credibility with the degree of stakes that major powers have in various regions of the international system, to assess their willingness to retaliate (see also Zagare and Kilgour 1993). Therefore, credibility is related to the objective indicator of state's interests or stakes in particular crisis areas and it is assumed that each actor is aware of its opponent's regional stakes.

This conceptualization of credibility implies the notion of immediate deterrence "where at least one side is seriously considering an attack while the other is mounting a threat of retaliation in order to prevent it" (Morgan 1983:30). It is, hence, different from the notion of credibility in a general deterrence theory, which explores situations in which opponents "maintain armed forces to regulate their relationship even though neither is anywhere near mounting an attack" (Morgan 1983:30). As Huth and Russett (1984) point out, immediate deterrence is not only easier to identify, but it is also more important to analyze because it reveals why deterrence fails or succeeds in situations where the credibility of threat is of utmost importance.

Past research is instructive here. Bueno de Mesquita and Lalman (1992) analyze the relationship between the stakes and the costs of war, but they define stakes in terms of dissatisfaction with the status quo. There are some studies of the characteristics of the "issues at stake" and the probability of crisis escalation (Gochman and Leng 1983), but the concept of issue salience has not been used in terms of regional salience as it is here. Namely, the issues at stake are usually observed according to their importance for the state's vital interests, such as control of one's own or adjacent territory and political independence. The present analysis suggests the regional stakes of major powers as an indicator of their credibilities to retaliate.

Formal Model

Building on previous work (Zagare and Kilgour 1993, Zagare herein), the present study offers a model of power transition theory that is convenient for empirical testing in the future. Rational choices for each player are specified according to the probability each player has of shifting from one outcome to another. All three variables—status inconsistency (SI), relative capability (CAP), and credibility (CR)—are postulated to be important in making rational choices, though not with equal weight for each possible outcome.

Four possible outcomes of the interaction between two major powers, faced with cooperation (C) and defection (D) as options, may be presented within a game-theoretic framework. Figure 7.1 reveals the sequence of decisions that this interaction entails. In what follows, SI_{Ch}, a measure of the Challenger's status inconsistency, indicates Challenger's incentive to upset the status quo. Defender's and Challenger's threat credibility is given by the

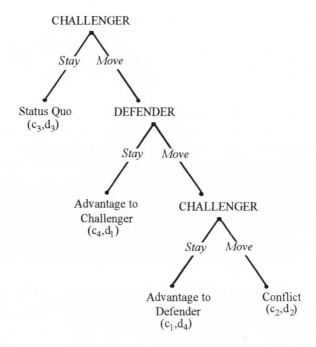

Fig. 7.1. The sequence of decisions.

probabilities CR_{Def} and CR_{Ch}, respectively. Defender's and Challenger's capabilities are indicated by CAP_{Def} and CAP_{Ch}, respectively.

The Challenger's incentive to upset the status quo is measured by the index of status inconsistency SI_{Ch}, and it ranges between 0 and 1. The lower the value of this index, the less underrecognized the status is. If SI_{Ch} is greater than 0.5, Challenger's status is underrecognized and, hence, the Challenger is dissatisfied with the status quo. Therefore, this measure meets the following requirements:

if Challenger's status is not underrecognized, then $0 \le SI_{Ch} \le 0.5$;
if Challenger's status is underrecognized, then $0.5 < SI_{Ch} \le 1$.

Therefore, all three factors—actor A's dissatisfaction with the status quo (SI_A), relative capability (CAP_A), and credibility (CR_A)—yield certain probabilities and their values range between 0 and 1. Operational definitions should thus specify empirical measures in terms of probability values.

The Challenger's utility from initiating a conflict may be summarized as follows:

$$U_{Ch} = (SI_{Ch})[(1 - CR_{Def})(1 - CAP_{Def})(c_4)$$

$$+ (CR_{Def})(CAP_{Ch})(c_{2+}) + (CR_{Def})(1 - CAP_{Ch})(c_{2-})$$

$$+ (1 - CR_{Def})(1 - CAP_{Ch})(c_1)]$$

In the case that the Challenger moves from the status quo (*CC*), with incentive SI_{Ch}, the probability for the next outcome (*DC*), which gives an advantage to the Challenger (with the payoff c_4), depends upon the Defender's credibility and capability. More precisely, the Challenger would have an advantage from upsetting the status quo without further escalation as long as the Defender is not sufficiently credible and capable to retaliate, that is, with the probability $(1 - CR_{Def})(1 - CAP_{Def})$. However, if the Defender's threat to retaliate is credible, conflict (*DD*) occurs, which yields the twofold utility for the Challenger: c_{2+} is the payoff from winning the war with the probability $(CR_{Def})(CAP_{Ch})$ and c_{2-} is the payoff from losing the war with the probability $(CR_{Def})(1 - CAP_{Ch})$. The last two pairs of probabilities may also be interpreted as probabilities for success or failure in the case that the Defender retaliates.

The Challenger may back down after the Defender's initial retaliation if it is not capable of continuing the conflict. This last possible outcome of the game (*CD*), which yields the payoff c_1 to the Challenger, also depends upon Defender's willingness to accept cessation of conflict and not to punish the Challenger with more costs. Therefore, this outcome will occur with the probability $(1 - CAP_{Ch})(1 - CR_{Def})$. Finally, the Challenger's preference to stay at the initial outcome (*CC*), with a payoff of c_3, equals $(1 - U_{Ch})$.

So far I have explained the Challenger's utility from upsetting the status quo. The final outcome will also depend upon the Defender's decisions. The Defender's expected utility from a conflict is as follows:

$$U_{Def} = (SI_{Def})[(1 - CR_{Def})(1 - CAP_{Def})(d_1)$$

$$+ (CR_{Def})(CAP_{Def})(d_{2+}) + (CR_{Def})(1 - CAP_{Def})(d_{2-})$$

$$+ (1 - CR_{Def})(1 - CAP_{Ch})(d_4)]$$

The probability for war (P_w) can be derived from the Challenger's utility (U_{Ch}) and the Defender's utility (U_{Def}) from conflict:

$$P_w = (U_{Ch}) + (U_{Def}) - (U_{Ch})(U_{Def})$$

The equation indicates that war is more likely if either the Challenger has high utility from initiating conflict or the Defender has high utility from preempting or defending, or both.[1] An addition rule is used because U_{Ch} and U_{Def} are neither mutually exclusive nor independent. They are not mutually exclusive, because both the Challenger and the Defender may have high utilities from initiation and preemption respectively. The values of U_{Ch} and U_{Def} are not independent since they also reflect the probability that an actor will defend if its opponent attacks. Hence, P_w indicates that war is more likely if both actors have high utilities from conflict. War should be expected if $0.5 < P_w \leq 1$ and, inversely, peace is expected if $0 \leq P_w \leq 0.5$.

To make predictions, each player's payoffs must be specified as well. To incorporate the model into the theoretical framework, it is assumed that a player's payoff depends on the extent to which an outcome accrues gains or losses to the player's power position (capabilities) and diplomatic prestige. The values of potential gains and losses increase as the player interacts with stronger and more prestigious opponents. Inversely, they decrease if an opponent is weaker and less prestigious. Thus, the outcome that gives the advantage to a player without costs imposed by the opponent (*DC* for Challenger and *CD* for Defender from fig. 7.1) yields the following payoffs:

$$c_4 = (CAP_{Def}/CAP_{Ch})(PR_{Def}/PR_{Ch})$$

$$d_4 = (CAP_{Ch}/CAP_{Def})(PR_{Ch}/PR_{Def})$$

In other words, the payoffs are determined by the ratios between the opponent's and actor A's own capabilities (CAP_A) and prestige (PR_A), wherein $PR_A = 1 - SI_A$. It is worth remembering that these payoffs do not reflect real changes in capabilities and prestige (diplomatic status), but rather they indicate values that players attach to each outcome.

Winning a war is another outcome that brings an advantage to the player, minus the costs of fighting. Since the gains included in c_4 and d_4 would be reduced significantly if the opponent is a more powerful and prestigious actor, c_{2+} and d_{2+} may be calculated as follows:

$$c_4/c_{2+} = 1/R$$

$$d_4/d_{2+} = 1/R$$

R is defined as the ratio between the opponent's and the player's own capabilities and diplomatic prestige. The equations show that the ratio between c_{2+} and c_4 equals the ratio between R and 1.00 (the same rule applies to d_{2+} and d_4).

This stipulation for payoffs means that $c_{2+} = (c_4)^2$ if $c_4 < 1$ (i.e., the opponent is weaker and less prestigious), and $c_{2+} = 1/c_4$ if $c_4 > 1$ (i.e., the opponent is stronger and more prestigious). In other words, each player's gains from a peaceful status quo change to its own advantage would be higher than the gains from winning the war (in either case, $c_4 > c_{2+}$ and $c_{2+} < 1$). Also, the costs of fighting are much higher for a weaker player than for a stronger player. If the Challenger faces, for instance, a two times stronger (and more prestigious) Defender, its payoff from a peaceful reversal in their positions (c_4) would equal 2. The Defender's payoff from upsetting the status quo (d_4) would equal 0.5. If they decide to fight, the payoffs from winning the war would be $c_{2+} = 0.5$ and $d_{2+} = 0.25$ for the weaker player and the stronger player, respectively. That is, the war would be costlier for the Challenger (the payoff c_4 is reduced by 75 percent) than for the Defender (d_4 is reduced by 50 percent). In short, what this payoff specification suggests is an asymmetry of payoffs for antagonists who are asymmetric in power and prestige.

Since both c_{2+} and d_{2+} incorporate the costs of fighting (and each is less than 1), a more uniform procedure can be applied for specifying the payoffs from backing down before the war escalates, and from losing the war. The payoffs from these less preferred outcomes would exponentially decline (i.e., $c_1 = (c_{2+})^2$, etc.), so that the following requirements are satisfied:

$$c_4 > c_{2+} > c_1 > c_{2-}$$

$$d_4 > d_{2+} > d_1 > d_{2-}$$

There is one exception to the last requirement. Namely, if two powers have equal capabilities and prestige, their payoffs from all outcomes equal 1. These conditions define the situation of high uncertainty (both players have equal chances to change the status quo to their own advantage). Players' choices would depend upon the level of their credibility as well as upon the nature of their prestige—whether both powers are underrecognized, overrecognized, or both of them have consistent status.[2] However, as the power gap between them increases in the same direction as the prestige gap this uncertainty decreases. In short, the model reflects the assumption that war is *least* likely between a satisfied preponderant actor and a dissatisfied, but weaker opponent. Inversely, war is *more* likely as actors begin to approach each other in power potential and credibility.

Note that the model is consistent with the original version of power transition theory as long as the more capable actor is credible and vice versa. Deviations from the theory arise when these variables are inversely related. The likelihood of war depends, then, upon the magnitude of the differences in players' capabilities and credibilities and the extent to which the challenger is dissatisfied with its status.

Conclusion

The formal model presented here improves upon previous power transition efforts by suggesting the importance of the factors of credibility and status inconsistency to measure dissatisfaction. However, absent empirical results, the claim of improvement cannot be made very strongly.

My model suggests that instead of relying simply on power as the only determinant of conflict, researchers should examine how power interacts with other factors that might contribute to the onset of war. Contrary to the balance-of-power theory, the power transition model goes beyond a simple power argument. The original empirical examination of the model, however, does not explicitly incorporate dissatisfaction with the status quo, despite the fact that it is included as a part of the theoretical conceptualization (Organski and Kugler 1980). I attempt here to correct for this omission as well as to amend it with a third factor, namely, credibility (conceptually revised in terms of regional stakes). The combination of the two additional conditions—satisfied diplomatic status and low stakes (credibility) on the part of the challenger—might explain why some transition periods are peaceful. Previous research efforts (Bueno de Mesquita 1981a, 1985; Bueno de Mesquita and Lalman 1992) encourage us to combine formal models with rigorous empirical testing. This is clearly the next step for the model developed here.

NOTES

I am indebted to Frank Zagare for his insightful comments. An earlier version of this paper was presented at the 1992 annual meeting of the American Political Science Association. I am grateful to Jacek Kugler and panel discussant James Morrow for their valuable comments.

1. While the model includes the possibility for the Defender's preemptive strike (i.e., the reverse order of moves in the U_{Def} equation), it must not allow for this possibility to arise due to the Defender's dissatisfaction with the status quo. Namely, the theory assumes that the Defender (i.e., status quo power) is a satisfied nation. "The dominant nation is the main architect of the international order and is assumed to be satisfied, while the challenger must be dissatisfied" (Kugler and Organski 1989b:186). If an empirical test measures actual values of each nation's attitude toward the status quo, then presuming that a nation is the Defender, its SI_{Def} may sometimes reflect its dissatisfaction with the order. Since this would be inconsistent with the assumption about the Defender, it should be specified that SI_{Def} is a constant value that does not exceed a certain threshold. The most adequate threshold seems to be around 0.30 because it rules out the possibility for war between two satisfied actors, notwithstanding other factors. That is, war between two satisfied actors is not likely even in an extreme case when their relative capabilities and credibility may yield them very high utility from conflict. In this case, P_w will not be greater than 0.50 only if each actor's SI_A is less than 0.30. Thus, the empirical analysis should examine each nation as if it were the Defender and, then, as if it were the Challenger. When a nation is presumed to be the Defender its SI_{Def} equals 0.30. However, when the nation is presumed to be the Challenger, its SI_{Ch} is variable, reflecting the extent of its actual satisfaction or dissatisfaction with the status quo.

2. The last condition regarding players' prestige is important since the Challenger's incentive to upset the status quo would vary according to these circumstances.

CHAPTER 8

Relative Power, Rationality, and International Conflict

Daniel S. Geller

Two basic issues involving capability distributions and power transition theory are examined in this study: first, the nature of the dyadic capability balance— parity or preponderance—most likely to be associated with the *occurrence* of great power wars; second, the question of the interaction effect between the static power balance, differential growth rates in capabilities, and conflict *initiation* among contenders. The analysis includes the empirical extension of transition theory to secondary major-power wars not associated with system leadership and to subwar militarized disputes among contenders. This study will focus on the issue of the relevance of power transition theory to a broader set of great-power wars, and on the question of the relationship between relative capabilities and the identity of the conflict initiator.

The theoretical and empirical work on power transitions by A. F. K. Organski (1958) and by Organski and Jacek Kugler (1980) constitutes one of the most important contributions to an understanding of the relationship between power distributions and war. According to the general thesis advanced in both *World Politics* and *The War Ledger*, a hierarchy of power determines and supports the international order at any given time, in the sense that the rules of the system and the division of values reflect the interests of the dominant state and its allies. However, the power relations among states are not permanent, and differential or uneven national capability growth rates ensure that the international distribution of power will shift and undermine the foundations of order. A growing disjuncture between a changing power distribution and the hierarchy of prestige produces a disequilibrium that, if uncorrected, results in crisis. In this way, shifts in power relations favoring dissatisfied challenger states rather than the dominant nation create the necessary conditions for transition wars. In short, the theory identifies a mechanism for war in the operation of power balances that lead away from preponderance and toward parity. More specifically, the thesis suggests that both static and dynamic power balances are associated with the occurrence of conflict, with points of contention developing over the relative importance of power parity versus actual transition, as well as the relevance of political capacity, geography, alliance capabilities, and third-party interventions in major-power wars (e.g., Moul 1985; Kim 1992; Kim and Morrow 1992; Geller 1993).

Other critiques of power transition theory and its empirical test involve its application to wars that do not impinge on system leadership. For example, Organski's (1958:322–23) original thesis dealt only with war between dominant and challenger states. However, *The War Ledger* reports confirmation of the theory based on an empirical test of power balances from the Franco-Prussian War, the Russo-Japanese War, and World Wars I and II. The questionable relevance of the Franco-Prussian and Russo-Japanese Wars to the test of power transition theory in *The War Ledger* has been noted by Bueno de Mesquita (1980:376–80), Thompson (1983a:103), and Levy (1985:353–54). (See, also, Lemke and Kugler herein.)

Another issue in dispute revolves around the question of conflict initiation. Specifically, most theories of war and lesser forms of international conflict rest on assumptions of rational choice. For example, Morgenthau (1948), Blainey (1973), Waltz (1979), Bueno de Mesquita (1981a), and Gilpin (1981), to cite but a few, all argue that rational calculations guide decisions of war and peace. Moreover, all of these explanations of interstate conflict explicitly incorporate decision maker estimates of relative power among competitors. As Small and Singer (1982:194) note, it is " . . . likely that the relative prospects for . . . victory or defeat will play a key role in such a decision." In fact, data linking the frequencies of war initiation and subsequent victory often have been used as evidence in support of the assumption of rational choice. Connections between power transition theory and rational choice by Kugler and Zagare (1990) question the original version of power transition theory (Organski 1958) postulating initiation by the weaker contender. Thus, the issues of relative power and conflict initiation bear heavily on theories of war and on assumptions of rational decision making.[1]

Interlevel Power Dynamics

The applicability of certain test cases in *The War Ledger* (i.e., the Franco-Prussian and Russo-Japanese Wars) to power transition theory in its original form as an explanation for wars that determine system-level leadership (Organski 1958:299–338; Organski and Kugler 1980:19) is debatable.[2] Britain—the dominant global power of the nineteenth and early twentieth centuries—was not a combatant in the conflicts noted above. However, Houweling and Siccama (1988a) apply power transition theory to lesser wars between major states and report empirical evidence in support of the extended transition thesis: capability shifts among the set of all major powers (and not just contenders) are associated with war.

There may be a closer connection between the relatively few and isolated instances of war for system-level leadership and the more numerous but lesser conflicts among major powers than only the dyadic balance of capabilities between combatants. Specifically, shifts in system-level power structure may have an *interactive* effect with dyadic capability distributions and the occurrence of warfare. The structural theories of Organski (1958), Gilpin (1981), Modelski (1983), Wallerstein (1984), and Thompson (1988) all suggest that systemic stability and world order are associated with the dominance of a single state. As the systemic distribution of power changes from a unipolar concentration of capabilities to a more diffuse multipolar balance, they argue that the disintegration of the hierarchy will lead to a challenge against the dominant state and result in global war. However, it also may be postulated that as the international system moves from a high concentration of resources in the leading nation toward multipolarity (power diffusion), the conflicts among the *entire set* of major states will become increasingly frequent, due to the weakening of the principal defender of the order. Movement toward power equality within these nation-dyads may trigger violent interactions that—though not related to system leadership—are still of considerable magnitude. This suggests that the erosion of the system-level power structure may be linked to lower-order wars among major powers, as well as to system-shaping global wars. Hence, systemic power deconcentration/diffusion and the reordering of the hierarchy among the strongest states may result in secondary great-power wars that precede the system-shaping global wars discussed by Gilpin, Modelski, Thompson, and Organski. In this way, power distributions at both the systemic and dyadic levels of analysis may interact synergistically to produce both global and secondary wars among the set of major powers.

Traditionally, theories of war have been formulated at either the state, dyadic, or systemic levels of analysis, and attempts at unification have been relatively rare.[3] However, Doran (1983a) and Thompson (1983b) discuss a possible connection between system- and state-level power dynamics and the occurrence of global war. In a subsequent study Doran (1989) presents empirical evidence of this linkage in his analysis of the causes of World War I. Vasquez (1986) also notes the possible interaction of systemic- and dyadic-level power dynamics as responsible for the different forms of warfare found in the nineteenth and twentieth centuries. Related arguments by Goertz (1992) and Bremer (1993) suggest that systemic-level attributes should be viewed as "contextual" factors that may influence conflict patterns at lower levels of analysis. This position has been reiterated by Rasler and Thompson (1992a) who note that studies of war that focus on the structural context of interaction should consider the more complex possibility that smaller structures may exist within larger ones, as in the case of dyads within a broader system. They argue (1992a:11) that one of the key problems in producing consistent results in studies of capability distributions and warfare may " . . . stem from attempting to analyze the contextual influence of *one* structure on international behavior when more than one type of structure is crucial."

A recent empirical example of the interaction of power dynamics at two levels of analysis may be found in Houweling and Siccama (1991, herein). Their examination of the interaction of dyadic power transitions with critical points in the (national-level) power cycle indicates a synergistic effect: the occurrence of a power transition in a dyad where at least one member is passing through a critical point in its power cycle is a sufficient condition for war. Houweling and Siccama conclude that system-level forces are "channeled" by dyadic- and national-level factors in producing their effects. A similar finding has been reported (Geller 1992a) regarding the interaction of systemic- and dyadic-level power distributions. The empirical results of that study indicate a synergistic effect between systemic capability deconcentration, dyadic power transitions, and warfare: power transitions among the entire set of strongest states are most likely to result in war under conditions of systemic-level capability diffusion (cf. Houweling and Siccama 1993). Regarding conflict initiation patterns among great power rivals, Huth, Bennett, and Gelpi (1992) produce findings of an interaction effect between system- and national-level variables. Specifically, they argue that both system uncertainty (a structural attribute) and risk propensity of national decision makers affect conflict initiation. The results indicate that the impact of systemic factors is mediated by unit (national) level variables.

In line with this reasoning, three hypotheses on the parity/preponderance arguments found in power transition theory are tested. Two of these hypotheses involve linkages between systemic- and dyadic-level power dynamics and war occurrence.

H1. Parity in capabilities is more likely to be associated with the occurrence of war than is preponderance.

H2. Under a condition of increasing systemic-level power concentration (stable hierarchy), dyadic capability differentials (parity/preponderance) will be unrelated to the occurrence of war.

H3. Under a condition of decreasing systemic-level power concentration (unstable hierarchy), parity in capabilities is more likely to be associated with the occurrence of war than is preponderance.

Rationality, Capability Distributions, and Conflict Initiation

Most theories of international conflict and strategic interaction are based on assumptions of rational choice. These explanations also generally incorporate decision maker estimates of the relative power balance between disputants. Based on early Correlates of War (COW) data, Singer (1972), Bueno de Mesquita

(1978, 1981a), and Small and Singer (1982) note the strong empirical association between war initiation and victory, and discuss the implications of this association both for capability distributions among disputants and for rational choice assumptions in conflict theory. For example, Bueno de Mesquita (1978:252) states that since war initiators are assumed to act to maximize expected utility, it should be anticipated that most war initiators possess a capability advantage over their targets. He subsequently (1981a:21–22, 29–30) argues that given the record for victory, war initiators either have maintained some " . . . systematic advantage, or they have been extraordinarily lucky." He holds the more plausible explanation is that they possessed a capability advantage and employed it, and he concludes that war initiation is calculated on a rational basis.

Providing extensive evidence from their COW project data, Small and Singer (1982:194–95, 199) report that from 1816 through 1980:

> . . . initiating forces emerged victorious in 42 (or 68 percent) of the 62 cases [of interstate wars], suffering 20 defeats. . . . There were 22 wars initiated by major powers against minor powers, and the majors were successful in all but two, or 91 percent, of the cases. . . . From [this] viewpoint, it could be urged that foreign policy decision makers are indeed quite competent in evaluating capabilities. . . . Furthermore, . . . one might go on to conclude that the prewar process is indeed a highly rational one (in the limited, problem-solving sense). . . .

When data on subwar international disputes are combined with the data on wars, the evidence looks remarkably similar: more powerful nations initiate conflicts against weaker nations. Gochman and Maoz (1984:597) examine the distributions in their Serious Interstate Dispute database (for the years 1816 through 1976) and note the strong tendency for major powers to be initiators rather than targets in conflicts with minor powers.[4]

A curious anomaly involves the pattern of conflict initiation and victory among the set of strongest states. Specifically, Small and Singer (1982:195) report that from 1816 through 1980 " . . . when a major power initiated war against another major power, the initiator was victorious in only 3 of 9 such wars, or 33 percent. . . ." This finding is disconsonant with the more general patterns of war/dispute initiation and victory, and implies either that there is something about great-power conflict that affects the rational-choice abilities of decision makers for those states, or else that conflict initiation among major powers occurs under conditions different from those of other nations.

One possible explanation for this anomaly may be found in the work of Organski and Kugler (1980) on the power transition. According to the thesis advanced by Organski and Kugler (1980:55), two factors account for major conflicts between the most powerful nations in the system: (1) the relative capabilities of the states; and (2) the growth-rate differential of power between the challenger and the dominant nation. The theory suggests that major wars are produced by unstable power balances—war occurs when a dominant nation has its advantage in relative capabilities eroded due to the rising power trajectory of a challenger.[5] Houweling and Siccama (1988a) maintain that the theory holds for the entire set of major-power wars and not merely contender wars and produce evidence in support of this contention. Regarding the identity of the conflict initiator, Organski's (1958:333) original description reads as follows:

> It might be expected that a wise challenger, growing in power through internal development, would hold back from threatening the existing international order until it had reached a point where it was as powerful as the dominant nation . . . , for surely it would seem foolish to attack while weaker than the enemy. If this expectation were correct, the risk of war would be greatest when the two opposing camps were almost equal in power, and if war broke out before this point, it would take the form of a preventive war launched by the dominant nation to destroy a competitor before it became strong enough to upset the existing international order. In fact, however, this is not what has happened in recent history. . . . [World Wars I and II involved challengers attacking] the dominant nation and its allies long *before* they equalled them in power, and the attack was launched by the challengers, not by the dominant camp.

Bremer (1980:69) reports empirical evidence in support of the hypothesis that it is the weaker of the great powers that begins the conflict. Specifically, Bremer notes that the set of second-ranking nations (over the 145-year span from 1820 to 1964) had the highest relative rate of war initiation, whereas the set of first-ranking nations (of the top five ranks) had the lowest war-initiation rate. He concludes that:

> To the extent that initiation is a valid indicator of who "forces" a dispute into war, this finding would be consonant with the theoretical positions of those like Organski . . . who emphasize the inherent conflict between the [dominant state] and the number two nation in the hierarchy.

Other evidence adduced in support of the contention that the power transition aggressor will be the weaker state is provided by Maoz (1982:160). His analysis of dispute data indicates that conflict initiators tend to be dissatisfied states undergoing significant increases in capability growth rates, rather than satisfied nations experiencing a relative capability decline.

However, Bueno de Mesquita (1980:379–80) points to a problem in this aspect of Organski's early notion of power transition logic: the hypothesis of a weaker challenger initiating a transition war against a more powerful nation can be true only if the challenger's decision makers are extremely risk-acceptant. Moreover, the transition theory stipulates that alliances are not critical among the factors in contender wars. In short, according to Bueno de Mesquita, " . . . it is irrational for a weak state to initiate a war against a stronger state when the war is expected to be bilateral and the goal is to win. . . ."

Levy (1987:83–84), in his analysis of preventive war, also notes the lack of a theoretical basis for the presumptive identification of the transition aggressor as the weaker state. Faced with a faster-growing challenger, Levy argues, preventive military action by the dominant power would appear to be a plausible foreign policy choice. Gilpin (1981:191, 201) similarly discusses calculations that might make "preemptive war" an "attractive" option for a dominant power whose capabilities are declining relative to a rising challenger. In *The War Ledger* Organski and Kugler (1980:57–60) reexamine the war-initiation hypothesis in terms of the power balances prior to World Wars I and II and note that in each case the challenger had passed the transition point before the conflict erupted. Nevertheless, they indicate that "[w]ith such small numbers [of cases] one has, at best, traces of trends . . ." (see also Lemke and Kugler herein).

In sum, power transition theory holds that capability distributions are involved in war decisions and that the specifics of these distributions determine the identity of the conflict initiator. In line with this reasoning, the following general hypothesis will be tested.

H4. Among great power dyads the distribution of capabilities is associated with the identity of the conflict initiator.

Measure Construction

The data used to test the hypotheses on power distributions,[6] conflict initiation, and war were drawn from the COW database.[7] The National Capability data set includes six measures covering three dimensions of national attributes. Two measures involve *military* capabilities (military expenditures and military personnel); two measures assess *industrial* capabilities (energy consumption and iron/steel production); and the final two measures involve *demographic* variables (total population and urban population).[8]

Since this study focuses on the war behavior of a restricted class of major powers, a Composite Index of National Capability (CINC) was developed to reflect the *relative* capability scores of only this *subset* of nations (cf. Bremer 1980:63–66). Major powers were identified on the basis of COW coding procedures (Small and Singer 1982:44–45). CINC scores were created by first obtaining the sum of major-power values on each of the six capability variables. The percentage of the total value for all majors that was controlled by each nation on each of the six measures was calculated. A summary score was then computed indicating the percentage share of the *total* capability pool possessed by each major power. For example, in 1816 England controlled a 33 percent share of the total capabilities possessed by all major powers, whereas France had a 20 percent share of the pool. These summary capability scores (CINCs) combining the six variables were computed for all major powers annually from 1816 through 1976. A further distinction was then drawn between those major powers (i.e., contenders) that controlled at minimum a 10 percent share of the capability base and those major powers whose CINCs were below 10 percent. This narrowed the set of major power/contenders to between three and five states in a given year. The nations that met these criteria are shown in table 8.1. A final procedure involved the computation of average CINC scores for each contender for five-year periods beginning in 1816.

The measure of structural power distribution was constructed to reflect changes along a concentration/diffusion continuum, involving calculation of the average change in the lead nation's share of the capability pool of all contender states over ten-year periods beginning in 1820. An increase or decrease in the lead nation's average CINC score was noted between each successive period. The lead nation's CINC share for the years 1816 through 1819 constituted the initial baseline. A summary statement of systemic concentration phases (combining contiguous ten-year test periods) is shown in table 8.2. The individual capability trajectories of the contenders and the shifts in systemic power concentration are as indicated in figure 8.1.

The next step in the research design involved the determination of relative power within every contender dyad. The ten-year periods used to identify shifts in systemic power concentration were established as the temporal boundaries within which dyadic power relationships were measured. Dyadic power relationships were defined to best permit testing of the different hypotheses. For

TABLE 8.1. Contenders

Year	Countries
1816–59	England/UK, France, Russia, Austria-Hungary
1860–64	England/UK, France, Russia
1865–99	England/UK, France, Russia, Prussia/Germany
1900–34	United States, England/UK, France, Russia/Soviet Union, Germany
1935–45	United States, England/UK, Soviet Union, Germany
1946–49	United States, Soviet Union, England/UK
1950–76	United States, Soviet Union, People's Republic of China

H1–H3, dyadic capabilities among contenders were determined to reflect either *preponderance* (the relative capabilities differed by 20 percent or more) or *parity* (the relative capabilities differed by less than 20 percent). Given the set of identified contenders, there were eighty-five possible dyadic combinations for the ten-year test periods between 1820 and 1976. Each of the eighty-five dyads was then classified on the basis of the presence or absence of war during the period according to the interstate war list in *Resort to Arms* (Small and Singer 1982:82–95). For H4, following the Organski and Kugler (1980:49) power transition categories, dyadic balances were defined as follows: *unequal* (the relative capabilities differed by 20 percent or more); *equal without overtaking* (the relative capabilities differed by less than 20 percent); or *overtaking* (the passing of one nation's CINC score by the other nation's CINC score one or more times during the ten-year test period). The COW Serious Interstate Dispute database generated eighty-four conflict dyads involving the identified set of contender nations: thirteen warring contender dyads (in six wars) and seventy-one subwar dispute dyads (in sixty-four disputes). The identity of each contender as conflict initiator or target was then recorded.[9]

Data Analysis

Table 8.3 displays the frequency counts used to determine the association between dyadic power distributions and the incidence of war. The Phi coefficient indicates that the variation in dyadic capabilities is related to the occurrence of war. Specifically, dyadic balances characterized by parity in capabilities are associated with war in 25 percent of the cases (eight of thirty-two), whereas paired distributions characterized by preponderance are associated with war in less than 10 percent of the cases (five of fifty-three). This finding confirms H1: among great-power dyads, parity in capabilities is more likely to be associated with the occurrence of war than is preponderance.

Table 8.4 displays the distributions used to determine the effects of dyadic power balances under systemic conditions of increasing and decreasing capability concentrations. Power transition theory suggests that systemic stability and world order are associated with the dominance of a single state. However,

TABLE 8.2.   Phases of CINC Concentration/Deconcentration

Decreasing Concentration	Increasing Concentration
1820–29	1830–69
1870–1914	1920–29
1930–39	1946–55
1956–76	

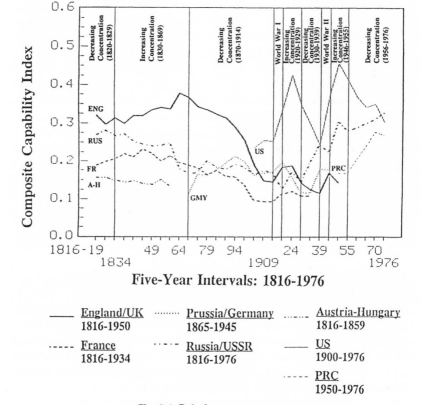

Five-Year Intervals: 1816-1976

| ──────── | England/UK | ········ | Prussia/Germany | ······· | Austria-Hungary |
| | 1816-1950 | | 1865-1945 | | 1816-1859 |

France ‒‒‒‒‒ 1816-1934   Russia/USSR ·‒·‒ 1816-1976   US ──── 1900-1976

PRC ‒‒‒‒ 1950-1976

Fig. 8.1. Relative power scores.

it is postulated that as the systemic distribution of power shifts from a unipolar concentration of capabilities to a more diffuse multipolar balance, the weakening of the principal defender of the order will lead to increasingly intense conflicts among the set of major powers and, ultimately, to global war. In short, increasing systemic power concentration reinforces the existing hierarchical

TABLE 8.3.   Dyadic Power Distributions and War, 1820–1976

	Preponderance	Parity
No War	48	24
War	5	8

Phi = 0.20953 $p = 0.0534$ $N = 85$ dyads

structure and should minimize conflict among the strongest nations, whereas a condition of systemic power diffusion weakens the existing hierarchy and will encourage violent conflicts to occur. Hence, parity in dyadic capability distributions should be most strongly associated with warfare under conditions of decreasing systemic power concentration.

Table 8.4a supports H2. Dyadic power balances among contenders are not associated with war in international systems where the concentration of power is increasing. Dyadic capability distributions characterized by parity are associated with war in about 11 percent of the cases (one of nine), while balances characterized by preponderance experience war in 10 percent of the cases (three of thirty)—virtually identical findings. Table 8.4b also displays the effects of dyadic power balances under systemic conditions of decreasing capability concentration. Once again, parity in dyadic balance is related to war in about 30 percent of the cases (seven of twenty-three), whereas war occurs in dyads characterized by preponderance in less than 8 percent of the cases (two of twenty-three). The Phi coefficient (standardizing the relationship across tables with different populations) of 0.274 is the strongest of the set. This finding demonstrates a substantive empirical linkage between systemic- and dyadic-level power dynamics and the causes of contender wars and confirms H3. In short, a stable international hierarchy (where systemic capability concentration is increasing) tends to minimize violent conflict among the set of strongest states, irrespective of dyadic power balances. However, an eroding international hierarchy (where systemic capabilities are diffusing) presents the greatest potential for war, with dyadic power balances characterized by parity more than three times as likely to experience war as dyads characterized by preponderance.

TABLE 8.4. Systemic and Dyadic Power Distributions and War, 1820–1976
Increasing Systemic Concentration

a	Preponderance	Parity
No War	27	8
War	3	1

Phi = 0.01543 $p = 0.9232$ $N = 39$ dyads

Decreasing Systemic Concentration

b	Preponderance	Parity
No War	21	16
War	2	7

Phi = 0.27400 $p = 0.0631$ $N = 46$ dyads

TABLE 8.5.   Capabilities and Conflict Initiation, 1820–1976

	Conflict Class	
Initiator Capability	Wars	Militarized Disputes (subwar)
Initiator Inferior	5	35
Initiator Superior	8	36
	$Z = 0.8320$	$Z = 0.1186$
	$p > 0.1000$	$p > 0.1000$

The computation of the Z statistic here employs the binomial test used by Bueno de Mesquita 1981, 22.
$N = 84$ dyads

Tables 8.5, 8.6, and 8.7 involve the question of contender power balances and conflict initiation. Organski's theory (1958) specifically posits that transition conflict will be initiated by the weaker contender. H4 states the relationship more generally. Table 8.5 depicts frequencies for the balance of static capabilities and war/dispute initiation among contender dyads for the years 1820 through 1976. The distributions indicate that contender states with inferior capabilities are just as likely to initiate wars and lesser disputes as are contenders with superior capabilities. There is no relationship between capabilities and either form of conflict initiation. These findings regarding contender capabilities and conflict initiation provide an explanation for the Small and Singer (1982:195) anomaly that in wars between major powers, the initiator wins in only 33 percent of the cases. The frequencies show that—for contender dyads— the static distribution of capabilities is not associated with conflict initiation. Thus, among contender states, war and dispute initiators are as likely to be inferior to their opponents as they are to be superior in the static balance of relative capabilities.

TABLE 8.6.   War Initiation, 1820–1976

	Dyadic Power Condition		
Initiator Capability	Unequal	Equal, No Overtaking	Overtaking
Initiator Inferior	2	1	2
Initiator Superior	3	2	3

Tau C = 0.0 $p = 0.5$ $N = 13$ war dyads

TABLE 8.7.   Subwar Dispute Initiation, 1820–1976

	Dyadic Power Condition		
Initiator Capability	Unequal	Equal, No Overtaking	Overtaking
Initiator Inferior	23	7	5
Initiator Superior	14	12	10

Tau C = 0.276 $p = 0.01$ $N = 71$ dispute dyads

Table 8.6 depicts the distributions for contender war initiation from 1820 through 1976 under shifting dyadic power conditions. In other words, the table indicates initiator identity under differential rates of capability growth. The Tau C statistic reveals that, again, no meaningful pattern is evident: the frequencies show that war initiators tend to have superior capabilities under all three power conditions, but these frequencies are not markedly different from those for contenders with inferior capabilities. In short, the thirteen contender war dyads reveal no significant pattern relating initiator capability to shifting dyadic power conditions (i.e., differential rates of growth). These frequencies are consistent with the results reported by Organski and Kugler (1980:57–60) regarding the identity of the war initiator. In their discussion of World Wars I and II, they point out that the conflicts were begun by the challenger *after* the transition occurred.

Table 8.7 provides the set of distributions for all contender militarized disputes below the level of war covering the years 1820 through 1976. Here, the Tau C indicates the presence of a pattern between initiator capability and shifting dyadic power conditions (i.e., differential rates of growth). Under unequal power conditions, stronger nations initiate fewer disputes than their weaker counterparts, perhaps due to their ability to secure their goals without militarized conflict. However, changing power relations alter the likely identity of the aggressor. Specifically, as the relative power condition shifts from inequality to transition, the probable dispute initiator increasingly tends to be the contender with the capability advantage.[10] This finding is consonant with *The War Ledger*—although it does not conform to the initial Organski (1958) postulate—and suggests that the actual passage of the dominant state by the challenger may lead this newly stronger nation to attempt to utilize its capability advantage. Kugler and Zagare (1990) note the differences in power transition logic as presented in Organski (1958) and Organski and Kugler (1980). They correctly conclude that *The War Ledger* version is consistent with an expected-utility model of rational calculation. Thus, not every transition results in war (Geller 1992a; Houweling and Siccama 1993), but dispute initiation is prompted by the interaction effect of static and dynamic power balances.

Conclusion

This study has examined two fundamental issues involving capability distributions and power transition theory: (1) the nature of the dyadic capability balance—parity or preponderance—most likely to be associated with the *occurrence* of great power wars; and (2) the relationship of power balances to the identity of the conflict *initiator.* The first question required the extension of transition theory to contender wars not associated with system leadership. Specifically, the relevance of certain test cases included in *The War Ledger* to the original thesis in Organski (1958) has been questioned in the past. However, transition theory has been applied to the broader set of major-power wars and empirical support for this extension has been reported (Houweling and Siccama 1988a, herein; Kim 1989, 1991, herein).

As a means of examining the relationship between the broader set of great-power wars and the capability balances associated with conflict, a possible linkage was explored between variations in systemic power concentration (and the strength or weakness of the international hierarchy) and dyadic balances characterized by either parity or preponderance. The findings indicate that a stable international hierarchy (where systemic power concentration is increasing) tends to minimize conflict among the set of strongest states, irrespective of the dyadic power balance. However, a disintegrating hierarchy (where systemic-level capabilities are diffusing) presents the greatest war potential, with dyadic power balances characterized by parity more than three times as likely to experience war as dyads characterized by preponderance.

In sum, this study indicates that capability balances do affect the occurrence of great-power conflicts and provides evidence regarding the interaction of power distributions at multiple levels of analysis. More generally, the findings reported here suggest the broader applicability of the basic power balance thesis presented in *World Politics* and *The War Ledger.*

The analysis presented in this study also demonstrates that, in terms of conflict patterns among the strongest nations, an interactive relationship exists between the static power balance, differential growth rates in capabilities, and conflict initiation. Under unequal dyadic power conditions, stronger nations initiate fewer disputes than their weaker counterparts. As capabilities converge, pressures to exploit transient power advantages make the stronger state the more probable conflict initiator. The results of this study are consistent with the findings of *The War Ledger,* and are thus consonant with an assumption of rational calculation in decisions involving conflict initiation. In addition, these findings may account for the Small and Singer (1982) anomaly of fewer relative initiator victories in great-power conflicts than are found in the general set of interstate wars. Moreover, all of the findings reported in this study reinforce the basic contention of transition theory that a shifting power balance is of critical significance among the causes of war and peace.

NOTES

1. The risk orientation of decision makers is an element that logically impacts on the relationship between power distributions and war (Bueno de Mesquita 1981a, 1989). When dealing with conflicts between major powers, for example, Bueno de Mesquita (1981a:124–25) assumes decision-making risk-neutrality or risk-acceptance. Recently, prospect theory (Kahneman and Tversky 1979) has been suggested as an alternative to expected utility as a theory of foreign policy decision making under conditions of risk (Levy 1992a,b). Prospect theory differs from expected-utility theory in positing that risk orientations vary according to estimates of outcomes as deviations (gains or losses) from a reference point. Specifically, it is hypothesized that individuals tend to be risk-acceptant with respect to losses and risk-averse with respect to gains. Regarding both theories, the introduction of the risk-orientation variable creates extraordinary measurement problems (Levy 1989:234), especially in studies dealing with large populations over extended periods. Within the expected-utility context, a simplifying assumption for large population studies is to posit that risk orientations conform to a normal distribution.

2. In the course of the analysis reported in *The War Ledger,* the Russo-Japanese War is reclassified as a center-system major-power war rather than as a contender war (Organski and Kugler 1980:51–53).

3. Notable exceptions are Houweling and Siccama (1991, 1993), Geller (1992a), and Huth, Bennett, and Gelpi (1992). Rasler and Thompson (1992a) make a related argument based on the interactive power dynamics of global and regional structures. They suggest that the world's most serious wars have resulted from the confluence of global power diffusion coupled with regional capability concentration. It should be noted that multilevel explanatory frameworks in the area of foreign policy analysis are less unusual (e.g., Geller 1985).

4. Maoz (1983:215,221) has argued that capability advantages determine victory or defeat only in contests between major powers. In all other dispute classes, conflict initiators win disproportionately not because they are stronger than their targets, but rather because they possess higher relative levels of resolve.

5. War as a result of a changing power balance due to uneven growth rates is a theme consistently found in the literature of international relations. For example, power shifts due to differential growth rates are explicitly or implicitly noted as a source of conflict by Gilpin (1981), Howard (1983), Modelski (1983), Wallerstein (1984), Thompson (1988), and Midlarsky (1990a). Moreover, all of these works incorporate assumptions of some form of rational calculation (e.g., cost-benefit/expected utility) in war decisions.

6. The terms "power" and "capability" are often employed interchangeably, although they are not identical (e.g., Organski and Kugler 1980:4–8). The measures and index utilized in this study reflect both the resource capacity (or power potential) and a crude es-

timate of the military strength of nations. The comparative scores generated here are representative of national *capabilities,* but the term "power" will be used in places as a synonym. Alternative measures of power may tend to produce different estimates of relative capabilities (e.g., Geller 1992b). Organski and Kugler (1980:33–38) use GNP as an indicator of power for its parsimony, reliability, and theoretical relevance; however, in a later study, Kugler and Organski (1989b:119–26) find stable and consistent rankings for the United States, whether hierarchical position is measured by GNP, exports, or relative political capacity. Arguments for the use of multiple indicators are presented in Singer, Bremer, and Stuckey (1972) and Bremer (1980). Various approaches to power measurement are discussed in Moul (1989), Merritt and Zinnes (1989), and a comparison of the COW index and GNP is provided in Kugler and Arbetman (1989a).

7. Capabilities were determined by the COW National Capabilities (1816–1985) data set, war was measured by frequency counts from the Small and Singer (1982) *Resort to Arms* list of interstate wars (1816–1980), and militarized conflicts were drawn from the COW Serious Interstate Dispute database (1816–1976).

8. The National Capabilities data set has entries on the six variables for all nation-states from 1816–1985. The capability indicators are as follows: military personnel, military expenditures, energy consumption, iron and steel production, urban population, and total population.

9. The wars and warring dyads in the analysis are:

War	Dyads
Crimean	Russia/UK
	Russia/France
Italian Unification	Austria-Hungary/France
Franco-Prussian	France/Prussia
World War I	Germany/Russia
	Germany/France
	Germany/UK
	US/Germany
World War II	UK/Germany
	Germany/USSR
	US/Germany
Korean	US/PRC
	UK/PRC

For a description of the militarized conflict data set, see Maoz 1982 and Gochman and Maoz 1984. By definition, "a serious interstate dispute is a set of interactions among states involving the explicit, overt, and government-directed threat, display, or use of force in short temporal intervals" (Maoz 1982:7). The dispute initiator is the state that begins the military confrontation with the first codable action. The target is that state against which the initiator directs its action. Most wars begin as militarized disputes—although the dispute initiator and war initiator are not necessarily the same. Subwar militarized disputes are included in the analysis under the reasoning that these nations are likely to engage in multiple conflicts before the actual outbreak of war. More specifically, as the capabilities of contender states converge over time, hostilities may grow more frequent. This point is made explicitly by Organski and Kugler (1980:28).

10. When war frequencies are added to those of lesser militarized disputes (i.e., combining the frequencies from tables 8.6 and 8.7, producing eighty-four conflict dyads), the pattern depicted in table 8.7 remains the same with a Kendall's Tau C of 0.2415 and a p of 0.0189.

CHAPTER 9

Power Shifts and the Onset of War

Frank Whelon Wayman

What made the war inevitable was the growth of Athenian power and the fear which this caused in Sparta.

—Thucydides, 5th century B.C.

This chapter throws fresh light on perhaps the oldest proposition in the profession of international studies—namely, Thucydides's hypothesis that a shift in the relative strength of states produces war; a central part of Organski's (1958) power transition theory. The chapter presents several heretofore unspecified reasons why power shifts do increase the risk of war. It then marshals evidence on how often power shifts have been followed by enduring rivalries and war in the past two centuries.

Several important contemporary authors, including Organski and Kugler (1980), Doran and Parsons (1980), Gilpin (1981), Keohane (1982), Kennedy (1988), Levy (1987), and Houweling and Siccama (1988a), have examined the relationship between war and change. By far the first of these voices was that of Organski (1958, 1968), who argued that power transitions were a crucial cause of major modern wars. Organski contended that certain moments of equality, when one side was surpassing another, bred warfare. Organski and Kugler's (1980) formulation, which dominated the early empirical investigations of the subject, is extended in this chapter.[1]

The purpose of this chapter is to improve conceptualization of this important topic. Whereas Organski and Kugler (1980) emphasize power transitions as a cause of war, I argue that their concept of power transition is actually too narrow. Not only power transitions, in which one state surpasses another, but also rapid approaches short of transition, in which one state closes in on, but does not surpass, another, increase the risk of war among major powers. While Organski and Kugler and Houweling and Siccama are correct in reporting an important association between power transitions and war, this chapter integrates their work with that of Levy, Thucydides, Gilpin, and Keohane. This latter group of writers have emphasized power shifts—a genus of which power transitions and rapid approaches are species—as the more fundamental cause of war. I conclude that this broader approach is more accurate, both on theoretical and empirical grounds.

Why Shifts in Relative Capability Can Cause War

When one state is growing in capability relative to another, the growing state gradually becomes more powerful. This is not to say that there is a magical moment when, after a "power transition," it now has more capabilities than its rival and therefore can prevail in all of their struggles. For one thing, there is a limited domain and scope of the power conveyed by material capabilities (Deutsch 1978); many issues of concern to the declining state will be beyond the reach of the growing state's power. Second, there is a loss-of-strength gradient (Boulding 1962), such that the material capabilities of a weaker state may be sufficient to prevail in combat over a more capable state, when the latter is fighting far from home. Instead of one moment of power transition, there is a continuous tendency for a state's power to decline as its relative capabilities decline.

A power shift proceeds as follows: a shift in material capabilities alters the relative power of the two sides and also alters their perceived power. This chain of events has five important consequences. First, when power shifts, there is an adjustment of loss-of-strength gradients, so that the geographic locus and issue areas in which the two sides are approximate equals changes. This means that there will be new conflicts of interest that are salient to the two sides. This conflict salience can be thought of as the product of "intensity" and "pragmatic relevance" (Wayman 1975), where intensity is the difference between the value the nation-states put on winning and losing the conflict and pragmatic relevance is the degree to which the contest between the two sides is evenly enough matched that choosing the right strategy (for example, either negotiating or fighting) can reasonably be expected to make a difference in which side wins or loses. When a conflict is over important stakes (i.e., is "intense") and can be decided one way or another by what moves the parties make (i.e., is "pragmatically relevant"), then the conflict will be salient. Shifts in relative capability will change the agenda of relevant and salient conflicts. For instance, when Britain was strong and China was weak, salient issues included British acquisition of extraterritorial rights in China and stationing of British troops in Peking after the Boxer uprising. A century later, as Britain declined relative to China, the relevant and salient issue changed to whether Britain could gain meaningful concessions from China as Britain gave up her last vestige of former Chinese territory, Hong Kong. Thus, capability shifts will affect which issues are pragmatically relevant at a given geographic locus, as relative capabilities will change the issues at the margin of each state's "scope" of power (Deutsch 1978). If this occurs over a century, adjustment can be gradual and peaceful. Rapid capability change makes negotiation of differences vexing, in that neither side has experience in negotiating over the novel matters that have become salient. No modus vivendi has been worked out on these conflicts; no habits of living with disagreement on these matters have been acquired. In such an atmosphere, both sides may make unintentionally provocative statements and moves.

Second, a power shift will produce appetite in the gaining state and apprehension in the declining state. As a state acquires the capability for power, it often acquires an appetite for the goods that are now within its reach.[2] Insofar as such goods are often seen as competitive, in which one state's gain is another state's loss, such appetites are threatening to other states. A modicum of apprehension comes simply with awareness that the other state is gaining. Today, for instance, Americans are wary of the growth of Japanese power, even though Japan is a trading state with no substantial ability to project military force and is one of America's closest allies. The concern in such situations is that economic gains will gradually give the rising power the capability to distance itself from and eventually threaten the leader. Gains in military capability cause much more intense apprehension. Because military capability reflects a political choice to allocate scarce resources to a war machine, changes in the military balance cause worries that the rising state is gaining militarily because it has the intention of threatening the leader. When apprehension stems from a reassessment of both capabilities and intentions, as it does in such circumstances, tensions are likely to rise very fast. Such apprehension can also result when gains in economic capability are accompanied by words or deeds that can be interpreted as hostile.[3]

The danger of this pattern, in the context of power shifts, is that the rising and declining states will become trapped in a vicious cycle of growing hostility, such as an arms race, or such as the escalatory sequence described by Leng (1983) in his study of recurrent crises. If this were the end of the matter, one might predict that power shifts would lead to arms races, and that these arms races would lead to crisis escalation and war. One might think of an example such as World War I, which followed a power shift between Britain and Germany and a subsequent naval arms race between them, and one might hope that war might be avoided through a policy of appeasement conciliation.

Third, however, to respond to a growing appetite with conciliation may produce war by convincing the rising power that the declining power is weak (Holsti 1967). In such an interaction, the declining state would make war more likely by failing to deter the rising power from going after the vital assets of the declining power (Levy 1983b) and by inducing a growth in the rising power's appetite with each concession. It is therefore very difficult for the declining power to respond properly to a power shift, because either an excessively hawkish or an excessively dovish reaction to the rising challenger may bring on war.

Kagan (1987), for example, argues that before World War I, when Britain and France were challenged by the rise of German power, it would have been better for Britain and France to respond in an even more forceful manner than they did. A strong military buildup by them might have convinced Germany that it had no chance of winning a war and thus might have tamed the German expansionists.

Fourth, in a power shift it is difficult to calculate who is ahead. Blainey (1973) has argued persuasively that one cause of war is disagreement over the pecking order. One state is usually more powerful, in the sense that it has the resources, including material capabilities, for winning a war. If both sides agree on who is likely to win, the weaker will usually be wise enough to submit, rather than fight and lose. But, as Blainey points out, nations often differ in their perceptions, and war is more likely when two sides each think they would win a war (see also Stoessinger 1974; Levy 1983b). Since only one side will in fact win, anything that increases the error in the weaker side's estimation of relative capabilities is a factor that makes war more likely, because it increases the likelihood that both sides will think they can benefit from war.

When the relative capabilities of states are not changing, they are more likely to have an agreed sense of who is ahead than if relative capabilities are changing. When relative capabilities are changing, insofar as people are optimists, the gaining state is likely to overestimate the rate at which it is gaining. And the declining state, also likely to paint a brighter than real picture, is likely to underestimate its rate of decline. This means that a moment may arrive when both think they have a military edge on the other. If the cost of war is not too great relative to the spoils of victory, this differing conviction over who will win may push them over the brink into combat.

Fifth, a power shift may lead to forecasts that the trend is going to continue. Such a forecast, if made by the declining state, may lead to a calculation that war is better now than later, because declining power will lead to a weaker bargaining position and war-winning capability. This is the logic of preventive war (Levy 1987). The declining state may initiate war or push the gaining state into a position in which it becomes the initiator.

One thing that restrains all these tendencies toward war is a large expected cost of war. If both sides realize that war is going to be devastating (Levy 1983a:148–49; Mueller 1988), then war becomes a much less likely outcome of these calculations. Thus, two nuclear superpowers experiencing a power transition will almost certainly avoid preventive war. For one thing, the more destructive a future war would be, the less likely that the overtaking state will start one when it is stronger. This perception reduces the temptation to start a preventive war by reducing the expected cost of remaining at peace. Also, the more destructive a current war will be, the less likely the declining state is to start a preventive war. Preventive war begins because of the calculation that it is better to have war now than later, and if war now means the end of civilization, it would be madness to start a preventive war unless the probability of future war seems very near certain—a possibility that we have already ruled out by noting the high cost of future war. Obviously, these calculations exclude the possibility of a leader who believes that something like a Star Wars shield will shortly make war a low-cost affair. (For different perspectives on the constraining effects of expected high costs, see Kugler herein; Morrow herein.)

In short, the anticipation (by both sides) of a costly war reduces the strength of the causal linkages between power shifts and war. One other variable also affects the strength of these causal linkages. When the power shift is occurring between two states that have already been longtime enemies, the likelihood of war is much greater. To the degree that two states sense that they share the same preference concerning international outcomes, a power shift between them will not be too worrisome. The declining state may even view the advent of such a rising state as cause for relief, since a state in decline relative to one country may be in decline relative to several others. Such a state (i.e., the Ottoman Empire in its last two hundred years, the Austro-Hungarian Empire from 1848 to 1914, or the British Empire in the twentieth century) often attracts a swarm of enemies and is in dire need of a growing, vigorous ally.

Analytic Framework

In moving from such general principles to disconfirmable hypotheses, it is necessary to be more specific about precisely which capability distributions will lead to war. The simplest of these hypotheses are static and focus on the *level* of capabilities of each side. It is important to test these static, "balance-of-power" hypotheses against power shift hypotheses because, as shall be clear shortly, the results bear on the proper interpretation to be given to empirical confirmation of a power transition hypothesis. Scholars who emphasize such a static framework themselves disagree. As discussed in chapter 1, balance-of-power theorists argue that a relatively equal distribution of power will insure peace or at least prevent large wars. These theorists suggest that under conditions of parity, neither side can be confident of prevailing in a war, and neither side will recklessly escalate a dispute to the point of war. According to them, it is *in*equality that is dangerous. Inequality tempts the powerful with dreams of easy dominance, while inducing fear and perhaps desperation in the weak: "Domination . . . is not enough for some states to demand but is too much for most states to tolerate. . . . At worst, the decisive preponderance of the Soviet Union would produce the peace of a prison camp; at best, it would precipitate war hopeless or not, in defiance of its drive for universal empire" (Claude 1962:62).

Taking the opposite power transition perspective, other theorists contend that preponderance (i.e., inequality) is more conducive to peace, while equality leads to war: "an equilibrium increases the danger of war by tempting both sides to believe that they may win; but in a situation marked by the preponderance of one side, the weaker side dares not make war and the stronger side does not need to" (Claude 1962:57).

The above arguments deal with the relative *levels* of power; here I deal with dynamics. Organski (1968:338–76) has, of course, argued that the *rates of change* in those levels must also be considered. In his view, wars between extremely powerful nations, during the era of industrialization, have tended to occur when one of the nations is passing the other in capabilities. This hypothesis, encapsulated in the opening quotation of this chapter, suggests that major wars occur under a combination of conditions: (1) two sides are nearly equal in levels of capabilities, and (2) one is passing the other because they are unequal in their rates of change of capabilities. Organski and Kugler (1980:52), having examined twenty cases, conclude that contenders for world leadership do indeed tend to fight each other only when one is passing the other.

This chapter examines the dynamic relationship between power transitions, speed of growth, and conflict, with the dyad-decade as the unit of analysis; that is, the focus is on whether the presence or absence of war between a pair of nations in a decade is associated with a power transition (or related change in capabilities) between them in that (or the immediately adjacent) decade.

The hypotheses to be examined here, suggested by the results presented in *The War Ledger,* can be summarized as follows:

H1. War between two states will be more likely to occur in the decade *after* they have a power transition than in other decades.

In H1, in which the power transition precedes the war, the only plausible direction of causality is that the power transition produced the war. Furthermore, H1, if confirmed, would be useful for forecasting, since it predicts war based on developments a decade before the outbreak of the war.

In a simple statistical test, H1 might be confirmed for any of the following reasons, some of which are potentially misleading:

1. If equality causes war and power transitions do not, and if a large proportion of the time that nations are approximately equal they also experience a power transition, then there will be a positive, and spurious, correlation between war and power transitions.
2. If power transitions cause war and equality does also, then there will be a positive correlation between war and power transitions.
3. If not only power transitions, but any substantial change in relative capabil-

ities, cause war, then there will be a positive correlation between war and power transitions.

Therefore, in order to determine more accurately whether equality, change, power transitions, or any combination of these produce war, it is important to consider more fully the relationship between capabilities and war. A first step is the following hypothesis:

H2. War will be more likely to occur between two states in decades when (a) they are relatively equal in capability, but (b) they are not undergoing a power transition, than in other decades.

For example, from about 1870 to 1910, Austria-Hungary and Italy were approximately equal in capabilities, but did not have a power transition—a condition that we can call "prolonged parity." If the first of the above three reasons is correct (equality breeds war), power transitions and prolonged parity should both produce war. H1 and H2 would both be confirmed, since the association between power transition and war and prolonged parity and war would be about equally strong. In contrast, if H2 is rejected (as suggested by the Austro-Hungarian/Italian example, which did not produce a war), but H1 is confirmed, then equality would not be a sufficient condition for war; the two relatively equal states would have to have a power transition for war to become more likely.

Finally, there is the possibility that mere approach in relative capability would be sufficient to account for war, even without a power transition. For example, Sardinia (later Italy) and Austria-Hungary seem to have fought repeatedly, about mid-nineteenth century, when their relative capabilities were moving together but still far apart; they seem to have ceased fighting during the period later in the nineteenth century when their relative capabilities were no longer converging. Organski (1968) seems to lend some support to the convergence thesis when he argues that wars associated with power transition will tend to occur before the actual moment of transition, that is, during the phase when two countries' power shares are becoming more equal.

A major logical problem for Organski's (1958) argument, incidentally, is that if he is right and wars do occur before the power transition has been completed, and the wars lead to the defeat of the challenger, then it will be impossible to know, even in post hoc historical investigation, whether a power transition was going to occur. The war will so diminish the challenger's power that a transition that would have happened in a peaceful era will be delayed or stopped permanently by the lost war. The only way to escape this logical trap is to argue that wars are followed by a Phoenix Factor in which the loser is soon able to catch up to its prewar growth pattern. Organski and Kugler make this argument in another article, but what they find true for nations losing to the benevolent United States (e.g., post-World War II Japan and Germany) may or may not be the case in other situations or in general (Doran 1971; Organski and Kugler 1980; Wheeler 1980; and for reinforcement of the logic of the Phoenix Factor, see Arbetman herein). Another issue is that if wars are prematurely started by the challenger, the challenger is acting at a moment before the transition in which the decision makers (since, as we have known at least since David Hume, no one knows the future) have no idea whether they are going through the beginnings of a power transition or a rapid approach that will fall short of transition. Logically, then, what motivates challengers to start a premature war during a power transition should motivate them to start such a war in a rapid approach that is mimicking the early phases of a transition.

The question, does an approach toward equality alone produce war, or must the change involve equality and power transition? may be examined by testing a third hypothesis:

H3. War involving a dyad is more likely when there is a power transition than when there is a rapid approach.

Measuring Power Transitions and Major-Power War

These hypotheses are tested using data on relations between major powers. The major powers, which are defined on the basis of a reasonable degree of scholarly consensus, are Austria-Hungary (1816–1918); Prussia (1816–70) and its successor, Germany (1871–1918 and 1925–45); Russia (1816–1917) and the Soviet Union (1922–present); France (1816–1940; 1945–present); England (1816–present); Italy (1860–1943); Japan (1895–1945); the United States (1898–present); and China (1949–present). This list of powers and dates of inclusion (from Singer and Small 1972:23) is followed with only minor exceptions.[4] The wars fought between these major powers (Singer and Small 1972:60–69) were the Austro-Sardinian War, the Crimean War, the War of Italian Unification, the Seven Weeks' War, the Franco-Prussian War, the Russo-Japanese War, World War I, the Manchurian War of 1931–33, the Sino-Japanese War of 1937–41, a second Russo-Japanese War (in 1939), World War II, and the Korean War.

The capabilities of the major powers are measured by the Correlates of War (COW) Composite Capabilities Index (Singer, Bremer, and Stuckey 1972). This index assesses the military, industrial, and demographic capacities of each major power, with each of these factors measured by two different indicators and all six components weighted equally in computing the total capability of the nation-states. For each of the indicators, the total "holdings" of all the major powers are calculated and then each nation's percentage share of that total is calculated. These six percentage shares are then averaged for each nation, to estimate its percentage of total capacities.

Power transitions were operationalized as cases in which one state, previously behind in percentage of total capacities, moves ahead and then remains ahead. Given that the data are measured at five-year intervals, a power transition therefore occurs when one state moves ahead over a five-year period and is still ahead at the end of the next five years. A reversed, or double transition, on the other hand, involves one state temporarily surpassing another over an initial five-year period, only to fall back behind in the next five years.

Prolonged parity was operationalized with two goals in mind: to keep the maximum gap between nations A and B small enough so that the cases of prolonged parity had about the same degree of equality as the cases of power transition, but to keep the gap large enough so that there would be enough cases for meaningful analysis. In the end, these two goals were achieved by averaging the gap between A and B at the start of the decade with the gap between A and B in the middle of the decade and counting as power parity those cases in which the average gap was 10 percent or less. In this way, seventeen cases of power parity were identified.

Approach-without-transition was operationalized with a similar pair of conflicting goals: to only look at cases of rapid approach, but to insure an adequate number of cases. The cutoff eventually selected was that the gap between A and B at the end of a five-year period had to be at least 40 percent less than it had been at the start of the period. In this way thirty-two cases of rapid approach (without a power transition occurring) were identified. Five of these cases were also cases of prolonged parity.

Thus, the 319 dyad-decades of the study have been apportioned into six conditions: single power transition ($N = 28$), double power transition ($N = 7$), rapid approach ($N = 27$), rapid approach and also prolonged parity ($N = 5$), prolonged parity ($N = 12$), and other ($N = 240$). Comparing the probability of war under each of these six conditions will allow one to test H1, H2, and H3. But to test these hypotheses under better conditions, one additional variable needs to be introduced.

The Role of Enduring Rivalries

Organski (1968) makes clear his conviction that only rivals will fight during a power transition. His argument is based on the assumption that national leaders calculate the costs and benefits of war and see no benefit in fighting friends. I agree with this argument but believe it can be transformed into an empirical question, so that one can test how strong an influence rivalry has on the relationship between power transition and war. To do this, I distinguish between relatively conflict-prone and relatively conflict-free pairs of states.

The first set are those pairs that are relatively conflict-prone in their relationship with each other, as indicated by their having more than one militarized interstate dispute with each other over the course of a decade. The militarized interstate disputes (MIDs) are those identified by the COW project (Gochman and Maoz 1984). Such a dispute involves the explicit threat to use force, a display of force, or the use of force by one state against another. While my relatively conflict-prone dyads are those experiencing at least two MIDs in a decade, the relatively less conflictual dyads are those that have entirely avoided militarized disputes with each other, or have had only one such clash in the course of any ten-year period. I call the relatively conflictual group "enduring rivals." I call their conflict an "enduring rivalry."

The concept of rivalry is broad and multidimensional, similar to the concept of social class.[5] Just as social class encompasses more than income, enduring rivalries encompass more than just repeated armed confrontations. Rivalries, in the full sense, are potentially multidimensional, involving (1) competition over some stakes; (2) approximate parity; (3) possible arms races, economic boycotts, and other efforts to overcome the opponent; and (4) repeated clashes. This chapter operationalizes enduring rivalries by measuring events (namely, militarized disputes, which are aggregations of specific incidents). Hence, this chapter focuses on the fourth dimension of the broad concept of rivalry.

For measurement purposes, each dispute was assumed to have a one-decade impact on relations between the two sides. For example, a dispute that started in 1902 would be assumed to have affected relations adversely until 1912. If a second dispute occurred in that interval or in the following year (1913 in the current example) then the two disputes are linked together as an enduring rivalry. And, to be consistent, the persistent disputatiousness is assumed to end one decade after the last dispute in the chain.

Among states that are not persistent disputants, I hypothesize that a pair undergoing a power shift will be more likely to have a war than a pair not having a power shift. The logic is that states that are not entangled in repeated militarized conflict with each other will still have some disagreements and some residual reasons to view each other with suspicion, so if one side begins to gain in capabilities, the other will be apprehensive. There will be some risk that the power shift will trigger a set of perceptions that degenerates into a conflict spiral, but the risk is less than if the two had been persistent disputants.

To test these power shift hypotheses, it was possible to make use of the COW serious dispute list and select those pairs of nations for study that were persistent enemies of each other. The COW serious (militarized) interstate disputes (Gochman and Maoz 1984) include a total of 966 such disputes in the period from 1816 to 1976. While this list is undergoing constant revision, especially because of questions regarding the categorization of minor-power disputes, the data set is relatively stable and complete with regard to major-power disputes. This chapter is concerned with those disputes that involved at least one major power on each side.

Any such compilation of enduring militarized quarrels, generated from a rule such as the ten-year rule that dichotomizes cases, will obviously produce some close calls. On the one hand, Britain and Germany *do* have an enduring militarized quarrel starting in 1900 because the dispute they had in 1911 occurred just in time to be linked to a dispute they had in 1900. On the other hand, France and Germany do *not* have an enduring militarized quarrel between 1924 and 1935 because the thirteen-year gap between their disputes is a couple of years too long to qualify them as persistent disputants. However, changing the ten-year interval by a few years (e.g., down to six years or up to twelve years) does not have a major impact on the span of enduring militarized quarrels. The disputes tend either to come in clusters only a few years apart or to be spaced across wide intervals. The ten-year rule was selected because it provided a certain simplicity and clarity of presentation.[6] For purposes of this chapter, the coding rules adopted seem to have produced a list of enduring militarized quarrels that has reasonable face validity and that is sufficient for the partitioning task at hand.

Testing the Power Shift Hypotheses: Power Shifts and Wars

By controlling for the impact of enduring militarized quarrels, the hypotheses listed above can be expanded. In addition to H1, that power transitions are associated with war, it will be possible to see whether power transitions are associated with decades of persistent disputatiousness and whether power transitions that occur between persistent disputants are more highly correlated with war than are power transitions between nonrivals. These same elaborations will be explored in the cases of H2 and H3, so that the effect of prolonged parity and of rapid approach on persistent disputants and the less disputatious can be contrasted.

As may be seen in table 9.1, H1 is supported. A power transition will double the likelihood of war in the subsequent decade (from the 10 percent background rate to 19 percent). The reversed transitions, as might have been anticipated, do not have as strong an effect as the single transitions, but do increase war by a modest amount (to 14 percent). Contrary to H3, the rapid approaches seem even more war-prone than the transitions. And prolonged parity, contrary to H2, is no more war-prone than the background rate. Grouping the transitions and approaches together (as in table 9.3), one can more clearly see their effect, which is to double the risk of war.

These findings are related to the literature on war and peace in several ways. Weede (1976) found that overwhelming preponderance (10 to 1 superi-

TABLE 9.1.   War and Capability Shifts, with Wars in the Subsequent Decade

	Single Power Transition	Rapid Approach	Rapid Approach and Prolonged Parity	Reversed Transition	Prolonged Parity	Other
Total *N*	26	23	4	7	10	213
Number of wars	5	7	1	1	1	22
% wars	19%	30%	25%	14%	10%	10%
			Among Rivals			
N	18	10	3	1	2	77
Number of wars	4	5	1	0	0	11
% wars	22%	50%	33%	0%	0%	14%
			Among Nonrivals			
N	8	13	1	6	8	136
Number of wars	1	2	0	1	1	11
% wars	13%	15%	0%	17%	13%	8%

ority) virtually eliminated interstate war. Overwhelming preponderance, he established, is a pacifying condition, compared to approximate equality or mild preponderance. The present study, which has no cases of overwhelming preponderance, finds that mild preponderance and approximate equality are equally war-prone. Neither is what Weede would call a pacifying condition. This finding could be interpreted as consistent with Blainey's (1973) hypothesis that wars are only possible when each side thinks it can win. The argument to establish this interpretation would be that overwhelming preponderance is pacifying because it establishes so clearly who would win. Mild preponderance is not pacifying because under mild preponderance there will be theaters and circumstances under which both sides see an opportunity for victory, just as surely as this will be the case under conditions of approximate equality. This argument eliminates one possible challenge to the validity of Organski's statistical evidence in favor of the power transition hypothesis. A critic could have argued that Organski's finding of a correlation between power transitions and war was due to the fact that all power transitions involve approximate equality and that any approximate equality—either prolonged parity or power transition—breeds war. The present study, like earlier ones (Organski and Kugler 1980:50–52; Houweling and Siccama 1988a), shows that prolonged equality does not breed war and that something about the transition process other than mere equality produces war. The evidence that rapid approaches also are associated with war, and are associated with war to the same degree that power transitions are, provides the crucial additional evidence to say that either kind of power shift—transition or approach—breeds war. This means that Organski and Kugler have merely established a specific instance of a more general pattern. The general pattern is that change—any change in relative capabilities, not just power transitions—produces war. My findings are consistent with those of Houweling and Siccama, that this is not a deterministic law fitting every relevant case, as Organski and Kugler argue, but a broader pattern of major-power interaction, a pattern that often is true but is neither a necessary nor a sufficient condition for war. War is a very general and complex phenomenon. Such conditions are influenced by a host of contextual variables (for example, geographic contiguity and power shifts). In settings with a multitude of such variables, any one variable only increases the probability of disaster, but no one of such contextual variables is a necessary or sufficient condition.

War may go through stages of development. One way of conceptualizing the stages on the road to war is to examine the escalation process up the ladder of threat, display, and use of force. The role of capability shifts in the escalation process may be seen in the lower portions of table 9.1 and in tables 9.2 and 9.3. First, the single transitions may be prolonging existing rivalries or producing new ones, for decades of power transition are a leading indicator of decades of rivalry (see table 9.2). Second, a single transition or rapid approach during an enduring rivalry increases the chance that the rivalry will escalate to war (see the middle rows of table 9.1, on rivals). Third, transitions, rapid approaches, and prolonged parity increase the likelihood of war, even among nonrivals (bottom rows of table 9.1).

So that this evidence can be examined at a higher level of generality, it has been grouped in the same two categories used previously—power transitions and approaches together, versus all other cases (table 9.3). Consistent with the theoretical argument, the association between capability shift and war is

TABLE 9.2. Enduring Rivalries and Capability Shifts with Rivalries in the Subsequent Decade

	Single Power Transition	Rapid Approach	Rapid Approach and Prolonged Parity	Reversed Transition	Prolonged Parity	Other
Total *N*	26	23	4	7	10	213
Number of rivalries	18	10	2	0	2	86
% rivalries	69%	44%	50%	0%	20%	40%

stronger among enduring rivals than among nonrivals (as may be seen in the lower portions of table 9.3). If the states are not rivals, and there is no power shift between them, the chance of them having a war in the subsequent decade is only 8 percent. If these nonrivals have a power shift, the risk of war approximately doubles, to 14 percent. That 14 percent is the same risk of war as for enduring rivals who are not undergoing a power shift. In other words, a power shift among nonrivals can create as much of a threat of war as among rivals who have not experienced a power shift. This finding is intriguing because rivals presumably have a reason to fight, or at least a record of threatening to fight. Such a background has often bred war (Leng 1983). Strangely, non-rivals seem to lack a track record of brinksmanship.

If the rivals have a power shift, their risk of war approximately doubles, from 14 percent to 31 percent. So a power shift doubles the risk of war among rivals, but also, to a slightly lesser extent, increases the risk of war among nonrivals. And, since all the results in this section have involved power shifts affecting future wars, the timing clearly suggests that the power shifts may be *causing* this increased risk of war. Also, the delay between the power shift and the war means that power shifts give ample early warning that there is a higher than usual risk of war.

All these arguments and findings are consistent with and supportive of the

TABLE 9.3.   War, Power Transitions, and Approaches with Wars in the Subsequent Decade

	Power Transition or Approach	No Power Transition or Approach
Total *N*	60	223
Number of wars	14	23
% wars	23%	10%
	Among Rivals	
N	32	79
Number of wars	10	11
% wars	31%	14%
	Among Nonrivals	
N	28	144
Number of wars	4	12
% wars	14%	8%

Note: For the entire data set ($N = 283$), gamma = 0.45, Tau B = 0.16; for the rivals ($N = 111$), gamma = 0.48, Tau B = 0.20; for the nonrivals ($N = 172$), gamma = 0.29, Tau B = 0.08.

following model: power shifts cause a subsequent increase in militarized inter-state disputes and consequently help create enduring rivalries. Power shifts produce an even greater increase in war. By implication from the above two results, power shifts produce an increase in the probability that an enduring rivalry will escalate to war. Finally, there is a weak interactive effect, so that power shifts have a slightly stronger effect on the outbreak of war when they occur in the context of enduring rivalries than when they do not.

General Findings about Power Shifts and War

The specific power transitions that go to war are ones in which both states tend to be located on the same continent (namely, Europe). This might suggest that geographic proximity is an important factor in triggering war from power transition. This peculiar European effect of power transitions on war may, however, be viewed more narrowly as a German phenomenon. A majority (60 percent) of power transitions that are related to war in the subsequent decade involve Germany as one of the pair in transition. What might have appeared to be produced by proximity (all pairs that are in Europe) may really be related to proximity to Germany. Another related caveat involves the interdependence of cases in dyadic analyses. Three power transitions related to Germany mean two cases of war: 1914 and 1939.

While a number of interpretations of the results are thus possible, it is perhaps appropriate to close the presentation of the data with a general overview. Of the twenty-six power transitions, five (19 percent) are associated with war in either the same or the following decade. And out of the thirty-seven dyad-decades of war, five (14 percent) are associated with power transition in the previous decade. As for rapid approaches, they have a stronger delayed association; with 30 percent of them followed by war, they are associated with 19 percent of the wars.

Replication with Doran and Parsons Data

To see if the results were an artifact of the COW data set, the analyses were re-done using the Doran and Parsons (1980) indicators of the relative capability of nations; the same data set used by Houweling and Siccama (1988a). This analysis indicates that the results are consistent, approaches again being a slightly better early warning indicator than power transitions, but with both predictors associated with war at about double the background rate. In the Doran and Parsons data set, there are eighteen single-power transitions and twenty-nine rapid approaches. The single-power transitions are associated with three wars in the subsequent decade and the rapid approaches are associated with seven. Thus, 17 percent of the power transitions and 24 percent of the rapid approaches are associated with war. Thus, power shifts account for a third of the wars (ten out of thirty-two) in the Doran and Parsons data set.

Conclusions

Power transitions and rapid approaches both seem to be early warning indicators in international affairs, since they both are associated with increases in war in the next decade. This finding alters the earlier literature in several respects. The earlier evidence, except for Houweling and Siccama's, was anecdotal or limited to a subset of the three biggest world powers. The earlier writers saw neither the analytic potential of distinguishing power transitions from rapid approaches nor the synthetic potential of combining them into the concept of power shifts. They did not have evidence on the rapid approaches that have been incorporated in the present investigation. And (in part because Organski [1958] predicted war would occur before power transitions), they overlooked the role of power shifts as a leading indicator of war. This work might best be described as part of the tradition of empirical theory about world politics. Other authors, such as Gilpin (1981), have worked in another tradition, constructing theories of world politics that are less constrained by, and less oriented toward, empirical testing (see Gilpin 1981:2–3). I believe these two traditions are complementary; indeed, the title of Gilpin's book, *War and Change in World Politics,* could easily be the title of this chapter. Robert Keohane, in assessing not only Gilpin's work but the overall state of theory in world politics, closed by emphasizing a classic issue, dealt with by both traditions:

> The problem of peaceful change is fundamental to world politics. . . . The question remains for us to grapple with: Under what conditions will adaptations to shifts in power take place without economic disruption or warfare? (Keohane 1982)

The findings of this chapter show that change has tended to produce war, at least in the past. Keohane, however, criticized "structural realist" theory (i.e., most of the major international relations theorists of the past) for being too deterministic at the systemic and dyadic levels and thus suggesting that individuals and nations had little influence over events. In contrast to the studies he criticized, these findings have been weakly probabilistic and thus leave ample room for national strategists to manipulate events.

One way to realize why structural forces are not deterministic is to think again about the evolution from peace to war as a series of steps (Vasquez 1987). If war is like going over the brink of a cliff, there is a sequence in the many steps to the brink of war and over the edge. The power balances between states are important in the walk up to the cliff tops; shifting power balances create suspicion, rivalry, and tension. But the steps past the warning signs and over the brink of the precipice involve the bluff, blustering calculations, and miscalculations made by decision makers in crises. As Kagan (1969) put it, Thucydides went too far in saying that the growth of Athenian power made the Peloponnesian War inevitable. The decision makers on both sides failed to foresee the enormous destruction the war would bring and failed to think as clearly as they should have in the councils that produced the decisions to go to war; if the decision makers had reasoned better, they could have avoided war. Kagan's criticism of Thucydides's determinism should not be taken too far, either. The Athenians and Spartans, in making their decisions, had to be influenced by the Greek power distribution and the direction in which it had been changing.

On the one hand, it is true that leaders do have choices. The perceptions and beliefs that shape leaders' decisions during crises are powerfully influenced by the relative capabilities of the two sides, and especially by the way those capabilities are shifting. For instance, concerning the onset of the Peloponnesian War, Kagan is right that reasonable men could, if they had been in power, have avoided war between Athens and Sparta. But, in defense of Thucydides's position, the rise of Athenian power may have produced enough arrogance in Athens and enough fear in Sparta that such reasonable views were in disfavor. When views are in disfavor they can still prevail, but need to be defended most articulately and forcefully. The odds are against them. Thus, power shifts do not make war inevitable, but they do increase the odds against the peacemakers.

NOTES

1. The full scope of articles on war and change is too broad to cite fully in this chapter; consequently, I focus on the power transition literature that I wish to redirect incrementally. Recent additions to the power transition literature include Organski and Kugler's restatement of their power transition analysis, plus novel perspectives such as Morrow's model of the role of rational choice (1989), Kim's treatment of the role of alliances (1989), and the Doran (1989) and Houweling and Siccama (1991) debate on the role of power cycles.

2. These goods include the ability to restructure the system, as well as resources (Gilpin 1981).

3. To anticipate the statistical portion of the paper, it should be noted here that, while it is possible to distinguish in theory between apprehension from economic and military power shifts, there is far too much multicollinearity to do so in empirical tests. That is to say, because economic and military capabilities are in fact very highly correlated, it is statistically next to impossible to sort out the differential impact of military and economic power shifts. Except for a few salient cases, such as the rise of the United States at the turn of the century or of Japan in our own time, economic and military growth go

hand in hand, and what one must focus on in testing hypotheses is a general growth in material capabilities.

4. These modifications are to include China in the list of analysis after 1900, Italy and Japan after the end of postwar occupation, and Sardinia after 1830. The rule underlying these changes is that members who become major powers are included for the duration of their system membership. This procedure is consistent with that of Singer, Bremer, and Stuckey (1972), from which the data have been derived. Singer and Small (1972) terminate Soviet, French, English, U.S., and Chinese major-power status in 1965—the terminal year of data analysis in their study. In this chapter, these nations are treated as major powers continuously, through the last year of the data analysis (as, indeed, I believe they should be treated up to the present).

5. The concept of an enduring rivalry was first operationalized in print by Wayman (1982) and Diehl (1983a).

6. A future researcher should be able to construct a more subtle indicator should research objectives require it. For instance, for the purposes of studying arms races, Diehl (1983a) needed a measure that reduced the number of dyad-years of conflict. In contrast to Diehl, my research design can benefit from a larger number of dyad-years of conflict, because it is useful for statistical purposes to have equally numerous cases of relatively conflictual and relatively nonconflictual dyads. Of course, even then, a potential methodological problem involving dyads is that all the cases may not be independent.

CHAPTER 10

Balances of Power, Transitions, and Long Cycles

William R. Thompson

If a sample of people walking by in the streets were asked what international relations model first came to mind, the odds are that it would be some version of the venerable balance-of-power model. They might not be able to provide a detailed description of the model's features or say much about its historical applicability. They might not even agree that they were talking about the same model. Even so, the idea that one side needs to balance an opposing side's capabilities is so intuitively appealing and so pervasive in everyday speech that most of the people surveyed would be able to offer a crude definition of the model's central mechanism.

The balance of power may be conceptually appealing and its terminology pervasive, but is it accurate to say that power balances are conducive to peace and international stability? Or, as argued by power transition analysts, is it the other way around? Power transition arguments emphasize the destabilizing and conflictual implications of a challenger catching up to a declining leader. As the transition is accomplished, that is, when the dissatisfied challenger has caught up, matched, or balanced its capabilities with its dominant foe is when war between a rising challenger and a declining leader is most probable.

Balance-of-power and power transition arguments are highly contradictory. Both may be wrong, but both cannot be right. At least that is the way the two sets of arguments are usually portrayed. It is also the point of view put forward in the most often cited power transition analysis, Organski and Kugler's (1980) *The War Ledger* study. Yet the inherent antagonism of the two models, it will be argued in this chapter, is a matter of interpretation. Indeed, not only does the antagonism require careful interpretation, so, too, do the balance-of-power and power transition models. Interpreted from a long-cycle perspective, the intermodel antagonism disappears. In the process of theoretically integrating these different frameworks, the balance and transition models are also modified.

The outcome of this merger, it is argued, is a product with explanatory and predictive powers superior to any one of the models in a stand-alone capacity. To make this case plausible, some discussion of *The War Ledger* is needed. It

163

will be shown that Organski and Kugler's balance-of-power versus power transition contrast is a matter of interpretation and that the two models can be incorporated into the long-cycle framework. The way is thus paved for a new empirical test. The last section of the chapter will thus assess the predictability of war within the context of regional, global, and regional-global transitions since the sixteenth century.

Organski and Kugler's Balance-of-Power Model

One of the purposes of *The War Ledger* analysis was to construct an alternative to what Organski and Kugler call the respectably ancient balance-of-power model. This model is viewed as inescapably flawed in terms of its logical conception and completely wrong in terms of its prediction that an equal distribution of power will keep the peace better than an unequal distribution.

Organski and Kugler's version of the balance of power rests on several assumptions. At the very heart of the argument are the twin maxims that all states seek to maximize their power and that if one state achieves an advantage it will exploit it by attacking its now weaker rivals. Yet, historically, there have been strong limitations on how much a nation could do to improve its relative standing within its own borders. Therefore, alliance formation represented the best avenue to augmenting one's strength.

As long as coalitions could be made and unmade sufficiently fast to ensure no state or group of states gained the upper hand, the system's equilibrium should be maintained. If the natural equilibrating tendencies should break down, there was always a fail-safe mechanism to save the day. One state, the balancer (for example, Britain in the eighteenth and nineteenth centuries), remained aloof and sufficiently stronger than the other states in the system. The aloofness put the balancer in the position that it had no vital interests at stake that prevented it from intervening on behalf of whatever side seemed weakest. The sufficiently stronger characterization indicates that once the balancer intervened, its weight would be enough to restore balance and equilibrium to the system.

Thus, as long as a system homeostatically manages to maintain an equality or balance of capabilities, it should continue to be peaceful and stable. Only when the system breaks down and a condition of power inequality emerges is the probability of war and instability likely to increase.

Organski and Kugler pointed out that this model is neither very accurate historically nor particularly consistent. All states have not sought to maximize their power. All stronger states have not attacked their weaker adversaries and neighbors. And, even if they had, how is it possible that one strong state could somehow remain above the fray? Should it not also be expected to exploit its capability advantage until it can dominate the system? Most of all, Organski and Kugler suspected that an equal distribution of power was not a particularly good prescription for avoiding war. On the contrary, they thought the balance equation had it backward. Peace is more probable, they argued, when an unequal distribution of power prevailed.

I think their last conclusion is at least partially correct, but not necessarily for the same reasons that underlay their analysis. To elaborate this position, more attention must be paid to the construction of their alternative, the power transition model. It, too, has its problems, in my estimation. But it is also important to note at the outset that the Organski and Kugler interpretation of the balance of power is not the only assessment available. Indeed, one might argue that they have constructed or adopted the most ultrarealpolitik version of the balance of power conceivable and, in so doing, ensured its failure as a predictive model.

One could point out that a model that insists that individual states are not likely to grow stronger on their own does not seem to match the historical sequence of coalition wars against Philip II, Louis XIV, Napoleon, and Hitler. These successive enemies of the iterative defensive coalitions had allies too, but it was not their expansionary coalitions that aroused fear and a sense of acute threat. The perceived threat emanated primarily from a single source that was often personified in highly simplistic terms as an individual leader who had successfully galvanized his state's resources. The bandwagoning allies of the main source of threat were only rarely very impressive, either in capability or action.

Alternatively, the conclusion that equality of capabilities is the most desirable state of affairs is awkward in the sense that it does not necessarily follow from the model Organski and Kugler have constructed. If the system truly does require a strong balancer, some, albeit selective, inequality would then seem to be highly desirable. Could not one then argue that the system's homeostatic propensities are severely weakened to the extent that no strong balancer is available to save the day? If so, it is as the balancer's relative capabilities decay and the system moves away from inequality toward equality that war and instability become more probable. The implication here is that the stability outcome would seem to depend on whether the imbalance favored a balancer or a nonbalancer. If that is the case, the question is not one of inequality versus equality, but of one type of inequality versus another.

I do not wish to suggest that Organski and Kugler created a straw man balance-of-power model that is so inherently inconsistent that it could never succeed. So many interpretations of the balance of power exist that no one can ever be accused of manufacturing a weak version for purposes of analysis. At the same time, the version put forward by Organski and Kugler is not the only interpretation possible. Different assumptions are possible. Different conclusions are likely, too.

For example, if one strips the version depicted by Organski and Kugler of its most dubious components (power maximization, alliances as the main route to enhanced position, and the emphasis on equality), a leaner balance-of-power core statement remains. To wit, it states that there is some tendency for threatened states to coalesce in reaction to an ascending source of expansionary threat. In addition, this balancing tendency is greatly facilitated by an *external* balancing force. Interestingly, in this stripped-down version, it is a movement, threatened or otherwise, toward greater inequality due to the rise of a strong state inclined toward territorial expansion that makes destabilizing war more probable. But it is not the fabled maintenance of equality that keeps the peace. At best, equality and peace must await the successful suppression of the expansionary threat and the postwar restoration of some type of balance. At the same time, the balancer is conceptualized as *external* to the system in question. This means that the system and its equality/inequality calculations are more complicated than portrayed in the Organski and Kugler version. The balance of power works for some subgrouping of actors or a subsystem of the larger international system. Otherwise, there would be no place from which an external balancer could intervene. If the balancer is outside the subsystem, one then needs two different calculations of capability dispersal—one for the subsystem and one for the larger system. It also implies that there need be no guarantee that subsystemic and systemic inclinations toward equality/inequality will be perfectly synchronized. However, the most dangerous or most destabilizing situation presumably would be a movement toward inequality in the subsystem in the guise of an emerging subsystemic hegemon concurrently with substantial erosion in the larger system's inequality—as the potential balancer's capability to intervene decays.

This deduction can be derived from an integration of Dehio's (1962) reading of European balance-of-power politics and leadership long-cycle theory. It does not assume power maximization, an altruistic balancer, or the perpetual value of either equality or inequality. The stabilizing effect of equality/inequality depends on where and when it is located.

But there is also a very prominent role in this interpretation for Organski and Kugler's power transition argument. It is certainly not necessary to dismiss their alternative model simply because one disagrees with their reading of the balance of power. To do so would in fact be counterproductive because the balance-of-power conceptualization makes more sense with the power transition conceptualization than without it. Before this assertion can be pursued further, it is first necessary to introduce the power transition model and its 1980 examination by Organski and Kugler.

Organski and Kugler's Power Transition Model

The power transition model described in Organski and Kugler 1980 has two major features. First, systemic stratification is vital to the model. The focus is restricted to major or great powers, which are divided into two main groups. One group, the powerful and satisfied, is led by the system's dominant state. This group is the primary beneficiary of the prevailing world order's allocation of privileges and, indeed, was in all likelihood present at the creation of the order. The dominant state is not only the most powerful great power, it is also the chief architect and beneficiary of the systemic status quo. The other group, the powerful and dissatisfied, feel that they received less than their fair or proportional share of the available benefits. One reason for the discrepancy between what is received and what is desired is that dissatisfied great powers are said to be likely to emerge as major competitors only after the creation of the world order. The established beneficiaries are reluctant to surrender some portion of their privileges to newly arrived and arriving powers despite shifts in the relative positions of the great powers.

The second major feature of the model is its central dynamic of structural change. Position in the system is determined by relative power that, in turn, hinges on socioeconomic and political development. Uneven rates of domestic growth will enable some great power to overtake and surpass the positions of other great powers. The most dangerous situation is when a powerful, dissatisfied great power overtakes the leader of the powerful and satisfied group. When the dominant state and its principal challenger are roughly equal in power, the structural incentives encourage the outbreak of a major war. Both sides can hardly overlook the dramatic narrowing of the gap in respective capabilities. The challenger will be encouraged to attack in order to accelerate the opportunity to change the rules of the game in its own favor. The once-dominant state will seek to suppress the challenger's ascent before it is too late. Moreover, the more quickly this dangerous window of equality appears, the more destabilizing is the effect. The less expected it is, the more difficulties decision makers will have in dealing with its implications once it occurs.

Organski and Kugler (1980:61–62) have summarized their own argument:

> The mechanisms that make for major wars can be simply summed up. The fundamental problem that sets the whole system sliding almost irretrievably toward war is the differences in rates of growth among the great power and, of particular importance, the differences in rates between the dominant nation and the challenger that permit the latter to overtake the former in power. It is the leap frogging that destabilizes the system. . . . Finally, this destabilization and the ensuing conflict between giants act as a magnet, bringing into war all the major powers in the system, dependent as they are on the order established by their leaders for what they already have, or for what they hope to gain in the future if they upset the existing order.

While this transition model can be summarized succinctly, testing the model is a less straightforward proposition. Clearly, the model does not predict when any two states are most likely to engage in war or even when most of the possible dyads of states in the system are apt to become conflictual. Most states simply lack the capability to upset structural equilibria. Since only a few states possess the power to do so, an appropriate test must be confined to the suitable subset of the war-making state population.

Organski and Kugler (1980) introduce several important auxiliary assumptions at this stage of their analysis. They suggest that prior to the twentieth century, the system's great powers were exclusively European and that the basic issues of contention among them were essentially European as well, even if they were often manifested in, or focused on, the control of resources and territories outside of Europe. Only during World War II did the European oligarchy give way to the emergence of several non-European great powers. It was not that all of these non-European powers suddenly became politically significant in the 1940s. Two in particular, the United States and Japan, were classified as choosing to keep their distance prior to World War II. By doing so, they remained on the sidelines of the main arena of action.

A second auxiliary assumption further differentiates the great-power group. As implied in the model's emphasis on the transition between the dominant state and its principal challenger, many of the system's elite actors are unlikely to be powerful enough to overtake the system's dominant power. Thus, Organski and Kugler differentiated major powers according to still another categorization—this time between contenders and noncontenders. Contenders are those major powers who are strong enough to shape the future orientation of world order. Operationally, Organski and Kugler take this to encompass the system's most powerful nation and any other nation with 80 percent or more of the leader's capability. If no state meets this 80 percent qualification, the focus is shifted to the next two strongest great powers, thus allowing for unequal contenders.

The third assumption is that the best way to measure national capability is through a state's gross national product (GNP). All other possible indices are thought to be reflections of the more fundamental economic base.

Finally, a fourth assumption restricted the type of wars that might qualify as wars of structural transition at the apex. Qualifying wars must have major powers on both sides. To ensure that we are only looking at "all-out" efforts to win, moreover, a candidate war's battle deaths must exceed those of earlier wars. War losers must also face the loss of territory or population, or both.

The auxiliary assumptions yield a differentiated and rather idiosyncratic major-power elite for the post-1870 era. Nine major powers are identified for the period between 1860 and 1975 (the list is that of Singer and Small 1972). Each major power has a specific tenure as a member of the elite's center or periphery. Six of the nine (France, 1860–90; Germany, 1890–1945; the United Kingdom, 1860–1945; Russia/USSR, 1860–1975; Japan, 1950–75; and the United States, 1945–75) are also designated as contenders for certain periods of time.

Five wars satisfy the fourth assumption: the Napoleonic Wars, the Franco-Prussian War, the Russo-Japanese War, and World Wars I and II. However, the third assumption pertaining to GNP requires the first war be dropped for lack of data. The focus on the post-1860 era, so evident in their table 1 (Organski and Kugler 1980:43) is thus dictated primarily by the limited availability of information on the capability indicator.

Two more operational assumptions preceded the main test of the transition model. The model predicts that when a formerly weaker state catches up to and achieves equality with a stronger rival, war is most likely to erupt. Organski and Kugler defined "equality" as being achieved when the weaker state exceeds 80 percent of its stronger rival's position. Because relative growth rates are slow, Organski and Kugler use a roughly twenty-year time period as the unit of analysis (1980:48).

If one looks only at peripheral great powers, the equal and overtaking hypothesis proves to be a poor predictor. The same outcome characterized the central great powers, taken as a group. Only when the test was restricted to contenders did the historical record and the theoretical explanation begin to converge. Fifty percent of the equal and overtaking cases resulted in war while none of the nonovertaking cases were linked to war. Consequently, Organski and Kugler concluded that positional transition was a necessary but not sufficient factor in bringing about major wars among contenders.

Tests of the power transition hypothesis invariably appear to raise operational problems that make it awkward to endorse the consequent findings without major qualifications. Among the preliminary operational dilemmas that must be resolved, one has to make decisions about whose transitions should be examined and how best to capture relative capability or power positions. Neither one of these questions seems to have an obvious answer. For example, the original power transition theory concerned the likelihood of conflict between a rising, dissatisfied great power and the system's declining, dominant power. Along the way, this fairly specific theoretical focus has been generalized by some analysts to apply to any great-power dyad's conflict propensity. *The* system's dominant power and *the* leading challenger somehow evolved into *any* dyad's dominant power and challenger or overtaker. Thus, a once-specific theoretical statement about system structure has been translated into a fairly ad hoc hypothesis on interstate behavior by the manner in which the tests of power transition have been conducted.

Aggregating all major-power positional transitions and conflict, something that Organski and Kugler do not do but others have, does not tell us much about the relationship between the primary structural transitions of the system's leader and its leading challenger. To try to make some inference to a system's most important dyad, and therefore a systemic-level statement, based on dyadic behavior in general would of course amount to a fallacy.

Restricting the focus to exclude weaker major powers still falls short of corresponding to the original theoretical justification. Utilizing their GNP index, Organski and Kugler attempted to weed out what they called the "noncontenders." But identifying the state(s) with the largest GNP in a particular period of time, which Organski and Kugler have done, does not necessarily isolate the dominant status quo power and its leading dissatisfied challenger as stipulated by the original theory. It is, in fact, highly curious that Organski and Kugler chose never even to discuss which states, if any, approximated these roles that are so critical to the power transition argument.

Doubts may be raised as well about what the best way is to measure positional transitions. Organski and Kugler, of course, argued strongly for GNP as the superior capability index. How well GNP captures the qualitative (as opposed to the quantitative) dimension of economic capability is debatable. In an era increasingly emphasizing technological innovation over gross wealth and the double-edged advantages/drawbacks of large populations, the limitations of GNP are especially bothersome.

What is most clear, however, is that a focus on GNP is highly restrictive as to which actors and how many years of their interactions can be examined. As noted above, insisting on the GNP indicator limited the analysis to the post-1860 era. Macrosystemic transitions, after all, do not occur all that often. An operational procedure that so severely restricts the N size of any possible explanation deserves careful assessment.

Houweling and Siccama (1988a) have been able to extend the Organski and Kugler analysis back some forty-five additional years by using a multiple indicator capability index combining information on iron/steel production, population size, armed forces size, coal production, and urbanization. Depending on which major powers are scrutinized and how their positions are measured, overtaking can be seen as roughly doubling or tripling the propensity of major-power warfare.

There is certainly something to be said for hypotheses that are capable of surviving different types of tests. The Houweling and Siccama findings suggest, at the very least, that the Organski and Kugler findings were not strictly artifacts of their post-1860 focus or insistence upon the GNP index.[1] Even so, multiple indicator indexes, while often appealing because they are able to encompass several dimensions, are rarely problem-free themselves.

In particular, the employment of "bulk" indicators such as population or armed forces size (or GNP) can bias the examination in favor of huge but not always highly efficacious land powers at the expense of smaller, more sophisticated maritime powers. The use of Houweling and Siccama's data, for example, portrays China as passing the United States in 1960. The Soviet Union is recorded as accomplishing this same feat in 1970. Similarly, the entry of new major powers into the elite subset can also distort the outcome. When the Chinese data are included after 1950, for instance, China is credited with having passed Britain, France, Germany, the Soviet Union, and Japan in 1950. If these "transitions" lack something in face validity, the subsequent findings on aggregate transitions and their war propensities can hardly be accepted at full value either. The point to be made with these various observations is not that it is hopeless to attempt to test the power transition argument. Nor is there any intent to dismiss previous analyses as lacking in value. No matter how one does it, there will always be corners cut and liabilities to validity created. Rather, the less-than-novel point being made here is that how one goes about asking the question will assuredly have some impact on what answers are ultimately obtained.

Since the power transition arguments have considerable theoretical and logical appeal, and because some notion of transition is central to most if not all historical-structural interpretations of world politics, as well as several realist formulations (Levy 1987), we should be seeking to maintain as close a level of correspondence between the original structural theory and the empirical test as is possible. The underlying question of whether war is more probable under conditions of equality or inequality (the balance-of-power versus the power transition question) is also a fairly central theoretical issue in the study of world politics. Accordingly, we need to be especially careful that the way we treat the appropriate actors and their relative positions discriminates appropriately among the different types of actors in the major-power elite subset.

Leadership long-cycle arguments suggest a different approach to the question of major-power transitions. If nothing else, the data associated with this perspective make it possible to encompass many more years of positional flux. But it can also be argued that the leadership long-cycle approach is more compatible in many respects to the original power transition model (Organski 1958) than are most of the other empirical examinations of this question. The potential for maintaining a close correspondence between theory and empirical test, consequently, is unusually good.

Not only is the degree of theoretical compatibility high, the potential for a more "faithful" empirical test is also exceptional. The reason why the approach to be taken here can claim more correspondence is that it begins by first establishing which state occupies the peak of Organski's power stratification (the leading maritime power operating most successfully at the global political economy level) and which state is the leading challenger among a set of regional, predominately land powers. The principal question then becomes one of whether, or to what extent, conflict becomes more likely as the regional challenger overtakes the declining global leader.

A subsidiary, more ad hoc question is whether other major-power dyads engage in a similar process. If the transitions of all dyads are conflict-prone, one would think that this relationship is most likely to be found if due attention is paid to structured contexts. Ascending land powers might be expected to fight their way up their regional ladders just as seapowers would be expected to fight among themselves at the global or transoceanic level. Structured transitions might then be expected within the regional and global levels, as well as between them (the regional leader catching up with the global leader).

Before proceeding to an examination of these hypotheses, though, some elaboration of why we should be interested in distinguishing between global and regional actors and their respective theaters of operation is in order. To explain and defend this emphasis, a recent modification and extension of leadership long-cycle theory must be discussed. The ultimate aim of this next section is to demonstrate how the power transition model and leadership long-cycle theory can be integrated. It will also demonstrate that balance-of-power and power transition models can be reconciled within the leadership long-cycle framework.

Long-Cycle Assumptions and Dynamics

For the purposes of this analysis, leadership long-cycle theory can be restricted to two sets of dynamics, one global and the other regional.[2] Neither dynamic is viewed as a universal process. Both are conditioned by very specific temporal and geographical parameters. The decade of the 1490s is a convenient starting point for tracing the patterns of structural change that seem most important. The west European region, in addition, played a central role as the main arena for the regional dynamics, just as European states long dominated the global processes.

The global processes are focused on managing long distance trade problems that became particularly prominent after the Portuguese created an Atlantic path to the Indian Ocean. The management of intercontinental questions of political economy depends on the extent to which capabilities of global reach are concentrated in the control of the system's lead economy. Historically, one state, the "world power," has emerged from periods of intensive conflict in a position of naval and commercial/industrial preeminence. Naval power has served as one of the principal manifestations of global reach capability. It has been, and continues to be, critical for projecting military force, for protecting commercial sea lanes, and for denying extracontinental maneuverability to one's opponents. However, the ability to finance preeminent naval power hinges in turn on a commanding lead in economic innovation and the profits that ensue from pioneering new ways of doing things (Modelski and Thompson 1992, 1994).

Winning a global war, a period of highly intensive struggle over whose policies will prevail in governing the global political economy, creates a literally unrivaled opportunity for imposing new rules or reinforcing old rules of global behavior. The world power and its allies have defeated the leader's rivals and reduced their capabilities, while at the same time the world power has improved its own economic and military capabilities.

Such opportunities come and go. The relative lead in naval power tends to decay. The returns from economic innovation diminish and, frequently, are not replaced by a new surge of creativity, at least in the same place. Old rivals rebuild. New rivals emerge. The coming together of these various tendencies means that the leader's relative edge is gradually whittled away. The postwar concentration of global reach capabilities, the foundations for political leadership and order, is eroded and gives way to a period of deconcentration and deteriorating order.

Within this context, one or more challengers to the world power's leadership will appear. The most dangerous challenger tends to be a state that either has already ascended, or aspires to ascend, to a dominating regional position through the development of its land power. From the perspective of the incumbent world power, regional hegemony, especially in Europe, is the first step to a more direct attack on the global political economy. From the perspective of the ascending regional power, the world power and its penchant for maintaining the status quo is likely to thwart the prospects for further expansion. A clash becomes increasingly probable in order to determine whose preferences will prevail.

In the long run, then, global order is intermittent. Phases of relative order are followed by periods of relative disorder, punctuated by bouts of intensive combat to decide what type of order will be imposed. The regional processes of most interest overlap with the central global dynamic. Just as the global political economy has been characterized by successive peaks and troughs of capability concentration, so too has the European regional distribution of power.

Initially, regional power concentration in Europe was achieved by expanding agrarian economies and mobilizing large armies and increasingly bureaucratized states. Eventually the agrarian foundation gave way to industrialization, but the preference for large armies and autocratic states persisted. At the regional level, the pattern of expansion tended to be close to home. Ultimately, though, the decision makers of one of these continental powers would envision the possibility of expanding throughout the region and perhaps beyond.

Regional supremacy meant, among other things, control over adjacent or nearby zones of economic and commercial prosperity. If the global leader lacked some form of insularity from continental expansion, regional supremacy constituted a direct threat to its continued existence as an autonomous actor. Insularity did not eliminate the threat, but only made it less direct. Regional expansion and supremacy then meant that a large agrarian power would achieve control over resources and equipment designed for, and oriented toward, participating in the global economy. Regional expansion, therefore, would eventually lead to ambitions beyond the region.

Occasionally, then, a bid for regional supremacy would be made in western Europe. The attempt was always unsuccessful, due in large part to the nature of the region's geopolitical configuration. The aspiring regional hegemon would find itself confronted by eastern and western counterweights that were able to inject extraregional resources into the struggle over regional control. The east supplied large armies. The west supplied sea power. Depending on the strategies of the hegemonic aspirant, one or both counterweights were likely to intervene to maintain the balance of power. When both intervened simultaneously, the would-be hegemon found itself fighting a two-front war that it was unlikely to win.

This regional dynamic was unlikely to go on forever. Gradually, the ratio of constricted regional capabilities to expanding extraregional resources would become far too asymmetrical to continue encouraging attempts to dominate the European region. It seems that 1945 was a watershed in this respect, although more time may be needed to fully assess whether regional hegemony in Europe is truly an artifact of the past.

Prior to 1945, though, the main regional dynamic was a movement to and from the occasional concentrations of continental power that served as a foundation for attempts at regional domination and threatened directly or indirectly the status quo of the global political economy as well. Combining the central global and regional dynamics suggests that structural changes at the two levels were out of phase with one another and yet closely connected. Maritime power concentration was decaying at the global level while continental power concentration at the regional level was increasing. The global erosion no doubt encouraged the ambitions of the ascending regional power. The ambitions of the ascending regional power, similarly, encouraged the reformation of maritime power concentration in order to suppress the regional threat. The forced deconcentration of regional power thus led to, or at least greatly facilitated, the reconcentration of global power.

What does this theoretical argument imply for our initial concerns with balances of power and structural transitions? Clearly, a very specific interpretation of the balance of power as it worked in Europe between the 1490s and the 1940s is advanced. The emphasis of this interpretation is placed not on the virtues of equality, but on geopolitical configurations. European balances did not depend on defending coalitions that suppressed the threat to the status quo and restored equality to the distribution of power. The balance of power instead depended on the intervention of actors residing outside of

Europe with access to nonregional resources. Some semblance of regional equality may have been restored, but not without first reconcentrating the global distribution of power.

Power transitions take place at both levels. Regional powers rise and overtake earlier regional leaders. New world powers rise and overtake earlier global leaders. One might also say that there is considerable potential for power transitions between the levels as well. Regional powers rose and attempted to overtake global leaders in the process of seeking to dominate western Europe.

Thus, the general implication is that the conflicts among the major powers of the past five hundred years have been too complicated to ask simply whether equality or inequality is more destabilizing. Movements from equality to inequality in power distribution have proceeded at different levels and at different speeds. Similarly, there are at least two structures of leadership or patterns of stratification that structure the contestations of major-power rivals. Transitions within the regional or global arenas may lead to war, but the potentially most destabilizing transition is invoked by the clash between the ascending regional and the declining global leaders.

However, there are several historical caveats to this story. The dissynchronization of global and regional structures does not work the same way in all parts of the world. Nor has it always worked precisely the same way in the evolution of the European regional-global structural interactions. At the beginning (post-1494), the patterns displayed in the first half or so of the sixteenth century are not the same as the behavior demonstrated in the next four hundred years. Western oceangoing seapower was too new to play much of a role in suppressing the Hapsburg expansion of Charles V. The western and eastern balancer roles were assumed initially by France and the Ottoman Empire.

Western seapower was the most critical factor in blocking Philip II of Spain and Louis XIV of France. Coalitions of western seapower and eastern land forces were necessary to thwart Napoleon and the two German expansions of the twentieth century. But other differences and nuances are equally noteworthy. Prior to the end of the eighteenth century, the development of regional power concentration was gradual. The French bid for European dominance after 1792, however, was not preceded by a slow and premeditated buildup of French land power. The second French bid, unlike its predecessor, was explosive and short-lived. The first German bid that followed in 1914 was also out of pattern in the sense that German decision makers initially did not plan to conquer Europe; their war aims expanded only after the war had commenced. Consequently, we see no gradual and premeditated concentration of regional force prior to the outbreak of war in 1914. The second German bid, nevertheless, did revert to the older pattern established by Philip II and Louis XIV.

These historical caveats are inconvenient from a modeling perspective. We would prefer one pattern repeated over and over. But that is not the way world politics has worked in the past five hundred years. Yet there is still a pattern and there are similarities that can be exploited for explanatory and predictive purposes. There are also a few kinks that will need to be taken into consideration at the operational stage.

A New Test of the Transition Process

Hypotheses and Indicators

These expectations associated with the leadership long-cycle interpretation of modern world politics lead to several separate tests of the explanatory and predictive power of the power transition hypothesis. Specifically, the hypotheses are:

H1. Wars, and especially global wars, are more likely when a regional leader catches up to and passes a declining global leader.
H2. Wars are more likely when a new regional leader catches up to and passes a declining regional leader.
H3. Wars are more likely when a new global leader catches up to and passes a declining global leader.

The analysis of these hypotheses depends on several assumptions or auxiliary rules of procedure. The temporal unit of analysis is the five-year interval, beginning in 1490–94.[3] The examinations involve ascertaining, for each pertinent dyad and each five-year interval, whether one state's capability position overtook another and whether war was initiated. Levy's (1983a) great-power war information is the primary source used to determine the timing of war initiations. Five-year intervals in which war is ongoing but began in an earlier interval are excluded from the analysis. Following the convention utilized in the Organski and Kugler and Houweling and Siccama studies, overtaking requires that one state catch up to within at least 20 percent of the other state's capability position. Capability positions are measured in terms of either relative army (regional) or navy (global) size depending on the primary strategic orientation of the state and the hypothesis involved. Given the nature of the hypotheses, the actor focus is highly selective. Only regional and global leaders are pertinent, even though "leaders" are identified in terms of their proportional share of the pooled army or navy capabilities.

The identity of these global and regional leaders is stipulated in part on the basis of our theoretical interpretation of world politics, as opposed to simply determining which state had the largest capability share at any given moment.

The global leaders have been Portugal (1494–1580), the Netherlands (1609–1713), Britain (1714–1945), and the United States (1945–present). To attain global leader status, a global power is required to emerge as the winner of a global war and in possession of 50 percent or more of global power naval capabilities (Modelski and Thompson 1988). The regional leaders in western Europe have been Spain (1494–1644), France (1645–1870), and Germany (1871–1945). The ends of their leads are determined by the periods at which their share of regional army capabilities was eclipsed by the next regional leader (Thompson 1992). Once we know who the regional and global leaders have been, it is possible to identify the wars that involve the pertinent dyads (Spain versus France, 1494–97, 1501–4, 1511, 1514, 1515, 1521–26, 1526–29, 1536–38, 1542–44, 1552–59, 1589–98, 1625–30, 1635–48, 1648–59, 1667–68, 1682–83; Spain versus Portugal, 1580; Spain versus the Netherlands, 1580–1608; England versus the Netherlands, 1652–54, 1665–67, 1672–74; France versus the Netherlands, 1672–78, 1688–1713; Britain versus France, 1741–48, 1755–63, 1778–83, 1792–1815; France versus Germany, 1870–71, 1914–18, 1939–40; and Britain versus Germany, 1914–18, 1939–45).

The major regional and global powers that form the pools for calculating the leader's shares overlap to some extent. To qualify as a global power, a state must possess a minimum share of the world's naval capabilities (10 percent of total global power warships) and demonstrate oceangoing activity, as opposed to regional sea or coastal defense activities. If these thresholds are attained between periods of global war, the status is backdated to the conclusion of the preceding global war. The global power status is retained until the state is defeated or exhausted in global war and no longer qualifies as a global power in the post–global war era.

These global power membership rules are considered liberal in the sense that some states are introduced earlier and others retained longer than might normally be expected: Portugal (1494–1580), Spain (1494–1808), England/Britain (1494–1945), France (1494–1945), the Netherlands (1579–1810), Russia/USSR (1714–present), the United States (1816–present), Germany (1871–1945), and Japan (1875–1945). This approach is taken to minimize the amount of disturbance to the distribution of capabilities introduced by entrance and exit from the elite. When states are introduced relatively early, as most obviously in the case of the United States (1816), their capability shares are so small that the weight of their presence in the data set, as in real life, is minimal. Only as their foreign policy ambitions increase should we expect to see their capability shares begin to register.

The same philosophy, but a different approach, is used in developing the regional list: Spain (1490s–1800), England/Britain (1490s–1945), France (1490s–1945), Austria (1490s–1918, with the 1520–55 United Hapsburg exception), the Netherlands (1590s–1800), Sweden (1590s–1809), Prussia/Germany (1640s–1945), and Italy (1860s–1943). For the most part, Levy's (1983a) guidelines on identifying great powers have been followed, although in a few cases, again to minimize the impact of entrance, we have granted regional elite status earlier than might be thought of as conventional. We have also imposed a western European location as an additional requirement. Hence, the United States and Japan are in the global list but not in the regional list.

While these exclusions from the regional list should not be controversial, two others (the Ottoman Empire and Russia/the Soviet Union) may be seen as more so. There are several reasons for their omission. The most important reason is theoretical. In our Dehioan-influenced framework, these two states served as the external, eastern balancers. Occasionally, they chose to intervene in the affairs of western Europe. They were able to intervene because they could assemble very large armies drawn from manpower resident outside the region. As states located on the fringe of Europe, they also had more degrees of freedom in choosing when to become involved in western Europe. But due to their locations, they also had important and diverting interests in other regions of the Eurasian and African continents. Therefore, we do not regard the Ottoman Empire and Russia/the Soviet Union as indigenous members of the western European region.[4]

A second reason for the exclusion is a matter of comparability or incomparability. The Ottomans in the fifteenth and sixteenth centuries, and the Russians thereafter, were quite capable of putting together armies larger than any of the western European states. Even so, the numbers mobilized were always misleading. For much of the period with which we are concerned and certainly the first half of it, the Ottoman and Russian armies represented large hordes of fighting men who were often poorly armed and organized. Rarely were they able to mobilize and place in the field many of the men who might be counted as their troops. Both imperial states also had significant garrison obligations and non-European security concerns that further reduced the number of men actually available for combat. As a consequence, it would be very awkward to compare Ottoman and Russian army sizes with their western European counterparts who experienced organizational modernization much earlier.[5]

Finally, a third reason is more a matter of convenience. We have no serial information on sixteenth-century Ottoman army sizes. Even if one wanted to include Ottoman data, we would be hard pressed to do so. We do have a full series of Russian/Soviet army data, but anyone who insists that Russia/the Soviet Union must be considered a full member of the west European region would be forced to systematically discount Russian army sizes. It is not clear what the size of the discount should be and, for that matter, it is no more clear whether one discount would work equally well in the eighteenth, nineteenth, and twentieth centuries.[6]

In the absence of some discounting mechanism, the Russian numbers tend to overwhelm the size of west European armies, even though the same outcome was rarely realized on many European battlefields. Nonetheless, it is not clear that the exclusion of Russian/Soviet data at the regional level is as critical as one might imagine. If one correlates two versions of the regional series through 1944, one with Russia included and one without Russian data, the two series look very much alike. Not surprisingly, they are highly correlated ($r = .942$). Excluding the Russian data, therefore, need not pose too great a threat to the validity of analyses calculating regional leader shares.

Since the inclusion/exclusion of Russian data must alter the regional metric, it must make some difference to the size of the regional scores. And changing the nature of the regional-share scores could affect some of the transitional calculations. Exempt from this possible threat are the transitions between global powers and the regional-global transitions prior to 1715, the point at which Russian data would be introduced with a different identification of the west European region. Transitions between regional powers would not be substantively affected either, at least as long as one did not insist that Russia became the regional leader. The scores of the regional powers at their respective points of transition would simply be lower than they are under the current assumptions. That leaves two global-regional confrontations that might be affected in some way: Britain-France and Britain-Germany. The transitional coding associated with the first dyadic antagonism is not affected by reducing the French capability share. Only one of the twenty-three observations for the British-German dyad is influenced. The one five-year interval affected by the decision on Russia's location will be noted in the data analysis section.

It is also certainly true that relying exclusively on measures of military clout does constitute some movement away from the initial emphasis on "power potential." Unlike economic indices, military indicators are more apt to demonstrate manifested capabilities (as opposed to capabilities that need to be transformed somehow into military capabilities for war purposes). In this respect, the logic of ascending powers waiting until they had achieved something like battlefield equivalency with their opposite numbers presumably is more likely to be realized in the historical record. But utilizing a five-year interval (as opposed to the more customary twenty-year interval) restores to the examination a strongly conservative flavor. The army/navy distinction also corresponds very well to the types of distinctions that are being made about land/regional and maritime/global actors and their theaters of operation.

Another way of looking at this question of indicator validity is to ask whether different states would appear to be the most preeminent in western Europe if a different index were utilized. If we were to examine population size information, the possible alternative index for encompassing so many centuries, we should not expect to find exactly the same leadership intervals, since populations need to be mobilized for military purposes. There is also no reason to anticipate that all states of the same size are likely to mobilize similar proportions of their full populations. Military participation ratios are a function of such variables as the nature of perceived threats, the extent of foreign policy ambitions, and the ability of the economy to sustain armed forces. These factors will vary from country to country and time to time.

Still, we should expect to find some relationship between population bulk and army bulk. Large populations, other things being equal, should be able to produce and sustain large armies. To the extent that soldiers are used for maintaining domestic order, large populations will require large numbers of troops. To the extent that armed forces are required to expand state boundaries to encompass large numbers of people, some correlation between army and population can also be anticipated.

These expectations are borne out by the historical record. Spain, due to the temporary unification of the Hapsburg branches, assumed the population lead in the mid-sixteenth century, only to surrender that position once more to the French. France led in west European population size from the late sixteenth century through the first half or so of the nineteenth century. After 1871, a united Germany assumed the lead. Each of these shifts in population predominance is reflected in roughly similar shifts, subject to varying lags, in army predominance. This finding would seem to suggest that a reliance upon army size as the sole index of regional leadership, while it may greatly oversimplify the nature of regional predominance, does not distort too much the complexities of changing power distributions within western Europe.

Data Analysis

Table 10.1 portrays the outcome for H1 on the relationship between regional-global transitions and global wars.[7] Four trials are involved: Spain versus Portugal, France versus the Netherlands, France versus Britain, and Germany versus Britain. Table 10.1 contains two contingency tables. The first encompasses all four cases; the second drops the Spanish-Portuguese dyad from consideration. The effect of eliminating the sixteenth-century case is fairly dramatic. The first table reflects a significant but moderate association between war and overtaking behavior. The second indicates a much stronger association.

The Portuguese-Spanish interaction pattern in the early sixteenth century was fairly distinctive. There was, of course, a long history of conflict between Portugal and Castile, culminating in some respects in a brief war in the 1470s over the possession of colonial territory. What is most distinctive about the Iberian record was their agreement to essentially divide the world between them, with papal legitimation. The Portuguese sphere was centered on Brazil, Africa, and most of coastal Asia. The Spanish sphere extended from Latin America to the Philippines. With occasional deviations, the separate sphere arrangement

TABLE 10.1.   Regional-Global Power Transitions

	Equal/Unequal No Overtaking	Equal and Overtaking
No War	48	8
	(90.6%)	(66.7%)
War	5	4
	(9.4%)	(33.3%)

Dyads = Spain-Portugal (1520–80), France-Netherlands (1610–89), France-Britain (1715–94), and Germany-Britain (1820–1939). Cramer's V = 0.262* (* indicates statistical significance at the 0.05 or better level) (N = 65)

	Equal/Unequal No Overtaking	Equal and Overtaking
No War	42	2
	(91.3%)	(33.3%)
War	4	4
	(8.7%)	(66.7%)

Dyads = France-Netherlands (1610–89), France-Britain (1715–94), and Germany-Britain (1820–1939). Cramer's V = 0.522* (* indicates statistical significance at the 0.05 or better level) (N = 52)

worked reasonably well—no doubt due in part to the different principles on which Portugal and Spain operated. The Portuguese were interested primarily in a crude form of global commerce. The Spanish specialized in territorial control and extractive forms of exploitation. If both states had been genuine maritime powers, the history of their interaction in the sixteenth century might have more closely resembled that of Venice and Genoa in earlier centuries.

A second dimension associated with the Iberian dyad is that there were really two Spains in the sixteenth century. The first one was part of the unified Hapsburg drive for European dominance. The second one, most closely identified with Philip II, retained the continental aspirations while also posing more direct threats to maritime interests as represented by the increased value of American gold and silver, the Dutch Revolt, the absorption of Portugal, and the Spanish Armadas sent against England. It may indeed be inaccurate to regard Spain as a single actor between the 1520s and 1580s. In any event, it is clear that H1 mispredicts the initial Spanish overtaking in the 1530s and the conquest of Portugal in 1580.

H1 does much better in handling the cases involving France and Germany. There is a miss in the 1650s, due in part to the rapid increase in English naval expansion after its civil war (and thus a decline in the Dutch relative share). But the most obvious regional-global transition occurs in the 1660s when France and the Netherlands are at war.

In the 1740s, the French regional position moved toward the British global position but did not quite succeed in overtaking it. A second case took place in the first half of the 1790s as the French Revolution evolved into global war. Unlike the French-Dutch outcome, though, both of the French-British overtaking periods were also associated with the outbreak of war.

The first noticeable movement toward convergence of the German and British positions, in the 1870s, failed to qualify according to the 80 percent threshold rule. The World War I case does qualify even though the episode fell short of a full transition. Only in the late 1930s case did the German regional position finally surpass the declining British global position. Again, both cases of overtaking were linked to bouts of major-power warfare centering at least initially on the leading regional and global powers.

Only two trials are at stake (France vs. Spain and Germany vs. France) in the test of the second hypothesis on the relationship between regional transitions and war probability. The second hypothesis also receives some moderate empirical support in table 10.2. About 57 percent of the fourteen overtaking episodes are associated with war outbreaks, in comparison to the 23 percent linked to nonovertaking intervals. Nevertheless, the fact that two dyads could generate fourteen periods in which a challenger at least came close to catching up with the regional leader tells us something about European regional transitions. Even though the number of actors that were most directly involved were few (only three between 1490 and 1944), they obviously kept trying to improve or maintain their respective positions.

The lion's share of the regional transition attempts are found in the history of the Franco-Spanish dyad. The record is not one of regular spaced, repetitive contests. Rather, the true periods of most intense contestation were the Italian Wars in the late fifteenth-early sixteenth centuries and the mid-seventeenth century. The Spanish emerged as the dominant state in the dyad in the former, just as the French did in the latter. The numbers in table 10.2 are inflated somewhat by the number of short wars fought in the early period of Spanish ascendance,

TABLE 10.2. Regional Power Transitions

	Equal/Unequal No Overtaking	Equal and Overtaking
No War	36	6
	(76.6%)	(42.9%)
War	11	8
	(23.4%)	(57.1%)

Dyads = Spain-France (1490–1700) and France-Germany (1820–1944).
Cramer's V = 0.303* (* indicates statistical significance at the 0.05 or better level) (N = 61)

the protracted nature of regional transitions, and some indicator flaws. An example of a flaw in the indicator system is the apparent overtaking of Spain by France around 1610, which is entirely a function of relying on an army-share index. Spain demobilized rapidly during its truce with the Netherlands. France did not and therefore it looks as if the French position surpassed the Spanish one very early in the seventeenth century. A more accurate reading of the respective positions of this dyad, however, is shown when Spain entered the Thirty Years' War two intervals later.

The Franco-German dyadic record is not really much cleaner, even if it is shorter. Germany overtook France in the Franco-Prussian War (1870–71) but did not choose, or was not in a position, to establish the type of army numerical dominance enjoyed by the Hapsburgs in the sixteenth century. Franco-German army shares hovered in the same proportional range until Germany moved ahead in World War I. Germany's subsequent defeat and forced demobilization left France briefly in the dominant position by default, only to have the tables reversed again in World War II.

The third hypothesis is restricted to the leading seapowers of the global system. In view of the circumstances associated with the global transition in the sixteenth century, again only two trials are at stake: England versus the Netherlands and the United States versus Britain. Table 10.3 conveys the outcome. As indicated by the moderate measure of association in table 10.3, H3 has the least predictive success of the trio of hypotheses. Part of the problem is that, as in the case of H2, only two dyadic records are relevant. The first dyad fought somewhat inconclusively in a period of potential transition in the mid-seventeenth century. The second dyad, the Anglo-American one, fought too but their military combat ended long before their transitional period, centered around the years between the onset of World War I and the end of World War II. They were certainly engaged in fighting during some of these years, but not with each other.

Perhaps the simplest way to deal with this predictive failure is to invoke a ceteris paribus qualification. Both dyads have one or two common denomina-

TABLE 10.3. Global Power Transitions

	Equal/Unequal No Overtaking	Equal and Overtaking
No War	64	5
	(97.0%)	(71.4%)
War	2	2
	(3.0%)	(28.6%)

Dyads = England-Netherlands (1615–1794) and Britain-United States (1820–1939). Cramer's $V = 0.321^*$ (* indicates statistical significance at the 0.05 or better level) ($N = 73$)

tors. As they approached periods of transition, decision makers in the challenging (from Cromwell to Wilson through Roosevelt) and defending states were restrained by the perception that greater threats to their security lay elsewhere and that they could better deal with these greater threats in coalition than separately. Even so, these coalitions were slow to develop. Invariably, wartime conditions were necessary to fully crystallize the nature of the coalitions. And while joint attentions were focused on the mutual threat, the intraglobal transition occurred. The new global leader emerged from the war stronger than before. The old global leader emerged from the global war financially exhausted and in no position to contest the transition. Quite literally, the old leaders had worn out their sinews of war in the process of winning the global war.

Thus, despite the variable outcomes associated with the four tests of the hypotheses, the different transitional processes that they attempt to isolate remain inherently intertwined. Regional transitions can be protracted affairs. Yet posttransitional dominance, after the sixteenth century, was cut short by global-leader intervention. Old and new global leaders became increasingly less likely to fight one another due to the timely appearance of a mutually perceived regional menace that served as a serious distraction. In the absence of such a threat, as suggested by the extended Dutch-Portuguese combat of the seventeenth century and the first and second Anglo-Dutch Wars, old and aspiring global leaders would fight one another too.

Conclusions

Regional, global, and regional-global transitions, therefore, interact in ways that are difficult to readily capture in two-by-two matrices. The salience of structural transitions and the intermittent fusion of regional and global structures, nevertheless, are graphically and empirically demonstrated. All of the outcomes were correctly anticipated and statistically significant. Without the concept of transition, the major-power history of the sixteenth through twentieth centuries resembles the anarchic and seemingly random conflicts enshrined in realist assumptions about world politics. With the concept, and some theoretical rules about different theaters of operation, the apparent anarchy is reduced to a relatively structured order—or, more exactly, relatively structured regional and global orders. It is also in the intermittent fusion of these regional and global orders that the balance-of-power concept, or at least its classical interpretation, can be reconciled with the concept of power transition.

NOTES

1. Other published, empirical work on the power transition thesis that for the most part is highly supportive of the basic idea include Thompson 1983a; Doran 1989; Kim

1989, 1991, 1992; Kugler and Zagare 1990; Houweling and Siccama 1991; Geller 1992a, 1992b; Kim and Morrow 1992; and Huth, Bennett, and Gelpi 1992. A variety of perspectives, time frames, and operationalizations of both independent and dependent variables are employed in these works.

2. Core leadership long-cycle arguments are summarized in Modelski 1987; Modelski and Thompson 1988, 1994; and Thompson 1988. The emphasis on the interaction between regional and global structures has always been a feature of leadership long-cycle arguments, but its importance is highlighted in Thompson 1992 and represents an explicit incorporation of the Dehio 1962 interpretation. The current analysis transcends an earlier and incomplete investigation of the power transition hypothesis within a long-cycle framework (Thompson 1983a).

3. The customary interval for calculating transitions is twenty years, but this seems overly favorable to the hypothesis, especially when one is analyzing changes in military capability positions as opposed to relative economic wealth. A five-year interval is more demanding and, if successful, more impressive. Another deviation from usual practice is the collapse of the three dyadic conditions (unequal, equal but no overtaking, overtaking) into two (unequal and equal but no overtaking versus overtaking) because too few examples of the intermediate category were encountered to make the distinction worthwhile.

4. However, two western seapower balancers (the Netherlands and Britain) are found in the regional list. They also had some leeway in deciding whether to intervene in western European affairs but their locations, despite their watery insularity, marked them as members of the region. This was particularly the case for the Netherlands and certainly applies to England/Britain most strongly prior to 1714. Portugal, on the other hand, was never a great military power in the European region.

5. The maximum Ottoman army size in the late-fifteenth and sixteenth centuries was probably on the order of 200,000 men. In comparison, the United Hapsburg army size peaked at around 150,000 men in the middle of the sixteenth century.

6. See Wohlforth 1987 for an informative discussion of the problems associated with calculating Russian capabilities in the immediate pre-1914 era.

7. A more detailed examination of these hypotheses can be found in Rasler and Thompson 1994.

CHAPTER 11

Power Transitions and Military Buildups: Resolving the Relationship between Arms Buildups and War

Suzanne Werner and Jacek Kugler

The theory of power transition first developed by Organski in 1958 has been successfully modified and extended for the past three decades and has survived many rigorous challenges with few setbacks.[1] In fact, the consensus in the field has begun to converge to the notion that the probability of a major-power war is highest when one contending nation is at parity with a second contending nation.[2] Despite this success, an important question remains. While an imminent or a successful transition may be a necessary condition for major-power war and will thus increase the likelihood of war given this particular power arrangement, *which* of the fairly numerous occurrences of near or actual transition are dangerous?

The Transition Perspective

In *The War Ledger,* Organski and Kugler found transitions between contending nations to be a necessary condition for war; no wars occurred without a transition. However, five transitions and six additional cases of parity did transpire without war (see table 11.1). Therefore, the structural arrangement in the international system is not a sufficient predictor of conflict. The theory of power transition alone can specify the necessary conditions for conflict, but cannot satisfactorily determine *which* conditions of structural change will trigger a major-power war.

Organski and Kugler have explained the number of cases that meet the necessary conditions for war but do not in fact escalate to war by dividing the system into satisfied and dissatisfied actors. Transitions between satisfied actors, according to the logic of power transition theory, will not result in war. Although logically consistent, this argument reduces to tautology. They assume wars do not occur between satisfied countries. Thus, if a transition occurs without war, the rising country *must* have been relatively satisfied with the interna-

TABLE 11.1.   Relative Power Distribution. (From Organski and Kugler 1980:42–53.)

	Preponderance	Parity and No Transition	Parity and Transition
No Major War	4	6	5
Major War	0	0	5

$N = 20$, Tau C $= 0.5$, significance $= 0.01$

tional order. This assertion is impossible to falsify unless extreme conditions of preference polarity exist. In order to escape the tautology, it is necessary to consider the methods a nation-state could use to reflect its level of satisfaction with the international system. The division between satisfied and dissatisfied actors is then not simply asserted ex post facto but, instead, is an empirically testable proposition. By integrating the work on military buildups with the power transition theory, we are able to address this question, previously left unresolved, and determine under what conditions a situation of parity or transition will result in a major-power war.

The Arms Race/Military Buildup Perspective

The arms race literature also has quite an extensive history but, like the power transition theory, it has left unresolved many important issues. The original connection made between arms races and wars is usually credited to either Richardson or Smoker.[3] Richardson originally argued that the inevitable result of an unstable arms race, represented by the dominance of the defense coefficients over the fatigue coefficients, in his two well-known differential equations, was war. Similarly, Smoker asserted that the fear inspired by an arms race would lead to war unless a submissiveness effect kicked in and forced the two countries to back down.

Wallace's well cited articles in 1979 and 1982 provided empirical support for these earlier assertions.[4] Although he found a very strong relationship between arms races and the escalation of disputes to war, seeming to resolve the long-standing conflict regarding the relationship between arms races and war, his study was roundly criticized on methodological grounds.[5] The criticism has centered on two main points. First, he broke down multilateral conflicts into separate dyads, inflating the relationship between arms races and escalation to war. As Weede points out, nineteen of Wallace's twenty-three cases of arms races escalating to war are derived from World Wars I and II alone. Second, his arms race index not only places inordinate weight on the last two years of the arms race but also fails to adequately capture the interactive component that Wallace believed to be integral to arms races.

Since Wallace's original article, the empirical findings and the strength of the researchers' conclusions on the relationship between arms races and escalation to war have progressively weakened. Siverson and Diehl conclude after a thorough survey of the literature that, "If there is any consensus among arms race studies, it is that some arms races lead to war and some do not."[6] Like Siverson and Diehl, most researchers in the field now concur that while arms races may escalate to war more often than other disputes, the relationship is very weak.[7] While war is the final result of many arms races, numerous other military buildups have ended without the actors resorting to armed conflict.

We enter this heated debate in a peripheral manner. The arms race literature has focused on the empirical question of *whether* a relationship exists between arms races and war. We focus on the question of why. The limited theoretical development has encouraged ad hoc arguments that posit positive as well as negative relationships between military buildups and the escalation of violence. The weak empirical results confirm the inadequate level of theorizing. In this study we follow Morrow's advice and "Rather than asking whether arms races lead to war, we . . . ask *which* arms races lead to wars."[8] To this end, we simplify the often unmanageable phenomenon of arms races into a more parsimonious concept of military buildup. A military buildup reflects the most fundamental aspect of an arms race—an abnormal level of military expenditures—but eschews the additional criteria of interaction and acceleration. Simplification sharply clarifies the relationship between arms acquisition and security and avoids the difficult methodological problems that have plagued the arms race literature. Although parsimony is always a valid goal, we focus solely on the abnormal growth of military expenditures for additional theoretical reasons.

The distinction between arms races and military buildups rests fundamentally upon the inclusion of an interaction and acceleration effect. As Richardson originally specified his argument, an arms race required that the level of military expenditures of nation *A* be determined at least in large part by the level of military expenditures of nation *B* and vice versa. For every *action,* there would be a *reaction.* Likewise, the model of an unstable arms race suggested that the rate of acquisition would become increasingly faster. Two points regarding these criteria must be addressed. First, an interactive military response between countries *A* and *B* need not be present for the acquisition of arms to be a critical factor. If nation *A* begins to increase its military expenditures, nation *B* can match that threat in many different ways—only one of which is to match its expenditures to those of its opponent. To limit our study to the cases where both *A* and *B* are building and reacting to each other would ignore the many interesting cases where one of the nations has chosen an alternative strategy or found the growth rate of its opponent unthreatening. As will become clear later when we elaborate upon our criteria for military buildups, one nation's military buildup, or "abnormally" high level of growth, may still trail behind the "normal" buildup of a second nation. Similarly, to limit our study to the cases where both A and B are reacting to each other eliminates all of the cases where domestic rather than external forces determined the level of military expenditures. There is no theoretical reason to anticipate that the final relationship between war and military buildups need be related to the *source* of the buildup.[9]

Second, the focus on acceleration seems relevant only if war is viewed as a result of increasing internal crises or discontent rather than structural conditions. Acceleration may be a critical factor affecting the relationship between arms acquisitions and war if an ever increasing diversion of resources from consumption or investment to the military fuels internal discontent and thus increases the decision maker's incentive to launch an attack. If, on the other hand, arms acquisitions are considered conducive to war because of the increased threat to the other player, the rate of growth of military expenditures relative to the other player should be the key factor and not the rate of acceleration. The integrated logic of power transition and military buildups suggests that nations will choose to initiate a war when either threatened by the succession of a hostile government or presented with the window of opportunity to pursue their objectives. Thus, the overall pattern of growth and expenditures matters in this calculation; the growth rate of military expenditures of the rising country may dominate and threaten its opponent even if the high rate of increase has been steady throughout the decade and no acceleration has occurred.

The arms race literature has reached an impasse. Utilizing different assumptions, two very divergent hypotheses continue to receive support. A positive relationship between arms races and war is generally defended, as Singer did in 1958, on the basis of the "armaments-tensions" cycle to which arms races allegedly contribute.[10] Smoker and Richardson likewise support a positive relationship. Conversely, the peace through strength argument suggests that an arms race, or at least a military buildup, may actually diminish the likelihood of war.[11] Likewise, empirical tests have not been conclusive. It appears that some arms races enhance the prospects for war while others do not. The integrated logic of military buildups and power transitions contributes to this debate and overcomes the impasse. First, we focus solely on the effects of an abnormal growth in military expenditures and argue that interactive and acceleration components are not central to either the peace through arms or the armament-tension debate. Second, and more important, we argue that it is necessary to focus on the *context* within which the buildup is taking place in order to understand the theoretical connection between military buildups and war. Linking the structural conditions of power configurations with the choice by national leaders to initiate an abnormal period of arms acquisitions fills two important gaps in the arms race literature: it not only provides the theoretical foundation for but it also clarifies the empirical work on the relationship between military buildups and war.

Power Transition and Military Buildups

Power transition theory focuses on one primary condition: the configuration of and changes in the power relationships of the international system. The theory suggests that war is least likely when one country predominates over the other major powers. When the powers of the great nations are approximately equal, with one nation in a position to usurp the dominant position of a second nation, a major-power war is possible.[12] Parity provides the opportunity for the challenger to attack. A military buildup, on the other hand, reflects a conscious decision by political leaders to increase the average rate of growth of military expenditures.[13] The military buildup thus reflects the decision maker's choice to either challenge the system or to defend the status quo. Increasing the rate of growth of military expenditures thus prepares the two states for war, as well as effectively signaling their willingness to do battle. A power transition will only erupt to war if the challenger state has indicated its willingness to do battle; transitions between status quo states will not erupt. The choice to increase the growth of military expenditures is a useful and intuitive indicator of willingness and dissatisfaction. Likewise, the final outcome of a military buildup depends in large part on the *context* in which the buildup occurs. A military buildup that occurs under conditions of parity will be far more dangerous than one that occurs under conditions of preponderance. We have taken the first step in distinguishing which balances and which buildups will lead to war.

Parity has provided the opportunity for the challenger to attack. The military buildup has indicated willingness to either disrupt or defend the status quo. A final component of the integrated model thus must reflect the actors' relative commitment to their goals. The commitment to disrupt or to defend the status quo is directly reflected by the relative rate of growth of military expenditures. A challenger firmly committed to its purpose will not only indicate its dissatisfaction by increasing the rate of growth of its military expenditures, but will also increase the rate of growth so as to exceed that of the dominant state. The challenger then has indicated its dissatisfaction as well as its willingness to sacrifice to maximize its potential for victory. A dominant state firmly committed to the defense of the status quo, on the other hand, will likewise attempt to increase its military expenditures so that its growth rate exceeds that of its nemesis. Such an increase will clearly signal its feelings of threat as well as its commitment to defend its order; such commitment will clearly discourage the challenging state that has been unable to match the buildup.

If the rising nation's military expenditures are growing at a faster rate than those of the dominant country, relative equality is extremely dangerous to the stability of the international order. Perceptions of capacity and threat are significantly affected by the buildup of the most threatening component of a nation's power resources. The rising country has clearly signaled its intentions. In the language of Organski and Kugler, the rising country has identified itself as a member of the dissatisfied coalition intent on disrupting the status quo. A military buildup, in these circumstances, increases the fear of the dominant country, fuels the armaments-tension cycle, and, as a final result, increases the probability of war. The presence or absence of a military buildup is thus an indicator of a country's satisfaction with the international order, while the relative rate of growth of military expenditures is an indicator of the country's commitment. Thus, we anticipate a major-power war when the challenger state has the opportunity, willingness, and commitment to challenge the dominant order. Parity provides the opportunity. A military buildup indicates the willingness or dissatisfaction with the current system. The challenger's successful effort to exceed the dominant state's military effort indicates that its commitment to change exceeds its opponent's commitment to stability and thus promises future success. When each condition is met, a major-power war will be the final disastrous result of the changes in the international order and the tensions of the military buildup.

If any of the above three conditions are absent, a major-power war is averted. Parity may have provided the opportunity for a challenge and a military buildup signaled dissatisfaction, but if the dominant state can successfully increase its military expenditures at a faster rate than the challenger, the crisis is averted. The challenger backs off, for the dominant state has very successfully signaled its commitment and capacity to defend its order. The dominant country, at least for the time being, has reinforced its predominance and at least perceptually increased the gap between the two contenders. Recognizing the potential challenge to its interests, the dominant country embarks upon a military buildup that not only signals its firm commitment to the current world order but also buttresses its ability to defend the same. Just as the structural conditions influenced the decision making of the dominant country, the choices made affect the structure. Despite the opportunity and dissatisfaction reflected by conditions of parity and the presence of a military buildup, we would not anticipate a major-power war under these conditions.

Likewise, if countries are positioned in relative parity to each other with a transition in power a possible result, changes in relative power will occur without major conflict if neither country has initiated a military buildup. This situation can be thought of as similar to the transitions between satisfied countries depicted in Organski and Kugler's (1980) original work. The absence of a military buildup indicates that both countries are either satisfied with the status quo or not fearful of potential changes. Of course, it is unnecessary to assume that the policy preferences of the two countries are identical, only that the differences are not so significant or salient as to make either country endure the costs of escalation.

Finally, when the two nations are unequal, whether a military buildup occurs or not and regardless of which nation is building up at a faster rate, a major-power war would not be anticipated for the necessary condition outlined in the power transition theory is not satisfied. A dissatisfied country does not have the power to change the current order and, as a result, the dominant country is not threatened by the hostile signals it receives. Likewise, even if the challenger is building up at a faster rate, the marginal impact in the short term is minimal. Therefore, under conditions of predominance, the presence or absence of military buildups as well as the identification of the winner of the buildup should be unrelated to war. Even if the challenger is building at a faster rate than the predominate nation, the position of the leading country as well as its ability to defend its interests as reflected in the status quo remains secure. The power disparity generally ensures that the benefits of escalation do not exceed the costs for either nation unless the decision-making body of the nation is risk-acceptant.[14]

Integrating these formerly separate logics offers potentially high rewards. If the empirical results support our suppositions, the integration would have addressed and answered two of the most critical questions left unresolved by the power transition theory and the work on military buildups. Which potential or successful power transitions result in war? We expect that relative power changes accompanied by a military buildup, with the rising nation's military expenditures growing at a faster average rate, will result in major-power wars. Relative power shifts, however, will not result in war when either a military buildup has not occurred or the dominant nation is leading the military buildup. Which military buildups result in war? Military buildups that are led by the rising country under conditions of parity will escalate to violent conflict. Military buildups that are either undertaken when the two nations are relatively unequal or when the dominant country's military expenditures are growing at a faster rate than the challenging country's will not result in war. We will test these hypotheses by examining relations between a modified set of contender nations between 1816 and 1980. In the appendix, however, we replicate Organski and Kugler's findings in *The War Ledger* with slight modifications in order to accurately incorporate the phenomenon of military buildups.

Measurement

To test these rather involved hypotheses, it was necessary to make several critical decisions. Specifically, we needed to identify the contenders from the larger set of countries and to distinguish major systemic wars from the category of general wars. In addition, we had to decide how to identify dominant and challenging countries, equality, military buildups, and winners. As always, some of the choices may be questioned. However, throughout the process, we attempted to reflect the core concept we were trying to capture.

The Contenders

Which countries to include in the test proved to be one of the easier tasks due to the preliminary work done in *The War Ledger* and the limitations of space in this article. The theory of power transition developed and tested by Organski and Kugler suggests that the theory is only applicable to the top contending nations in the international system. It is those nations that will decide the international order and those nations that will fight for the privilege of choice.[15]

Organski and Kugler distinguished major powers from the rest of the world and contending nations from the rest of the major powers in the following way. Power resources easily separated the major players from the smaller actors.[16] In addition, alliances among the relevant actors distinguished between members of the central and peripheral international systems.[17] Only members within the central system are considered to be major powers. The choice of contenders from this already elite list relied upon a series of elaborate criteria. The most powerful nation in the world at any given time was always considered to be a member of the contending class. After that, a nation joined the elite ranks if its resources were at least 80 percent of the capabilities of the leading nation. If no nation met this criteria, however, Organski and Kugler considered the top three countries within the central system to be within the contending class.[18] This operationalization of the contending class is unnecessarily cumbersome and somewhat contrived. In addition, the top three nation rule has the unintended effect of biasing the structure of the system in favor of a tripolar system. Even if two nations towered above the rest of the system in strength, a third nation was still included. Likewise, if a fourth or fifth nation was only slightly smaller than the top three, it could be excluded from the elite set. Although we replicate Organski and Kugler's findings in the appendix to demonstrate the robustness of the model, we present here a slightly revised list of contending nations based upon an improved method of selecting the main players in the international system and an extended time span.

In order to differentiate the contending nations from the major powers, we focus on the natural gaps in power between groups of nations. All major powers within the central system are listed from most powerful to least powerful for each time span under consideration. The division is made between contenders and major powers by examining the largest percentile drop in power from one nation to the next. For instance, if the four largest countries in the international system, $A, B, C,$ and $D,$ have, respectively, 100, 75, 30, and 15 units of power, the largest percentile change in power is between nations B and $C.$

Therefore, *A* and *B* would be included as contenders, while *C* and *D* would not be. This new criteria for the selection of contenders seems to paint a more realistic portrait of the international system; Organski and Kugler's criteria would also have included nation *C* despite the fact that nation *C* does not have the capabilities to play in nation *A*'s or *B*'s ballpark. We retain Organski and Kugler's use of gross national product[19] to operationalize relative power positions due to its simplicity, availability, and ability to reflect the commonly accepted power positions in the international system.[20]

Within this elite group of contenders, it is necessary to identify the dominant country from the potential challengers in order to integrate the military buildup logic. The dominant country in the decade under analysis is that nation with the larger average GNP for that time period. Although this distinction appears straightforward, it becomes increasingly complex when relative power positions are changing. A state that has previously dominated the system retains its role up to the transition. Once the transition has occurred, however, the former hegemon becomes the new challenger. Likewise, a relatively weaker state may assume the role of leader once the transition has occurred, even without a war.

Major-power Wars

Just as the actors were carefully selected according to the criteria suggested by the theory, it is necessary to identify the cases of violent conflict that the theory attempts to explain. We are interested in the wars fought over the nature of the future international order—potentially system-transforming wars. To determine the set of conflicts to be studied, Organski and Kugler relied on three criteria. They argued that since the major powers alone can decide the fate of the system, major-power wars must by definition be fought with a major power on either side of the conflict. The entire population of wars since 1800 that meet this criterion include the Napoleonic Wars, the Crimean War, the Austro-Prussian War, the Franco-Prussian War, the Russo-Japanese War, World War I, World War II, and the Korean War. Although a major power fought on each side of these conflicts, the entire population of major wars should not be tested, for they have yet to meet the additional criteria that attempt to ensure that the war was fought over the nature of the international order. To this end, Organski and Kugler added two additional criteria. First, the battle deaths from the war must exceed the battle deaths of any previous war. Second, the war must present a real threat to the territory or population for the vanquished.[21] By these standards, Organski and Kugler distinguished those wars that could be classified as major systemic wars from the nine cases of major war. They conclude that the theoretical constraints reduce the number of conflicts to five: the Napoleonic Wars, the Franco-Prussian War of 1870–71, the Russo-Japanese War of 1904–5, and World Wars I and II.[22]

The set of major systemic wars is not, however, an accepted fact.[23] In fact, in this study, the criteria for distinguishing a major systemic war has been altered slightly from Organski and Kugler's definition. First, we argue that for the war to be potentially system transforming, a contending nation—not merely a major power as defined above—must fight on each side of the conflict. Although potentially disruptive, violent disputes between the major powers cannot fundamentally challenge the established order of the system. The contending nations alone have the capacity to set the rules and thus to change them if desired. Second, the necessity for the battle deaths to be larger than in any previous war is both unnecessary and irrelevant to distinguish major systemic wars. This criterion seems only to ensure that the war is extremely long and may, in fact, exclude the potential that through luck or strategy one nation or coalition won quickly despite the all-out effort of the vanquished coalition. The threat-to-core-territory condition established by Organski and Kugler sufficiently ensures that each nation will be putting all of its resources into the war effort. According to these criteria, the population of systemic wars reduces to four: the Napoleonic Wars, the Franco-Prussian War, World War I, and World War II. The Napoleonic Wars are excluded due to limited data. Obviously, the event under investigation is a very rare event, of interest not for its frequency, but for its potential effects.

Opportunity, Willingness, and Commitment

The condition of equality will also be necessary to operationalize in order to test the integrated theory. We have chosen to retain the criterion for equality utilized by Organski and Kugler. Thus, within the set of contending nations, if the GNP of the smaller nation is at least 80 percent of that of the larger nation, it is considered to be relatively equal. This large span compensates for the crudeness of the measurement of power as well as for potentially quick shifts due to qualitative changes in military technology or disruptions in the political systems. Nations not meeting this cutoff are considered unequal relative to each other.

The criteria used to identify dyads experiencing military buildups with each other will be the most controversial, due both to the absence of an accepted list of criteria and to the plethora of different ideas. As noted previously, we have referred throughout to the phenomenon under examination as military buildups and not as arms races. In order to identify which dyads are experiencing a military buildup a two-step process is followed. First, it is necessary to determine if either nation is experiencing an "abnormal" rate of military expenditure growth. Second, it is necessary to determine if the high rate of military growth of a country is directed at the second country in question. The latter problem is addressed by noting whether an extended dispute existed between the pair of countries during the time period.[24]

The former problem is far more difficult. What is an abnormal rate of growth in military expenditures? Although arms races and military buildups are generally accepted as common if not frequent events, the criteria to establish when a buildup has occurred is still not clear. The problem, however, has been resolved simply but satisfactorily by Michael Horn. We will use his operationalization to identify military buildups, with only slight modifications.[25] Thus we have operationalized an abnormal rate of growth necessary for a military buildup by comparing the overall average rate of growth of military expenditures[26] for a particular country (1816–1980) to the average rate of growth of military expenditures for each decade under examination.[27] If the decade average exceeds the overall average rate of growth, the country is coded as maintaining an abnormal rate of growth for that time period.[28] If one of the countries of a dyad has an abnormal rate of growth and an extended dispute exists between them, the dyad is considered to have experienced a military buildup. The "winner" of the buildup is the country whose average rate of growth of military expenditures for the decade exceeds the other's. The rate of growth is the proper unit of comparison for it reflects the changing relative power positions.

The Results

The results are extremely encouraging. When the two logics are combined into one interactive model, the ability to predict major systemic wars is vastly improved, for the integrated model differentiates dangerous conditions of parity and dangerous military buildups from those that are not. In order to demonstrate the utility of the integrated logic it is first helpful to examine separately the simple relationship between (1) structural conditions and war (table 11.2) and (2) military buildups and war (table 11.3). Although a weak relationship exists between equality and war and military buildups and war, it is clear that additional criteria are necessary in order to distinguish *which* conditions of equal-

TABLE 11.2.   Relative Power and War

	Not Equal	Equal
No War	17	23
War	0	5

$N = 45$, Tau B $= 0.28$, $p < 0.05$

TABLE 11.3. Military Buildups and War

	No Buildup	Buildup
No War	28	12
War	0	5

$N = 45$, Tau B $= 0.45$, $p < 0.05$

ity and *which* military buildups will escalate to war. Even if we had separated out stable parity relationships from those undergoing relative power changes, we would still find, as Organski and Kugler did in *The War Ledger,* that some transitions lead to war while others do not.

These results alone, however, are quite powerful. First, it is important to note that the inclusion of the additional element of military buildups will not *detract* from the results obtained from relying solely on structural arrangements. In other words, there were no instances where a war occurred without a military buildup. This is especially impressive since we excluded from the buildup averages those years in which intense warfare occurred so as not to inflate the buildup averages. Although the additional criterion could not improve upon the prediction of war, as the structural argument had already provided the necessary conditions for war, considering the effect of military buildup did not reduce the power of the power transition model. More important, however, the inclusion of military buildups reduces the error of the structural argument. Considering only the relative power relationship, there were twenty-eight cases that could have resulted in war, for they met the condition of relative equality. Only five cases did so. The potential cases for escalation, however, were reduced from twenty-eight to seventeen—a 39 percent reduction in error—by the additional consideration of military buildups (table 11.4). Thus, the overprediction of war, one of the key weaknesses of the power transition model, is significantly reduced by the inclusion of a test for military buildups. However, the tables

TABLE 11.4. Change in the Prediction Accuracy by the Inclusion of Military Buildups

		Predicted	
		No War	War
Actual	No War	$+11$	-11
	War	—	—

demonstrate clearly that despite their ability to accurately predict war, neither model explains a great deal of the variance due to the many cases that satisfied the necessary conditions but did not in fact result in war.

Although military buildups reduce the error in prediction of the structural argument, it still remains impossible to differentiate which buildups and which conditions of equality will result in war. The power of the integrated logic is clear when the three critical components of parity, military buildups, and identification of the winner are incorporated into a single test as shown in table 11.5. Military buildups in which the challenger is winning will be in bold print. The date following the dyad indicates the beginning year of the ten-year period.

With the exception of the United Kingdom and the USSR between 1906

TABLE 11.5.   The Integrated Logic: Opportunity, Willingness, and Commitment

	Not Equal		Equal	
	No Buildup	Buildup	No Buildup	Buildup
No War				
	(12)	(5)	(16)	(7)
	US-JA 1976	**US-USSR 1946**	UK-FR 1816	UK-FR 1836
	UK-FR 1876	**US-USSR 1956**	UK-FR 1826	UK-GER 1920
	UK-FR 1886	**US-USSR 1966**	UK-FR 1846	UK-USSR 1876
	UK-GER 1866	US-CHI 1976	UK-FR 1856	**UK-USSR 1906**
	UK-GER 1876	UK-USSR 1930	UK-FR 1866	UK-USSR 1920
	UK-GER 1886		UK-GER 1896	FR-GER 1876
	FR-USSR 1846		UK-USSR 1846	FR-GER 1886
	FR-USSR 1866		UK-USSR 1856	
	FR-USSR 1876		UK-USSR 1866	
	FR-USSR 1886		UK-USSR 1886	
	GER-USSR 1886		UK-USSR 1896	
	GER-USSR 1876		FR-USSR 1856	
			GER-USSR 1886	
			GER-USSR 1896	
			GER-USSR 1920	
			CHI-JAP 1976	
War				
	(0)	(0)	(0)	(5)
				UK-GER 1906
				UK-GER 1930
				FR-GER 1866
				GER-USSR 1906
				GER-USSR 1930

Note: Boldface entries indicate instances in which the challenger was "winning" the buildup.

and 1913, the results are exactly as anticipated.[29] No military buildups escalated to war except those that occurred under conditions of relative equality and when the challenging country was increasing its rate of growth of military expenditures at a faster rate than the dominant country. War did not occur unless each of the three necessary conditions was satisfied. When military buildups occurred when the nations were unequal or when the dominant country was winning, they did not escalate to war. Note also that who is winning the race appears relevant only when conditions of parity exist, as anticipated by the integrated logic. Note for instance the relationship between the United States and the USSR. In several decades, the USSR significantly increased its military expenditures, easily dominating the United States' efforts. Despite the tension and the obvious dissatisfaction of the USSR during this time period, the mutual buildup did not escalate to war. Despite the clear signal of dissatisfaction and obvious commitment to change, the structural arrangement ensured that the USSR did not have the opportunity to launch a successful attack. Furthermore, it is very interesting that the only dyad that defies the logic is the United Kingdom and the USSR in the period immediately preceding World War I. Although any explanation can only be supposition, it appears reasonable to assume that the war with Germany may have preempted the conflict brewing between these surprising allies. The resilience of this relationship can be demonstrated by a simple measure of association, Tau B, in table 11.6. The interactive variable, Buildup * Equality * Challenger, will only be equal to one when *each* of the variables are equal to one. When equality, military buildups, and challenger winning are simultaneously strictly positive, the interactive term will also be strictly positive and, as predicted, the binary war variable will also have a positive value. If *any* component of the interactive term is coded as zero due to inequality, no buildup, or dominant country winning the buildup, then the inter-

TABLE 11.6.   The Interactive Model

	Buildup * Equality * Challenger	
	0	1
No War	39	1
War	0	5

$N = 45$, Tau B $= 0.9$, $p < 0.01$

Buildup =   1 if military buildup occurred
            0 otherwise
Challenger = 1 if challenger led buildup
            0 otherwise
Equality = 1 if parity existed
            0 otherwise

active term will be equal to zero. As predicted, the dependent variable will similarly have a zero value under any of those conditions. Parity or transitions are not inherently conflictual. Similarly, military buildups need not escalate. It is the *interaction* of these conditions that fuels international conflict into major-power wars.

The power of the interactive model can also be demonstrated simply by noting the proportional reduction in error.[30] Lambda, a measure of the proportional reduction in error, provides a useful estimate of the ability to predict the dependent variable given knowledge of the independent variable. In Organski and Kugler's original findings, despite a statistically significant association between major systemic wars and an overtaking challenger, the independent variables were no better at predicting the occurrence of major systemic wars than would have been predicted by chance. The proportional reduction in error was zero. However, by considering the interactive variable presented above, the error of prediction is decreased by 80 percent, a significant improvement over the original model.

The importance of these findings can be expressed more clearly by considering subsets of the complete data set. We noted above that twenty-eight cases satisfied the necessary condition for war under the structural argument, although only five of those cases actually escalated. Likewise, seventeen cases satisfied the necessary conditions for war under the military buildups analysis. Again, only five of those cases escalated to a major-power war. When we finally consider the interactive effect of all three conditions, only six cases satisfy the necessary conditions for war. The overprediction error has been reduced almost to zero, for five of those six cases do result in a major-power war as shown in table 11.7. The importance of adding the third element, identifying the winner of the military buildup, becomes very clear by examining closely the second column in the table. In this case, the relationship between equality, military buildups, and war is completely confounded. For these six cases, a military buildup has occurred indicating dissatisfaction and parity has presented the opportunity to escalate the conflict. The dominant country, however, has

TABLE 11.7.   Military Buildups

	Not Equal	Equal Dominant Wins	Equal Challenger Wins
No Major War	5	6	1
Major War	0	0	5

$N = 17$, Tau C $= 0.76$, chi$^2 = 13.11$, $p < 0.01$

clearly signaled its commitment to defending the order by building up at a faster rate than the challenger. Its clear commitment averts the brewing crisis.

The integrated logic thus finally differentiates which military buildups and which conditions of parity will lead to war. Equality accompanied by a military buildup, with the rising nation's military expenditures growing at a faster average rate, results in major-power wars. Equality, however, will not result in war when either a military buildup has not occurred or the dominant nation is leading the buildup. Likewise, military buildups that are led by the rising country under conditions of parity will escalate to violent conflict. Military buildups that are either undertaken when the two nations are relatively unequal or when the dominant country's military expenditures are growing at a faster rate than the challenging country's, however, will not result in war. Both structure and choice matter.

Conclusion

Extensive research in the field of arms races has concluded somewhat weakly that while some arms races have escalated to war, others have not. No theory of arms races had previously been developed that identified *which* arms races lead to war. Likewise, while the theory of power transition offers the necessary conditions for war, it cannot predict *which* transitions will result in war. We have limited the definition of arms races so as to encompass only military buildups and have extended the concept of power transition to encompass the counterfactual or the "potential" transition so as to integrate the logic of military buildups with that of power transitions. The integrated logic would suggest that war is the probable result of three joint events: the existence of relative parity between two contending nations; a military buildup between the pair; and the average growth rate of military expenditures by the challenging nation exceeding that of the dominant nation. The condition of parity presents the challenger with the opportunity to successfully challenge the dominant state. The military buildup indicates the state's dissatisfaction with the current international order or with the growing challenge. Finally, the winner of the buildup clearly indicates the relative commitment and resolve of the two states. The military buildup thus provides a great deal of information to the leaders. The dominant state can determine the significance of the threat while the challenger can ascertain the difficulty of the coming battle. The buildup provides useful information that may encourage the challenger to alter its course. Two tests of the hypotheses strongly support the integrated logic.

The inclusion of the buildup conditions appears to be an important contribution to the power transition logic, as is the inclusion of the power transition conditions to military buildup research. The integration helps to identify when structural changes and military buildups will lead to war. Military buildups cannot be examined usefully in the absence of the context in which they occur. The structure of the system is an important element determining the outcome of a hostile buildup of arms. Similarly, power transitions cannot be examined without being cognizant of the signals and choices being made by the individual decision makers. Leaders of the rising nation determine whether it is worthwhile to challenge the status quo established by the dominant nation, while the leaders of the dominant nation must determine if they are willing to defend the status quo. Although relative power positions will eventually change and war may still be the end result, the addition of the military buildups to the power transition logic suggests that the timing and the tenor of the transition can be manipulated by skilled leaders. The state can thus choose whether to react to the possibilities that the power configuration presents by either building or not building up its military.

APPENDIX

TABLE 11.8. The Integrated Logic: Application to *The War Ledger* Cases

	Not Equal		Equal	
	No Buildup	Buildup	No Buildup	Buildup
No War				
	(16)	(4)	(11)	(6)
	UK-FR 1876	**US-USSR 1946**	UK-FR 1856	UK-GER 1920
	UK-FR 1886	**US-USSR 1956**	UK-FR 1866	UK-USSR 1876
	UK-FR 1896	**US-USSR 1966**	UK-GER 1896	**UK-USSR 1906**
	UK-GER 1856	UK-USSR 1930	UK-USSR 1856	UK-USSR 1920
	UK-GER 1866		UK-USSR 1866	FR-GER 1876
	UK-GER 1876		UK-USSR 1886	FR-GER 1886
	UK-GER 1886		UK-USSR 1896	
	FR-GER 1856		FR-USSR 1856	
	FR-GER 1896		GER-USSR 1886	
	FR-USSR 1866		GER-USSR 1896	
	FR-USSR 1876		GER-USSR 1920	
	FR-USSR 1886			
	FR-USSR 1896			
	GER-USSR 1856			
	GER-USSR 1866			
	GER-USSR 1876			
War				
	(0)	(0)	(0)	(5)
				UK-GER 1906
				UK-GER 1930
				FR-GER 1866
				GER-USSR 1906
				GER-USSR 1930

Note: Boldface entries indicate instances in which the challenger was "winning" the buildup.

This test replicates the findings in *The War Ledger,* utilizing Organski and Kugler's criteria for contending nations. The strong results suggest that the integrated logic is robust despite the change in operationalization. As above, the case of the United Kingdom and the USSR during the decade beginning in 1906 proves to be a false prediction. Otherwise, the model correctly differentiates the military buildups and structural arrangements that will lead to war from those that will not.

NOTES

1. See, for instance: Singer, Bremer, and Stuckey 1972; Gilpin 1981; Bueno de Mesquita and Lalman 1988; Houweling and Siccama 1988a; and Kim and Morrow 1992.

2. Bueno de Mesquita and Lalman 1992:chap. 6.

3. See Richardson 1960b, who showed that if the product of the defense coefficients $(k * l)$ exceeded the product of the fatigue coefficients $(a * b)$ in the differential equations representing the acquisition of arms by two nations $(dx/dt = ky - ax + g$ and $dy/dt = lx - by + h)$ no equilibrium would be reached. The acquisition of arms by one nation would spur the acquisition of arms by the second until resources were depleted or, more likely, a war settled the dispute. Also see Smoker 1969.

4. Wallace 1979, 1982.

5. See Lambelet 1975; Wallace 1980, 1983; Weede 1980; Altfeld 1983; and Horn 1987.

6. Siverson and Diehl 1989.

7. See, for instance, Diehl 1983b; Diehl and Kingston 1987; and Horn 1987.

8. See Morrow 1989.

9. Several scholars have examined the action-reaction component of arms races. For positive but weak findings, see Ward 1984 and Ostrom and Marra 1986. For negative findings, see Organski and Kugler 1980.

10. Singer 1958.

11. Morgenthau 1960. As Wallace (1982) points out, most of the evidence for the "peace through preparedness" doctrine is anecdotal and idiosyncratic.

12. In *The War Ledger,* Organski and Kugler distinguished between situations of parity where the power of the nations was roughly equal and transitions where the relative power of the key actors was changing. Although important, the distinction is problematic for the following reason. If a war occurs without a transition it does not necessarily refute their hypothesis, for it would be necessary to discover the counterfactual: would a transition have occurred without the war? The war may have been intentionally initiated, either by the dominant nation in order to avert the transition or by the challenging country in order to gain the advantage of surprise or simply due to miscalculation. Therefore, in this study we do not distinguish conditions of parity or potentially imminent transitions from actual transitions. Although this fact introduces its own problems, the addition of the military buildup criteria diminishes the bias that has been introduced. In addition, it allows us to examine the important cases where imminent transitions and

possibly major-power wars were avoided by the foreign policy *choices* of the actors. This greatly improves upon the purely structural theory of power transition, which, while not eliminating the potential for choice, did not explicitly incorporate it into the theory.

13. Some confusion may arise regarding our use of both GNP and military expenditures. As Kugler and Arbetman (1989a) demonstrate, utilizing both GNP and military indicators as *measures of power* would in effect be counting the same resources twice. It is important to note that we do not use military expenditures as an indicator of strength; GNP alone measures the relative power positions of the players. Military expenditures, instead, indicate the willingness and the commitment of the leader to initiate a war. Military expenditures reflect the choice, while GNP indicates structural constraints.

14. Bueno de Mesquita and Lalman (1992) demonstrate that the probability, *p*, of victory for either nation *i* or *j* need not be greater than 0.5 for it rationally to choose to attack if it is risk-seeking.

15. See the chapter by Lemke (herein) for a more developed and complex argument regarding the international order. He envisions the international system as a complex of intricately intertwined hierarchies. Within each hierarchy, battles over the rules of the order take place. Therefore, instead of having merely one set of contending nations battling over the order of the international system, several separate sets of contenders exist that struggle for the privilege of choice over smaller, more regionally centered questions. Preliminary tests of the linkage between power transitions and military buildups within the wider context envisioned by Lemke prove very encouraging. The logic at work in regional hierarchies appears to parallel the logic in the international system.

16. The identification of major powers by population, economic productivity, and military strength is available in Singer and Small 1968.

17. For a more elaborate discussion of the distinction between center and peripheral systems, see Organski and Kugler 1980:42–44.

18. Organski and Kugler 1980:44.

19. Angus Maddison (1991:195–222) provides an exhaustive list of all his sources and computations performed in order to derive the indices he provides for those interested. For this latest edition, Maddison slightly changed the base used for both France and Germany. While these changes affect the timing of some of the transitions, they do not affect the analysis when the condition of equality is examined instead. As previously noted, the empirical testing of transitions is exceedingly difficult, requiring knowledge of counterfactual events. The formulation used here reflects the logic in the power transition theory while avoiding this problem.

20. For a discussion of their choice of GNP and the advantages and disadvantages of alternate measures of power, see Organski and Kugler 1980:30–38, and Kugler and Arbetman 1989a.

21. Organski and Kugler 1980:45–47.

22. Despite Organski and Kugler's carefully defined criteria for the dependent variable, the set of major systemic wars does not meet their own standards. Kugler himself demonstrates (1990:201–13) that the Russo-Japanese War does not meet their second criteria regarding battle deaths; the total battle deaths were greater in the Napoleonic Wars than in the Russo-Japanese.

23. See, for example: Gilpin 1981; Levy 1985; Thompson 1988; and Midlarsky 1990a.

24. See Wayman 1989.

25. Horn 1987:29–39. As explained above, we will not incorporate his operationalization of acceleration. Nor will we incorporate his second measure of arms races.

26. As Horn (1987) notes, it is critical to identify only *peacetime* buildups. The results would be badly skewed if the high rates of growth that characterize most major wars were included in the overall average. Again following Horn's lead, we control for this problem by eliminating those years where battle-death rates exceed a critical amount. Horn examined several levels of intensity to determine which critical level of battle deaths to use. After carefully examining the data, we chose to exclude those growth rates during the years that battle deaths exceeded 7,500. This excluded all major-power war years but did not artificially deflate the averages. A major power may have been settling a small rebellion within its zone of influence and thus suffered minimal casualties while maintaining a military buildup against real opponents.

27. Generally, military buildups are examined over a relatively long period of time. The period examined, however, differs with each scholar. We chose ten years because it represented approximately the average length of time examined by other scholars. This choice, of course, will negate the identical replication of Organski and Kugler's test, for they used twenty-year periods. Despite this problem, it appeared to be the lesser evil. Military buildups examined over a twenty-year period not only fail to reflect reality but also minimize or exclude the many intense rivalries that have occurred over a shorter time period.

28. In computing the overall rate of growth for each country, it became obvious that in some instances one or two years of extremely high growth rates or outliers artificially inflated the country's overall average. The outliers dominated the results. As a result, decades that experienced high rates of growth were incorrectly coded as "normal." The only instance of this artificial inflation relevant to the test presented here is Germany. When the three outliers were replaced by the otherwise largest change in military expenditures (1933: 3.02, 1934: 1.57, and 1935: 2.27 were replaced by the rate of change 1.46 that occurred in 1890), the overall average was reduced from 1.073 to 1.047—a substantial and significant change. While retaining high rates of growth, the reduced overall average better reflected the trends in German spending. As a result, Germany was considered to have an abnormal rate of growth in the period 1866–75, where it had not previously.

29. Note that the Crimean War between the United Kingdom and Russia and France and Russia is not coded as a "war" due to our definition of the dependent variable— major systemic war. The Crimean War does not meet the criteria for a major systemic war. Therefore, within this context the relations between the United Kingdom, France, and Russia are coded as "no war" for the decade 1846–55: of course, if you change the dependent variable, the results will differ. If the Crimean War satisfied the criteria for a major systemic war and was thus coded, two cases would escalate to war without satisfying the necessary conditions outlined here.

30. Since the dependent variable is dichotomous and the error terms, as a result, are heteroscedastic, any regression analysis would require a discrete choice model. Therefore, we cannot report an ordinary least squares model. Similarly, the variance in the dependent variable is so small, due to the limited number of major systemic wars, that a logit model is not useful either. The measure of association presented with table 11.6,

however, successfully demonstrates the interactive component of the model, as well as accurately portraying the strength of the relationship. No major systemic wars occurred unless the necessary conditions were satisfied. In addition, only one case satisfied the necessary conditions for war but did not in fact escalate to war.

Part 3
The Phoenix Factor Revisited

CHAPTER 12

The Consequences of the American Civil War

Marina Arbetman

Originally, the Phoenix Factor presented the empirical observation that after major wars in developed countries, both winners and losers recovered within a relatively short period of time (Organski and Kugler 1977). It was as if the war had never occurred. This observation suggests that wars do not have as powerful an effect as popularly believed, and more important, that a nation's success in a war does not directly correlate with its economic recovery afterward. This finding implies that economic growth might be constrained by a ceiling imposed by the economic and political infrastructure of the specific nation. Conflicts motivated by the desire to improve economic position or destroy the enemy forever are then doomed to fail. The opportunity cost of conflict should not be overlooked, but the present value of wars is trivial compared to what leaders expect. The Phoenix Factor suggests that there is a natural path of growth and that it takes one generation to get back to this path if it is altered by a war (see Organski and Kugler 1977, 1980:chap. 3; Rasler and Thompson 1992b).

Contrary to this position, Olson (1982) argues that the fast recovery from wars is a function of the destruction of distributional coalitions during the war, because this destruction reestablishes competition among groups and enhances the productive use of resources. He contends that the rate of economic growth in the United States after 1865 was the result of the number of years the individual states enjoyed statehood and the trauma of defeat in the Civil War. Recent evaluations of the Phoenix Factor show, however, that recovery is more a function of economic than political destruction (Kugler and Arbetman 1989b).

The bulk of research that has been conducted to explain the Phoenix Factor focuses entirely on wars between nations. Yet, there is no theoretical reason to argue that the behavior of disputants within regions of a country should show different patterns of recovery than those of independent states. Therefore, the object of this inquiry is to extend previous research by including major conflicts within nations, namely civil wars. If it can be established that contenders in conflicts do recover within twenty years regardless of their status after the conflict, we can address the causes of this recovery in a more generalized manner. In addition, the argument that economic stagnation is due to having lost past wars will become unacceptable. As a consequence, causes of sluggish economic performance will have to be searched for elsewhere. Furthermore, waging wars as a strategy for economic development can be ruled out.

Concepts, Data, and Methodology

The intent of this inquiry was to include all major civil wars to determine whether internal disputes disrupt economic growth and are followed by recoveries similar to international wars—that is, whether the principles of the Phoenix Factor would apply. Yet, attempts at empirical research in this direction face the problem that economic pre- and post-civil war data are not readily available, and discrimination between winners and losers often is impossible. Such difficulties have thus far prevented the study of the factions involved in the Russian Revolution of 1917, the Spanish Civil War of 1936, the Biafra Civil War of 1967, and the Chinese Civil War of 1949.

As a result, this analysis is restricted to the American Civil War, for which simple economic data by state, twenty years prior to the war, are available (Easterlin 1960; Berry 1968). Other, more sophisticated studies do not go as far back as required by this study (for example, Balke and Gordon 1989) or focus on partial measures of economic performance such as commodity output, manufacturing production, agricultural output, consumption, and inflation, but not overall output (cf. Easterlin 1961; Engerman 1971; Goldin and Lewis 1975).

The Economic Costs of War

To calculate the costs of war, I adopted the trend-fitting methodology used by Organski and Kugler (1980) and also employed in several other studies (Houweling and Siccama 1988a; Kugler and Arbetman 1989b). This technique consists of selecting a base period previous to the war and determining the established trend by fitting a regression line. This trend is extrapolated and then compared with the actual performance. The base period of twenty years prior to the war was chosen to follow previously used methodology that has proven successful. In the same way, the starting point of the Civil War is the first year of full confrontation and the starting point of the recovery period is the first full year after the confrontation ended. The economic performance of the states is measured by the logarithmic transformation of gross national product (GNP) because it reduces the problem of underestimating and allows for more stable trend-fitting regressions than linear adjustments (Organski and Kugler 1980: 108–19).

This method ascertains the costs of war on the basis of percent change in constant GNP utilizing the following equation:

$$\text{TIME} = b_0 + b_1 * (\text{Log GNP}).$$

TIME = 0, 1, 2 . . . 20, where the last number in the series coincides with the year the war started.

The predicted values of this regression estimate the growth each country (or state) would have if no war had occurred. This estimate shows the proportion of GNP destroyed in terms of foregone years. Standardizing the measure of losses using time avoids the problem of comparing absolute losses among nations (or states) that grow at different rates. Using time as the common commodity allows for valid cross-time and cross-country comparisons, permitting us to disregard levels of development (Kugler and Arbetman 1989b).

The Phoenix Factor postulates that within twenty years of the end of the war, countries should once again be at their expected level of growth, having recovered from the costs of war. At that point the states should converge to the path of "normal" growth. These estimates of "normal" growth, the prewar trend, depend on the reliability of the regression estimates. The proportion of total variance explained, r^2, is used to evaluate the reliability of the estimates and the stability of the projections of costs given by the residuals.

The results for the American Civil War are shown in table 12.1. In the case of the American Civil War, the prewar estimates for most of the free states are very stable, with $r^2 > 0.80$ for eleven states. Trends for five states fall below this desired level of $r^2 = 0.80$, but are still stable ($r^2 \geq 0.69$). Results for Maine and Vermont are unstable. This instability is due to the lack of antebellum economic growth. As a result the slopes cannot effectively be distinguished from zero. In spite of the lack of significance of the variables in some cases, visual inspection suggests the trends are accurate. Similar unstable results are found for those slave states with minimal GNP rates of growth. Arkansas and Texas show prewar economic growth and stable estimates ($r^2 > 0.80$). Florida also experienced economic growth but has stable estimates ($r^2 = 0.69$). Finally, the border states, with the exception of Missouri ($r^2 = 0.93$), show growth but produce unstable results. Virginia poses unique problems. Even though the prewar estimates are good ($r^2 = 0.81$), the split between Virginia and West Virginia after 1863 makes the postwar analysis difficult.

The American Civil War

There has been a great deal of debate about the economic impact of the American Civil War. One point not debated is that the costs were very high for the entire country. Economically, all the states were devastated. But perhaps the most traumatic statistic is the loss of manpower. More American lives were lost in this war than in all other American wars combined: 618,000 (258,000 for the

TABLE 12.1. Estimates of the Rate of Growth in U.S. Pre–Civil War, 1840–60

State	r^2	b_0 (Significance)	b_1 (Significance)
U.S. (total)	.96	-23.62 (.000)	.96 (.000)
Free			
California	.81	-143.58 (.000)	32.58 (.000)
Connecticut	.69	-155.92 (.000)	38.51 (.000)
Indiana	.90	-103.14 (.000)	25.23 (.000)
Illinois	.96	-64.24 (.000)	16.63 (.000)
Iowa	.99	-14.59 (.000)	9.89 (.000)
Kansas	.81	-116.43 (.000)	32.59 (.000)
Maine	.17	-122.62 (.083)	31.44 (.063)
Massachusetts	.69	-196.89 (.000)	38.72 (.000)
Michigan	.96	-45.88 (.000)	15.44 (.000)
Minnesota	.81	-117.74 (.000)	32.58 (.000)
New Hampshire	.37	-142.06 (.006)	40.25 (.004)
New Jersey	.83	-131.17 (.000)	31.40 (.000)
New York	.69	-231.89 (.000)	38.76 (.000)
Ohio	.81	-164.22 (.000)	32.65 (.000)
Oregon	.81	-81.55 (.000)	32.59 (.000)
Pennsylvania	.77	-197.67 (.000)	35.31 (.000)
Rhode Island	.69	-126.70 (.000)	38.62 (.000)
Vermont	.09	-81.18 (.232)	24.13 (.181)
Wisconsin	.98	-15.82 (.000)	10.73 (.000)
Slave			
Alabama	.37	-164.71 (.005)	40.41 (.003)
Arkansas	.91	-65.13 (.000)	24.09 (.000)
Florida	.69	-81.03 (.000)	38.53 (.000)
Georgia	.34	-169.82 (.008)	39.56 (.006)
Louisiana	.003	-11.31 (.896)	4.75 (.806)
Mississippi	.39	-166.37 (.004)	40.76 (.002)
North Carolina	.03	-49.27 (.560)	13.31 (.484)
South Carolina	.008	41.67 (.615)	-7.40 (.701)
Tennessee	.21	-143.97 (.049)	34.05 (.036)
Texas	.81	-125.34 (.000)	32.58 (.000)
Virginia	.33	203.24 (.005)	-39.18 (.007)
Border			
Delaware	.68	-90.24 (.000)	38.82 (.000)
Kentucky	.42	-179.02 (.002)	41.21 (.002)
Maryland	.54	-170.04 (.000)	41.60 (.000)
Missouri	.93	-81.03 (.000)	21.56 (.000)
West Virginia[a]	.81	-90.54 (.000)	32.58 (.000)

Note: For all the states $N = 21$.

[a]Created in 1863 (Prewar trends state of Virginia)

Confederate States, and 360,000 for the Union; cf. Randall and Donald 1969). One of every six males of military age died in this conflict.

It is not clear what the effects of the war on the economic performances of the North and the South have been. Some studies argue that the Civil War was a strong modernizing force for the whole country and that it fostered industrialization (Thomas 1971; Luraghi 1972). Other studies conclude that the war had a devastating impact on the South (Sellers 1927; Engerman 1971; Goldin and Lewis 1975), while still other scholars debate its impact on the Northern states' economic growth (Gilchrist and Lewis 1965; Scheiber 1965; Engerman 1966; Andreano 1967).

Goldin and Lewis (1978), following Beard and Beard (1930) and Hacker (1940), indicate that without the war the rates of growth, at least for the North, would have been markedly reduced. It was the war that fostered industrialization, and these benefits offset the costs of the war itself. Some significant changes did take place after the war: commercial banks developed new types of assets, and the national banking system was created. Also, new modes of transportation, such as the railroad, contributed to the economy, but not as much as expected. All of these changes were supposed to have positive effects on the entire American economy. However, some authors state that the transfer of political power from South to North had a much more positive impact on the North than on the South, since after the war the North was able to impose its political agenda. For example, the North had the upper hand in passing legislation such as increases in tariffs, molding the institutions of the banking system, and issuing greenbacks (Beard and Beard 1930; Hacker 1940). This argument supports claims that conflict winners are able to take economic advantage of their victories. If this argument is correct, one should expect that the North would recover at a faster pace than the former Confederate states.

Although this chapter is not concerned with the causes of the war, some discussion about the economic situation of the South will shed some light on the economic consequences of the war. Some authors suggest that the economic failure of the system of slavery was one of the causes of the Civil War. The lack of southern growth is then attributed to the inefficiencies of the slave system in the allocation of labor, rendering low levels of productivity and economic stagnation in the antebellum period (Wright 1897; Beard and Beard 1930; Hacker 1940). A war could not but improve this state of affairs. The direct effect of abolishing the inefficient system of slavery would be a quick recovery after reallocating resources and changing the institutional arrangement. From this perspective the Confederacy should have recovered very quickly. But things did not work that way. The economic profitability of the former slaves did not improve after the war. Slaves were free to work in factories after the war, but they had to be paid. Furthermore, this change of jobs did not improve their productivity, since the labor was immobile (Phillips 1966, in Dacey 1976:113–16). The devastated southern economy was further strained because, as former slaves became factory workers, there was not enough labor available for agriculture, dramatically depressing agricultural output. This shortage of agricultural labor, coupled with the large dependency on cotton, are pointed to as the main causes for the poor postwar economic performance of the South (Ransom and Sutch 1977).

In contrast, Fogel and Engerman (1974) conclude that slavery was not moribund before the war and the economy in the South was growing rapidly. The rate of return of male slaves was similar to the rate of return of other investments, around 6 percent. Therefore, pushing the Confederacy into a war was economically unsound. But the consequences of abolishing slavery might have made the Confederacy's recovery very difficult. Ransom and Sutch (1977) look deeper into political variables to explain why the South never recovered. They find the lack of appropriate institutional arrangements was a major force behind the nonrecovery of the Confederacy in the postwar period. Olson (1982) also analyzes institutional arrangements as a key aspect of economic performance, but he reaches a totally different conclusion. He states that "[in] the Confederate states, the development of many types of special-interest groups has been severely limited by the defeat in the Civil War, reconstruction, and racial turmoil and discrimination." He continues, claiming that in line with his theory "these states should accordingly be growing more rapidly than other states," since distributional coalitions had been destroyed and competition among interest groups reestablished (Olson 1982:97). The South not only hosted the majority of the battles (thirty-nine), but also lost the war. The heavier combat that substantially destroyed much of its capital also affected the political arrangements. The Confederacy lived for about four years, centered around an entirely new government that was formed by coalitions that organized well before those states actually seceded from the Union. But when they were finally overthrown, "the political stability of these Deep South states was profoundly interrupted by the Civil War and its aftermath" (Olson 1982:101). If Olson is correct, the South should exhibit greater economic growth after the Civil War because it suffered the most damaging costs (Goldin and Lewis 1978).

A different alternative is presented by the empirical regularity of the Phoenix Factor. Kugler (1973) and Organski and Kugler (1977, 1980) find that for developed countries, it is not winners or losers that determine the recovery rate, but rather their prewar economic performance. Engerman (1966:372) reports that for the North, the per capita output grew at an annual rate of 2.1 percent after the war (1870–1900), in comparison to the 1.45 antebellum rate. He states (1966:375) "the higher postwar rate may merely reflect a 'catching-up' process induced by the decline in per capita commodity output during the Civil War decade." Cochran (1961) and Engerman (1971) claim that the process of

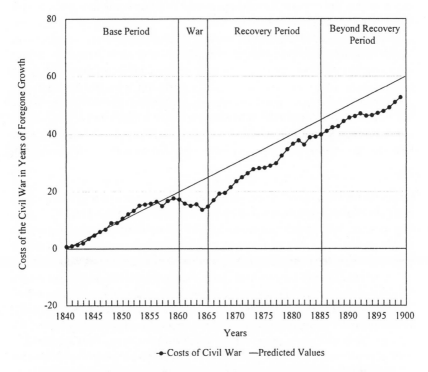

Fig. 12.1. The Phoenix Factor for all the states. Straight line is predicted values.

industrialization had begun before the war, therefore the postwar manufacturing boom was only a continuation of the trend. Furthermore, and coinciding with the results of the Phoenix Factor, Engerman (1966:373) finds that not "until almost twenty years after the war did per capita commodity output reach the level estimated by extrapolating the prewar trend, and it is at this time that the rate of growth returns to this earlier level." Following this logic, states that were not growing before the war should not be expected to grow afterward.

This chapter investigates the above set of conflicting arguments about the consequences of a devastating civil war.

The American Civil War and the Phoenix Factor

The overall results in figure 12.1 show very clearly that the United States as a whole recovered from the Civil War in a pattern very similar to that encountered in international conflicts of major magnitude. The recovery process was fast, but once the original losses were almost overcome, the nation reverted to pre–Civil War patterns.

For the United States as a whole, the Phoenix Factor clearly operates. The almost complete recovery from the costs of war takes place in less than twenty years, and lasting costs remain below three years in foregone growth. The postwar convergence reported in figure 12.1 is consistent with the Phoenix Factor phenomenon uncovered following major wars in developed countries (Organski and Kugler 1980). Generalizations are always dangerous with such restricted data, but these overall results suggest that the scars of major civil and international wars are eradicated within a generation and follow similar patterns.

Figures 12.2, 12.3, and 12.4 summarize the results of the Civil War but differentiate between the winners, losers, and neutral states in this conflict. The general patterns again coincide with the expectations of the Phoenix Factor. Much of the convergence results because many states did not experience economic growth in the antebellum period. Most of these nongrowth states were concentrated in the Confederacy, but some were in the North as well.

The general results show that states that stagnated before the war recovered and then reverted to the same stagnant trend as in the prewar period. This result is present whether one considers slave, free, or border states (see table 12.2 for classification of the states for the period of analysis).

The free states as a group similarly reveal the operation of the Phoenix Factor, recovering from heavy war losses within less than fifteen years (fig. 12.2). If analyzed separately, 85 percent of the free states behaved as expected by the Phoenix Factor. States such as Massachusetts, which were growing before the war, recovered from the losses of the war within twenty years. Maine, Vermont, and New Hampshire were exceptions, as they were not growing before the war, but after the war they surpassed expectations and within a decade were at their prewar levels. This result is what Olson expected from the South-

TABLE 12.2. Classification of American States
during the American Civil War

Free	California, Connecticut, Indiana, Illinois, Iowa, Kansas, Maine, Massachusetts, Michigan, Minnesota, New Hampshire, New Jersey, New York, Ohio, Oregon, Pennsylvania, Rhode Island, Vermont, Wisconsin
Slave	Alabama, Arkansas, Florida, Georgia, Louisiana, Mississippi, North Carolina, South Carolina, Tennessee, Texas, Virginia
Border	Delaware, Kentucky, Maryland, Missouri, West Virginia

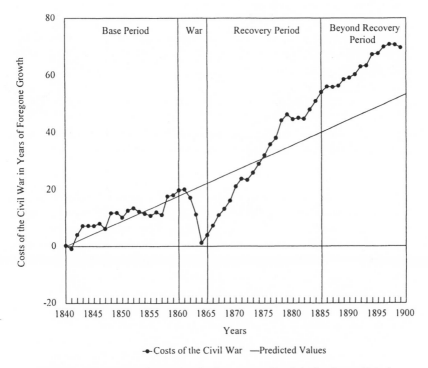

Fig. 12.2. The Phoenix Factor for the free states. Straight line is predicted values.

ern states! (See table 12.1 for the estimates of the rate of growth in United States pre–Civil War, 1840–60.)

As shown in figure 12.3, most of the slave states (70 percent) were stagnant prior to the war. In essence, the Civil War destroyed flat economies. In such scenarios it is expected that the states would have sluggish growth in the postwar period as well. The clearest examples are Georgia, North and South Carolina, and Louisiana, where postwar trends are as flat as predicted. Tennessee, Alabama, and Mississippi show a positive postwar trend, more in line with Olson's expectations than those of the Phoenix Factor. Virginia is something of an anomaly. Not only did Virginia not recover after the war, but its economy went into severe decline. This is understandable, given that one-third of its land was separated to create West Virginia in 1864. Other exceptions include Arkansas, Texas, and to a lesser degree Florida, showing profiles similar to those of the northern states: prewar growth and recovery within twenty years.

In summary, most states follow the expected Phoenix Factor prediction of

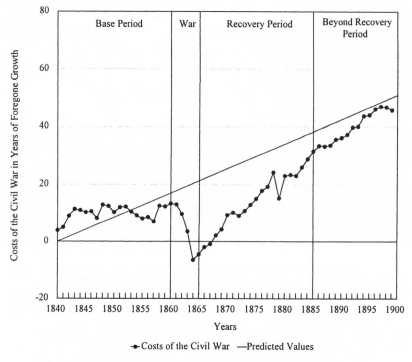

Fig. 12.3. **The Phoenix Factor for the slave states. Straight line is pre-dicted values.**

recovery within twenty years. The states that were not growing in the years prior to the war returned to their prewar sluggish patterns of economic growth. Three states seemed to have been revitalized by the war. From a flat economy, they went on to very positive growth. The states that were growing before the war continued to do so and recovered within the stipulated time. These results are counterintuitive. The South was devastated and defeated. Many of the political structures of the region were taken over by Northerners. One would antici-pate that patterns of recovery would be attenuated in this region. They were not. The South suffered far more because its economy was relatively backward than because it was devastated after the Civil War. Indeed, the potential long-term effects of this conflict seem to disappear across the United States. Such puz-zling results confirm the existence of the Phoenix Factor at the domestic level. The South, like Germany or Japan after equally devastating defeats, managed to recover from war. The difference in perceptions between the cases may be largely due to the fact that prior to World War II, both Germany and Japan were

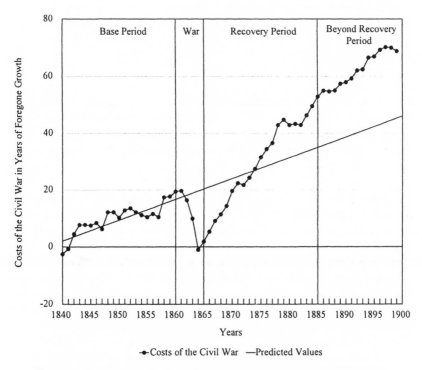

Fig. 12.4. The Phoenix Factor for the border states. Straight line is predicted values.

booming economies, and they not only recovered but maintained their impressive performances after the conflict. On the other hand, the South simply recovered and then reverted to a very slow growth pattern congruent with its preindustrial status. The long-term differences between the North and South that have only been overcome recently are seemingly due to the economic structures in these regions, and should not be attributed directly to the effects of the devastating Civil War.

Let us finally turn to the control group. While border states as a group recover in less than fifteen years, in accordance with the expectations of the Phoenix Factor, they display a mixed picture, individually (fig. 12.4). Missouri fully recovered to its prewar level. Kentucky, Maryland, and Delaware recovered, but it took them almost thirty years to get back to their prewar trends.

As stated earlier, controversy exists regarding the accuracy of some figures of economic performance. Goldin and Lewis (1975, 1978) argue that the costs of the Civil War have been underestimated, especially for the South. However, if the figures used in this chapter had been adjusted, they would only have rendered higher costs during the war, and the positive results would have been more pronounced.

Conclusion

The results of this analysis of the American Civil War strongly support the Phoenix Factor observation. The southern states whose economies were stagnating in the prewar period went back to their poor performance but the ones that were growing in the antebellum continued to do so. This finding seems to put aside the controversy over the efficiency of the slave system in the South and its demise. The states that were using the political and economic resources efficiently in the prewar period and were flexible to adapt to the postwar circumstances returned to prosperity. Confederate states with slow performance converged back to this pace. This characteristic of reverting to a sluggish rate of growth is found in two free states, Maine and New Hampshire, as well as some border states. Thus, the postwar economic recovery, or Phoenix Factor, is directly correlated with prewar economic performance, rather than with whether states emerged as winners or losers.

Goldin and Lewis (1975, 1978) state that there is "no evidence that the war benefited the United States or the North in a 'gross way.'" This statement is consistent with the present results and with the original Phoenix Factor argument. The southern states recovered, but lost in relative terms, not because the North prevented them from recovering, but because their prewar economic growth trends had been flat.

The implications of this study are fundamental. Nations involved in devastating civil wars recover in patterns similar to those that follow costly international conflicts. The scars of wars are eradicated in a generation. Long-term developmental trends are neither hindered nor accelerated by wars. Thus, war is neither a tool for development nor a method to prevent development. Why? The answer is not immediately apparent. We know that the patterns of recovery do not conform to standard expectations of industrial recovery through reconstruction of industrial capacity, because the South was an agricultural society and recovered following similar patterns to those in the North. We further question standard arguments about political intervention. For example, this work does not support Olson's insights that the destruction of political structures leads to recovery (1982). The South was politically devastated and occupied while the North was not, yet their patterns of recovery do not differ. It is evident that societies are resilient, yet we still do not know why. An answer to this critical question could shed light on the process of national development and should be vigorously pursued.

APPENDIX: THE PHOENIX FACTOR AND WARS BETWEEN DEVELOPING COUNTRIES

Originally, I intended to include the analysis of wars between developing countries with respect to the Phoenix Factor in this chapter. Yet, the analysis of the economic recovery from international wars in developing countries is quite limited in scope for three main reasons. The first one is the lack of historical data; one can only consider relatively contemporary wars. Even when focusing on these modern cases, data for the Vietnam War are unavailable. The second reason studies of developing-country wars are limited is that most wars waged among developing countries, in spite of being prefaced by loud political problems, are usually quite limited in terms of economic costs. For example, the Football War in Central America between El Salvador and Honduras in 1975, the Malvinas/Falklands War between Great Britain and Argentina in 1982, and the recent American-Iraq Gulf War do not qualify as major conflicts, since the severity of the wars did not measure up to standard definitional criteria.

The third and last constraint to this analysis lies in the fact that the effects of wars are often compounded either by subsequent additional major wars with developed countries or by other domestic conflicts. For example, it is impossible to analyze the recovery of Japan from the Sino-Japanese conflict (1937–41) because of the masking effects of World War II. Similarly, it is hard to see the effects of the Six Days' War of 1967 because within six years Israel was involved in the Yom-Kippur War of 1973.

In spite of these limitations, a few conflicts among developing nations will be analyzed. The data furnished by Angus Maddison in various publications (1973, 1982) provide a record of GNP fluctuations (and Net Material Product in the case of Russia) large enough to analyze the Manchurian War, the Sino-Japanese War, the Russo-Finnish War, the Sino-Indian War, the Second Kashmir War, and the Bangladesh Conflict. This sample is not comprehensive by any means, but I hope it can be extended in future research.

Obtaining and evaluating data for the Korean War proved to be more difficult, especially for North Korea. The enhanced difficulty originated from the question of how much of the prewar rates of growth were due to the then-industrial North, and how much to the agricultural South.

The prewar estimates for the international wars show very stable trends in most cases. Contemporary wars such as the Bangladesh Conflict, the Second Kashmir War, and the Sino-Indian War have $r^2 > 0.88$, except for China in the Sino-Indian War, whose r^2 falls below the desired value of 0.80, (here, $r^2 = 0.67$). The prewar trends for the Russo-Finnish War are also very stable ($r^2 > 0.94$). The results for the remaining two wars for China are $r^2 = 0.67$ for the Sino-Japanese War and $r^2 = 0.99$ for the Manchurian War. (See table 12.3 for coefficients and levels of significance.)

TABLE 12.3. Estimates of Prewar Rates of Growth for International Wars

War/Country	Year	N	r^2	b_0 (Sig.)	b_1 (Sig.)
Manchurian	1931–34				
Japan		—	—	— —	— —
China		3	.99	−371.08 (.029)	81.24 (.028)
Sino-Japanese	1937–41				
Japan		—	—	— —	— —
China		3	.67	−489.04 (.397)	104.87 (.389)
Russo-Finnish	1939–40				
Russia		17	.97	−45.99 (.000)	14.52 (.000)
Finland		17	.94	−89.66 (.000)	24.73 (.000)
Sino-Indian	1962				
China		16	.67	−88.64 (.000)	19.91 (.000)
India		16	.88	−146.96 (.000)	32.75 (.000)
Second-Kashmir	1965				
India		14	.95	−126.29 (.000)	28.16 (.000)
Pakistan		14	.95	−131.19 (.000)	29.74 (.000)
Bangladesh	1971				
India		18	.98	−123.19 (.000)	26.12 (.000)
Pakistan		18	.98	−85.67 (.000)	18.84 (.000)

Analysis

The case selection criteria was to choose international wars with more than 10,000 battle deaths and, when possible, to maintain continuity with the sources previously used in this kind of research for the sake of comparisons and future replications. The Manchurian War (1931–34), and the Sino-Japanese Conflict (1937–41) both qualify as major conflicts. Casualties were 60,000 in the first and as high as 1,000,000 in the second. I chose not to analyze Japan, because the effects of World War II were so devastating that they masked any recovery from the Manchurian War and the Sino-Japanese Conflict.

The Manchurian War was fostered by the slump in world trade in 1930–31. Japan's exports fell to almost half of their previous value and public pressure was increasing. Manchuria would be a way to resist the economic depression and also a show of force to deter China from expanding their infrastructure. Manchuko had the status of an independent country but behaved as a Japanese colony. The conflict between China and Japan lingered and in 1937 the Japanese decided that China should be brought together with Manchuko in a structure where Japan would be the nucleus. Contrary to Japanese expectations the war did not end quickly, and fighting continued until World War II ended.

Contrary to what one might expect, the Manchurian War did not halt China's economic growth at all, while the Sino-Japanese War imposed eight years of foregone growth. One problem to bear in mind when analyzing China is the discrepancy among the sources. Angus Maddison grants China a much higher rate of growth than do the World Bank or Summer and Heston (1978, for a full discussion of the discrepancies see Maddison 1983). In spite of the apparent relative overestimates reported by Maddison, I chose his source to maintain continuity with the sources previously used in this kind of research.

Unfortunately, it is difficult to see how long the recovery would have taken because eight years after the Sino-Japanese Conflict ended, the 1949 civil war imposed costs much higher than the earlier war. The civil war's losses were thirty-four years of growth! According to the Maddison data, it took China only four years to recover from these staggering losses. In the 1950s and 1960s the Great Leap Forward and the Cultural Revolution, coupled with the Sino-Indian War (1962), cost China fourteen years of foregone growth. But China again recovered from the ashes quite quickly. This long-standing border dispute over the Macmohan line was of little consequence. The most interesting point revealed by consideration of China is found by looking at the different effects of domestic upheaval as opposed to those of international conflict. The domestic problems seem to have as strong an effect, or even stronger, than international wars. However, recovery is twice as rapid, suggesting that domestic upheaval may leave fewer scars on the economy. In this case the Phoenix Factor emerges with a vengeance.

The Russo-Finnish War presents a problem similar to the previous case. As in the case of Japan, it would be deceiving to look at the recovery of the Soviet Union from this war, since the effects of World War II destroyed the traces of any possible recovery. The geopolitical position of the Karelian Isthmus was responsible for this war. In 1939 the Soviet army launched a colossal attack against Finland. For Finland this was a very consequential war: of the 90,000 casualties, 40,000 were Finnish—very large human losses in relative terms. When analyzing the costs of this war, it is apparent that the conflict did indeed affect Finland, but not the Soviet Union. Finland shows four foregone years of growth, but no recovery within the prescribed twenty years. The Phoenix Factor does not seem to be present following the Russo-Finnish War.

The Second Kashmir War (1965) and the Bangladesh Conflict (1971) are good examples of conflicts that entailed lots of political upheaval but very few economic costs. The long history of animosity between Pakistan and India is well reflected in the Kashmir territory. Pakistan opposed further integration of Kashmir with India because it would strengthen India's claims. But the war resulted in a stalemate, with both countries claiming some kind of victory and both accepting the U.N. cease-fire resolution of 1965. This war entailed 6,800 casualties but no economic losses for Pakistan, and only one foregone year for India—which recovered in less than two years. The Bangladesh Conflict was born of economic causes. East Pakistan was providing raw materials and agricultural commodities, but the income from these activities was being diverted into the industrializing West Pakistan. This dispute was exacerbated by religious reasons, since the Bengalis were regarded as latter-day Muslim converts. The active involvement of India fostered the dismemberment of Pakistan and the birth of Bangladesh. This conflict, with 11,000 lives lost, had no economic costs for either country. The economies of these countries continued to grow at the expected rate as if no conflict had ever taken place. The question of whether the Phoenix Factor exists after these wars becomes irrelevant. It is impossible for the Phoenix Factor to be present if there is no economic destruction to recover from. However, these cases neither confirm nor reject the hypothesis of the Phoenix Factor, since the trend of economic growth keeps the same prewar pattern.

Analysis of these wars is quite disappointing. The conflicts analyzed are either too small, or the lack of economic costs do not necessitate recovery, or the recovery is blurred by other circumstances. It may well be that since the size and scope of the wars studied is often quite small, the countries involved were not economically devastated by the conflicts. The question of whether the Phoenix Factor exists after these wars then becomes moot.

In other cases the conflicts were large, as in the costs of the Russo-Finnish War for Finland. Here, the Phoenix Factor should have been most clearly observed, but Finland never fully recovered. Thus, in this instance where the Phoenix Factor is given an opportunity to occur within the same framework of conflicts that includes developing nations, it is not present.

Finally, results are masked by other circumstances. Either a major war or domestic circumstances interfered, like the civil wars in China and Russia. In these cases, domestic problems superseded international wars. China's recovery seems to exceed any optimistic expectations of recovery, at least with the data at hand.

This leads us again to the explanations of the Phoenix Factor. Why did Finland never recover, while China recovered so quickly? It may be that previous results cannot be generalized to developing nations. Or, it may be that the Phoenix Factor can be accelerated by capable governments.

The above discussion suggests that there are very few wars among developing nations that lend themselves to the study of the Phoenix Factor. As explained above, the analysis for international wars is especially constrained by three main problems: lack of historical data, limited costs of the conflicts, and recovery masked by other national or international events. In general, most international wars in developing countries do not inflict costs large enough to necessitate recovery. Therefore, the analysis of the Phoenix Factor becomes moot. The one case where the Phoenix Factor was expected to be present was Finland in the period following the Russo-Finnish War. But Finland never recovered. This case reopens the question of whether other factors should be taken into account in analyzing the economic recovery of developing nations (Kugler and Arbetman 1987). Specifically, does the political capacity of different countries influence their rates of recovery?

NOTES

The assistance of Lorenzo de Guttadauro, who did most of the data collection, and of Douglas Lemke and Doris Fuchs, who edited the manuscript, is very much appreciated. I am indebted to Jacek Kugler, Bruce Bueno de Mesquita, and Ray Dacey, who have commented on this work.

Part 4
Power Parity in the Nuclear Context

CHAPTER 13

Beyond Deterrence: Structural Conditions for a Lasting Peace

Jacek Kugler

> The splitting of the atom has changed everything save man's mode of thinking: thus, we drift towards unparalleled catastrophe.
>
> —Albert Einstein

At the onset of the nuclear era, Albert Einstein advocated the rejection of war to reconcile man's political conduct with the devastation promised by the new physical reality. One response to Einstein's challenge is nuclear deterrence, defined as the ability to prevent attack by a credible threat of unacceptable retaliation. Historically, the legitimacy of deterrence is reinforced by the fact that nuclear weapons have not been used to wage war since the destruction of Hiroshima and Nagasaki in 1945 (Huntington 1982). Since the consequence of a nuclear strike has gone untested for almost half a century, the underlying assumption that nuclear weapons are directly responsible for the long period of peace among nuclear powers remains inconclusive. However, by exploring the logical integrity and empirical consistency of concommittal deductions, the stability of deterrence can be assessed. The query is simple but fundamental. Is deterrence stable? Does an increased capacity to kill enhance or diminish the prospects for peace? And does the proliferation of nuclear weapons elevate or lessen the likelihood of war?

Historical Evolution of Nuclear Deterrence

The goal of deterrence is the prevention of war. Bernard Brodie, the father of classical nuclear deterrence, proposed that nations armed with nuclear devices would refrain from initiating a nuclear attack to achieve any desired objective. Consider Brodie's concise statement following the end of World War II:

> The first and most vital step in any American security program for the age of atomic bombs is to take measures to guarantee to ourselves in case of attack the possibility of *retaliation* in kind. The writer in making that state-

ment is not for the moment concerned with who will *win* the next war in
which atomic bombs are used. Thus far the chief purpose of our military
establishment has been to win wars. From now on its chief purpose must
be to avert them. It can have almost no other useful purpose. (Brodie
1946:76)

In Brodie's view, the deterrer, to prevent war, consciously yields the initiative to its adversary and relies on the resultant fear generated by the potential consequences of retaliation. Prompted by the advent of nuclear technology,
this reliance on retaliation changed forever the normalities of political behavior. Indeed, even in a unilateral nuclear world, the deterrer, to avert war, will
choose to coexist with potential competitors whose policies are neither trusted
nor supported. Moreover, when nuclear weapons are held by more than one actor, nuclear powers still choose to accept the risk associated with a promised retaliatory response in contrast to a nuclear conflict so calamitous that the living
may well envy the dead. Brodie's "classical deterrence" postulate formed the
parameters that John Foster Dulles eventually utilized in outlining the specifi-
cations for Massive Retaliation (MR), the deterrence policy explicitly adopted
in the 1950s by the United States. Under MR, a preponderant nuclear defender
credibly threatens to devastate opponents when provoked. The expectation is
that nuclear terror can compel potential opponents to avoid challenges they
might otherwise initiate.

Brodie's stable deterrence conception reemerged when competing nuclear
powers achieved a second-strike retaliatory capability. Mutual Assured Destruction (MAD) postulated that mutual terror creates the conditions for lasting
peace. That is, stable and long-lasting deterrence results when potential opponents, fearing devastation, are dissuaded from attempting to impose their preferences on others.

After 1945, the concept of classical deterrence dominated the field. A con-
trasting trend argued that nuclear deterrence was a tenuous, potentially unstable
policy. Starting in 1950, Liddell Hart posed a thoughtful and direct challenge
to Brodie's stable notion of nuclear deterrence. Believing that conventional
wars could and would continue under the nuclear umbrella, Hart (1967) concluded that it was a gross fallacy to assume that nuclear weapons *alone* would
make an initial nuclear attack less menacing or more preventable.[1] Although
not totally dismissing the notion that a major nuclear war could be prevented
by the fear of nuclear retaliation, Hart did limit the political domain where nations could dominate others by terror alone.

In 1959, Wohlstetter (1959) challenged the classic treatment of deterrence
among nuclear powers by introducing the concept of the "balance of terror."
Wohlstetter proposed that if the defender did not have a sufficiently large and
well-protected nuclear arsenal, then the single threat of nuclear retaliation could
not deter a surprise attack. Thus, a nation could not avoid a preemptive attack

by threatening retaliation if most of the available weapons could be destroyed during the initial strike. To compensate for this "window of vulnerability" nuclear nations would have to retaliate when the warning of an impending attack was first detected. The expansion of nuclear arsenals that insured the survival of a retaliatory strike solved the immediate problem, but did not resolve the central issue of vulnerability. Wohlstetter concluded that nuclear weapons created terror, but for peace to succeed in the nuclear age, constant vigilance must prevail.

In the late 1960s, Wilkinson developed a similar argument to challenge extended deterrence.[2] In the shadow of the Sino-Soviet split, the discussion centered on the stability of deterrence. What would the Soviet Union do if China, after achieving nuclear parity, chose to attack by conventional means? Wilkinson argued that conventional superiority would determine the outcome, thereby nullifying the role of nuclear weapons. This imaginary scenario materialized in Europe during the 1980s, when NATO drastically reformulated its deterrence policy from retaliation to conventionality. The threat of mutual annihilation obviously failed to convince the United States that the Soviet Union would not press its presumed conventional advantage to challenge NATO in Europe.

From another perspective, Organski (1968) questioned stable deterrence, arguing that deterrence would prevent war under nuclear preponderance, but could lead to eventual conflicts under nuclear parity. While the first nuclear strike might be calculated to gain an advantage, Organski contended the retaliatory strike that could prompt an even higher retaliation offered no advantage to either side. Once a nation is threatened with obliteration, the defender must threaten to act irrationally and retaliate promptly to insure stability.

This sketchy outline of nuclear impediment illustrates that deterrence evolved from a clear preventive strategy to a concept of much greater complexity. In the process, deterrence lost much of its original meaning and evolved into a concept that encompassed a number of policies that offered dramatically different strategies for averting, minimizing, and in some circumstances even waging nuclear war. Hardin et al. (1985) identified three broad positions that delineate the multiple meanings of "deterrence." The main elements of three deterrence categories identified in his exhaustive work (1985:5–6) are paraphrased as follows:

1. The system of deterrence as presently constituted is extremely stable. (i) It is not likely to be significantly affected by changes in the levels of armaments in either direction for the foreseeable future. (ii) Leaders of nuclear nations, faced with the awful consequences of war, have become cautious and this behavioral transformation has reduced significantly the likelihood of any kind of war between the superpowers.
2. The system of deterrence is in delicate balance and the status quo is undesirable and tenuous. (i) New weapon developments threaten this tenuous stability. Thus, nuclear weapons should be curtailed, since elimination is im-

possible. (ii) Nuclear deterrence cannot extend beyond preventing the other side from using nuclear weapons. The behavioral transformation, if any, is limited to massive nuclear conflicts.

3. The system of deterrence is delicate and unstable. Thus, deterrence based on the threat to destroy large portions of each side's population is both morally untenable and unstable. (i) To assure stable deterrence nuclear nations must develop war-fighting, damage-limiting capabilities and strategies. (ii) Nuclear weapons cannot assure the prevention of nuclear war. Thus, nuclear nations must develop counterforce weapons and defense systems that insure success in case of war.

In the first two perspectives, the cornerstone of deterrence is nuclear retaliation, and the central theme is the prevention of overt nuclear conflict by the threat of nuclear war. The first perspective is congruent with Brodie's "classical" view that deterrence is ultrastable because the enormous costs associated with retaliation do not justify the risk of a nuclear attack. The second variant of deterrence resembles the conception of tenuous deterrence (Hart 1967). Doubts are raised that deterrence can prevent conflict at lower levels of intensity. When nuclear disparity exists, opportunities to achieve net gains from marginal challenges under the nuclear umbrella may still lead to conventional wars. It also reflects Wohlstetter's concern with windows of vulnerability created by technological changes. However, it does not address Organski's assertion that when nuclear parity is in place, persistent tensions may lead to challenges, and such challenges may escalate into full-fledged nuclear confrontations.

The third perspective redefines deterrence and shifts from preventing war to limiting escalation. In a dramatic departure from the previous two policies, the key to deterrence is no longer retaliation. Nor is the prevention of nuclear war the goal of deterrence policy. Rather, initiating and waging war are advocated for limiting potential losses. To gain the advantage, this perspective anticipates the need to exercise a nuclear option to avoid massive escalation. This war-fighting strategy reintroduces the first-use principle of nuclear weapons. The tenuous link to "retaliatory deterrence" is the assertion that a limited nuclear attack can "prevent" a nuclear conflict from rapidly escalating to massive proportions. With the "war-fighting" option, deterrence has now come full circle. Contrary to Brodie's direct challenge of the von Clausewitz dictum, nuclear weapons become the policy tools of war instead of the instruments to deter conflict.

Limiting the Scope of This Inquiry

There is a clear dividing line between retaliatory nuclear deterrence and first-strike nuclear deterrence. To many, retaliation is the essence of nuclear deterrence. Although war is still an empirical possibility, advocates who adopt a retaliation perspective mean not to wage, but to avoid, war. A number of distinguished scholars and practitioners including McGeorge Bundy, Morton Halperin, William Kaufmann, George Kennan, Robert McNamara, Madelene O'Donnell, Leon Sigal, Gerard Smith, Richard Ullman, and Paul Warnke, all architects of nuclear deterrence, have most forcefully stated the principle of retaliation:

> The United States should base its military plans, training programs, defense budgets, weapons deployments, and arms negotiations on the assumption that it will not initiate the use of nuclear weapons. ("Back From the Brink," *Atlantic Monthly,* August 1986:35)[3]

It is difficult to reconcile the record with the actions of nuclear powers since 1945 without accepting the impact of counterstrikes. Consider the empirical record. The United States' nuclear preponderance between 1945 and 1965 did not lead to war, even though the elites understood their temporary nuclear superiority. But after Hiroshima and Nagasaki the United States adhered to nuclear retaliatory deterrence and chose to bypass opportunities to strike first with nuclear weapons, (recall the reversals in Korea after China crossed the Yalu, or the grievous resolution of the 1956 Hungarian revolution; George and Smoke 1974; Freedman 1981).

The Soviet record is also consistent with the notion of retaliation. Recall that the USSR held nuclear preponderance over China during the Sino-Soviet dispute and accepted territorial threats and provocation without preemption.[4] Moreover, despite overt efforts by Iraq, Iran, and Pakistan to develop a potentially threatening nuclear capability, Israel has not used its reported nuclear advantage against its neighbors. Since World War II, no nuclear power has used nuclear weapons against a nonnuclear country, even in the face of serious military reversals or military defeats, such as those experienced by the United States in Korea and Vietnam, or by the Soviet Union in Afghanistan (for an exhaustive review of this literature, see Harvey and James 1992). This self-restraint suggests a remarkable commitment to retaliatory deterrence by nuclear powers even when the terror created by the risk of second-strike retaliation is absent.

The strategists that propose a war-fighting option deal with what will happen if a nuclear war is waged, and not with how nuclear conflict can be averted. Indeed, Colin Gray (1979), an articulate advocate of war-fighting "deterrence," proposed that when threatened a deterrer should initiate nuclear war to enhance the possibility of political "decapitation" of the enemy and minimize potential escalation. In a related effort, Huntington (1982) proposed that nuclear weapons, when necessary, be employed to advance policy goals rather than simply to deter massive conflict. However, regardless of specific tactics, once the use of nuclear weapons is advocated to gain a marginal advantage over an opponent or to decide the outcome of conventional conflict, the intent is not to prevent nuclear war, but to impose limits of escalation. Initiation, in sum, implies waging, not preventing, a nuclear war. Nuclear weapons are no longer unique tools to inhibit conflict for fear of the consequences. Instead, such weapons become military instruments for the continuation of policy by other means.

The emphasis of this chapter addresses the restricted concept of retaliatory deterrence, whose objective is to probe the structural stability of deterrence to prevent war. A separate, but hopefully avoidable topic is a deliberation on how to wage a nuclear war.

Models of Retaliatory Deterrence

Stable Deterrence

Brodie argued that the "unacceptable" costs of nuclear war and a commitment to retaliation secure the prospects for stable deterrence. Three policies, Massive Retaliation (MR), Balance of Terror, and Mutual Assured Destruction (MAD), characterize idealized approaches to deterrence.

Intriligator and Brito (1984), with impeccable logic, effectively formalized these ideas into a persuasive and consistent structure that transforms verbal statements into structured equations based on a variant of Richardson's arms race model. Their perspective also provides connections between these various policies.

Their structure directly links the period of Balance of Power, when conventional weapons were the only means of destruction, to the period of nuclear monopoly, when the United States could compel opponents by threats of MR, to the period of Balance of Terror, when limited nuclear arsenals were available, and finally to the period of nuclear parity, when nuclear nations can rely on a second strike to assure MAD.[5]

The anticipated costs of war differentiate stable "nuclear" deterrence from unstable "nuclear" and "conventional" deterrence. Like Brodie, Intriligator and Brito argue that the structural characteristics of the Balance of Power and MAD are identical in all respects but the number of anticipated casualties resulting from a serious conflict. They further argue that conventional power is unstable and falls in the cone of "war" while MAD is constant and falls in the cone of "peace." In other words, as one moves along the casualty plane for each nation the risk of destruction generated by a crisis increases dramatically.

The critical difference between conventional and nuclear deterrence is the risk associated with the magnitude of retaliation. Historically, conventional deterrence failed because the risks of confrontations were acceptable, while nu-

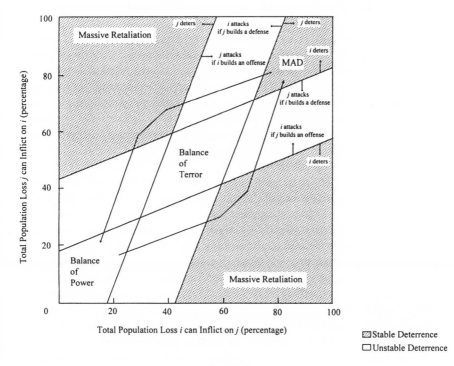

Fig. 13.1.

clear deterrence has succeeded because the risks associated with such a war are unimaginable (for alternate views of conventional deterrence see Quester 1966; Mearsheimer 1983).

The work of Intriligator and Brito demonstrates an effective analytical summary of nuclear deterrence since 1945. They supply the connections between each level of conflict and peace, explicitly indicating how stability can be enhanced or imperiled by the deployment of nuclear weapons (fig. 13.1). Total war can be waged anywhere in the region of conventional power formed below the lower thresholds for i_1 and j_1. This area establishes the limits within which competing actors will initiate war in pursuit of their policy goals, anticipating that a confrontation will produce "acceptable" losses. The distinction between this unstable Balance of Power or Balance of Terror and the stable region of massive retaliation and Mutual Assured Destruction is attributed to the potential destructive differences between conventional and nuclear weapons. As Brodie argued, nuclear war is no longer viable, not because it cannot be waged, but because it cannot be won. Thus, Intriligator and Brito show that evaluations based on conventional capabilities have limited significance in the nuclear era.

Consider the path of nuclear parity. Here identical distributions of power at low and high levels of destruction lead to two completely different outcomes. At the conventional level, parity or preponderance result in limited or total war. With limited nuclear arsenals, a Balance of Terror leads to probing, tenuous stability and the tendency for preemptive nuclear war. Once a second strike is credible, parity leads to MAD and secure peace.

Like Brodie, Intriligator and Brito argue that equality at *low* levels of destruction does not produce peace but instigates conflicts, where the outcomes are uncertain. However, in the cone of mutual deterrence, additional weapon deployments increase second-strike capabilities and, through redundance, further stabilize mutual deterrence. Contrary to conventional fears about an arms race, Intriligator and Brito's argument links nuclear arms buildups with increased stability. As overkill capacity steadily increases, competitors move further and further into the area of ultraequilibrium. Therefore, the unique contribution of nuclear weapons is that massive deployments are not destabilizing, but instead assure opponents, even in case of unexpected attack, that retaliation is still a strong deterrent.[6]

The Intriligator and Brito model is compelling because it goes beyond a simple description of specific conditions and provides paths that lead from war to peace. Consider the consequences of nuclear deployment from the perspective of actor i at the onset of the nuclear era. Starting in conventional power, i's ability to destroy j increases unilaterally along the horizontal axis. Nation i could initiate a conflict that j would resist in the region between thresholds one and two when the expected costs for j fall below those required for deterrence and the outcome is unknown. However, as i's ability to destroy increases, as i's nuclear arsenal expands unilaterally, the conditions for Massive Retaliation are achieved.

Under nuclear preponderance, when i makes demands, j, faced with absolute losses that exceed acceptable levels of suffering, yields. Paradoxically, under MR, i can deter war unilaterally, but cannot be prevented, other than by its own decision, from initiating a devastating conflict against any other nonnuclear opponents.[7] Under MR, only a commitment to retaliation prevents i from initiating war. This is the critical difference between deterrence and balance of power. The latter assumes that actors interact in total anarchy and choose to avert war only when the consequences are feared (Waltz 1979). Power preponderance, therefore, leads to war. On the other hand, a preponderant nuclear power does not initiate conflict despite an overwhelming military advantage unless forced to do so by prior aggression.

Note that Balance of Power has consistent rules for initiation and prevention of war. Deterrence has equally clear rules for the prevention of war but, in contrast, has inconsistent ones for the initiation of conflict. This is the murky area of "credibility" in deterrence. The escape clause is "acceptable threat" before a nuclear retaliation is launched. The question asked is, has this threshold been approached in the past? Would the United States have retaliated with nuclear weapons if Chinese forces had been successful in pushing American troops to the sea rather than the 38th parallel in Korea? Would the Russians have retaliated if China had pushed its claims against the Soviet Union beyond two small islands during the period of the Sino-Soviet Rift? Would Britain have resorted to nuclear retaliation if Argentina had succeeded in sinking key elements of the British fleet in the dispute over the Falkland-Malvinas islands? And if Arab states manage in the future to conventionally threaten one of Israel's cities, will such action merit nuclear retaliation?

As long as the "threshold" for nuclear retaliation is unspecified, the stability of deterrence, particularly unilateral nuclear deterrence, is indeterminate. Still, one must wonder, would these countries have acted otherwise if the ultimate outcome threatened their very existence and survival?

Returning to parity, Intriligator and Brito show one central and significant implication. Unlike the Balance of Power or Terror, MAD is ultrastable. Intriligator and Brito posit that an equilibrium of nuclear weapons leads to stability under conditions where the Balance of Power could generate war because the cost of war is higher, and initiators will avoid such risks. Departures from MAD, once achieved, are discouraged, since such nuclear disparities increase the likelihood of war. Note for example that if actor i chooses to move from mutual to unilateral deterrence by deploying weapons that allow for a preemptive first strike or new defenses, the possibility of nuclear conflict elevates. The reason is that limiting retaliation by j implies losing the ability to maintain deterrence. Thus, once the cone of mutual destruction is achieved, all participants would find it difficult to disturb stable mutual deterrence.

Among mature nuclear powers, changes away from MR to MAD or vice versa are potentially unstable. Therefore, the development of nuclear weapons to gain equilibrium with the existing nuclear powers, as the USSR did after 1945, China did after 1960, or the United States did with its Strategic Defense Initiatives (SDI), disturbs massive retaliation. There is an experience level with transitions from the area of MR, where the monopoly of nuclear arms ensures stability, to MAD where such assurances are mutual. Intriligator and Brito point out that the transition from MR to MAD is secure, provided the stronger party adheres to a policy of retaliation only. However, if the stronger party chooses to initiate conflict when it acquires nuclear weapons, a preemptive nuclear conflict could be waged. The weaker party still cannot deter the stronger.

The reverse is considerably more dangerous. Once two actors are locked in the cone of Mutual Assured Destruction, the party that believes new developments will destroy its ability to deter may be prompted to preempt. In this case, the results could lead to war because both sides have the capacity to overwhelm the opponent, and neither possesses the incentive to capitulate. For this reason, reversing mutual deterrence would be far more risky than challenging unilateral deterrence. Perhaps this is the reason Intriligator and Brito support the deployment of a "thin" defensive system to prevent small nuclear powers from threatening other nations, but oppose the development of a "thick" defense system that could threaten nuclear deterrence. Intriligator and Brito, unlike Brodie, emphasize the distinct capability differences of nuclear powers like the United States or Russia from the capabilities that could be deployed by Pakistan, Israel, South Africa, and a host of others, whose delivery vehicles are limited. The reason is that unilateral nuclear deterrence would continue to apply to most members of the international community, particularly if "thin" defensive systems were deployed.

Their model implies that large deployment of nuclear weapons under mutual deterrence enhances stability as the overkill capacity that each country acquires rises. Paradoxically, the "balance of terror" becomes increasingly stable as the losses that each contender expects from a confrontation continue to rise. Arms races that result in additions to nuclear arsenals improve MAD because they provide a cushion for each participant. Recalling Wohlstetter's concern, large arsenals minimize the potential disturbances produced by a technological breakthrough or the uneven deployment of new generations of nuclear weapons. As deterrence becomes MAD, it is ultrastable as war is no longer winnable.

However, the model has disturbing implications following effective deployment of defensive systems by two nuclear powers. If the defense structures of j match those of i, the deterrence capability of both actors declines. As both drift, first into conditions for unstable mutual deterrence, and eventually back into the cone of war, the specter of large war reemerges. Therefore, with the elimination of nuclear weapons, nations recreate the structural conditions for waging serious conventional conflict. This replaces the stable "assured destruction" with an unstable conventional situation. Finally, the proliferation of nuclear weapons improves both regional and global stability as MAD expands. Fewer nations can risk confronting one another when faced with the risk of nuclear war. Nevertheless, this well established but controversial implication that nuclear proliferation leads to stability is seldom accepted by practitioners (some do admit the connection; see Rosen 1977; Intriligator and Brito 1981; Waltz 1981; Bueno de Mesquita and Riker 1982; Berkowitz 1985).

Tenuous Deterrence

Liddell Hart's (1967) forecasts for the nuclear era have now been verified. Conventional conflicts continue to be waged by nations that wish to advance their marginal position but do not attempt to reverse the overall structure of the international system. A critical test of the validity of nuclear deterrence may be framed by the distinction between absolute versus marginal gains.

Consider the radical notion that nuclear weapons have not changed the calculus of war. For this argument to be true one must account for the absence of major war since 1945. One perspective, consistent with this view of international politics, is based on the notion of power parity and transition. This perspective was introduced in 1958 by Organski and empirically explored and expanded by a number of scholars. Power preponderance contends that nations do not initiate conflict when the outcome is clear beforehand. If they do, only limited confrontations are waged. However, when the opposing sides are equally matched and one threatens to overtake the other, the potential for massive confrontations exists. Indeed, under conditions of power parity the opposing sides have equivalent opportunities to succeed. When the participants are leading powers, the victor shapes the future of the international system. More important, the costs of such conflict are high, since both sides are evenly matched for mortal combat, and this situation could result in conflicts equal in devastation to both world wars.

Empirical analysis of conventional confrontations supports the plausibility of these structural preconditions. Since most major conflicts in the international system took place under parity, what remains is to determine if transition, territorial contiguity, alliance structures, arms competition, or other factors not yet disclosed are the real triggers for conflicts. However, the evidence at the nuclear level is absent.

In the same way that the proponents of MAD extended the logic of balance of power, it appears worthwhile to examine the implications of extending the parity logic to the nuclear period. Figure 13.2 summarizes dynamics and levels of conflict anticipated by the power parity perspective in the nuclear era.

Similar to Intriligator and Brito's perspective, the parity model anticipates that large disparity in power is linked to peace. Thus, under massive retaliation, characteristic of the early part of the nuclear era, the interaction between non-nuclear and nuclear nations is expected to be stable. As the capabilities of a non-nuclear power increase, so does its ability to resist demands by the preponderant nuclear actor. The conditions for a limited nuclear war increase as parity is approached, but a conflict is unlikely, because the satisfied status quo power has little incentive to take this initiative. Limited conflicts are frequently won by the dominant power. A few are decided in favor of the weaker side, but in such instances the relative costs endured by the challenger are much higher than those suffered by the dominant nation (i.e., Vietnam, Afghanistan). The most serious conflict is anticipated when the challenger overtakes the dominant nation in power and matches its ability to destroy.

Parity provides the necessary conditions for both peace and serious con-

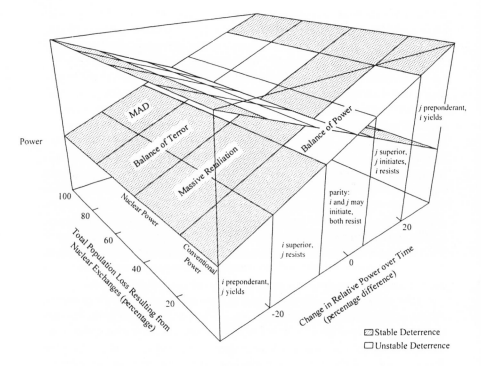

Power

Fig. 13.2.

flict. Therefore, deterrence under power parity is tenuous. The challenger makes demands with a probability of success that is equivalent to that confronted by the deterrer. The risk for such an action is high, but marginal probing by a risk taker may result in some gains that eventually, if repeated, could lead to escalation. Deterrence is determined by the joint decision of both nuclear actors not to act. If one chooses to break the impasse, the terror of nuclear weapons cannot deter the initiation and resultant escalation of such conflicts (Kugler and Zagare 1987; Powell 1990; Zagare herein).

It is shown here that, prior to 1945, decision makers faced with power parity considered marginal gains when deciding to initiate war. However, in similar circumstances, in an equal number of cases, conflict among major powers failed to materialize. Indeed, the conditions for stable and unstable mutual deterrence were present concurrently. Such results are consistent with the logical deduction that parity is a necessary but not sufficient condition for conflict.

Following this logic, under power parity the possibility of war is intensified. The challenger now dominates, and the old hegemon can only resist the dismantlement of its previous control of the international system. The condition of stable unilateral deterrence under preponderance is reimposed only when the disparity of capabilities makes it clear that the old dominant power cannot challenge the newly emergent dominant power.

In the nuclear era, so far, no country has traversed the path to both power parity and nuclear parity. Yet, under power parity, the Balance of Power, Balance of Terror, and MAD are all policies that share the one common condition of tenuous stability. The structural conditions for total war are present in all three situations, but with nuclear weapons the costs are higher. Under these conditions, nuclear proliferation begins to destabilize. This chapter does not investigate the specific nuclear levels that distinguish the peaceful from the conflictual outcomes. Rather, the intent is to show how the structural conditions implied by this model differ from the structural conditions anticipated by stable deterrence. It is not possible to test deterrence directly, but the implications of stable and unstable structures can be explored.

Consider the path of the United States and the Soviet Union since 1945. In the period between 1945 and approximately 1960, in spite of the cold war, the United States' preponderance of nuclear capability and power insured stability. Nevertheless, once parity in nuclear weapons was achieved in the mid 1960s, stability was still maintained despite the continuing disparity in power capabilities between the United States and the Soviet Union. The convergence in nuclear capabilities was never accompanied by conventional power parity overtaking. Employing this argument, the Soviet Union would have to initiate a conflict against an opponent whose power was far superior. As new information about the Soviet economy has become available, it is now apparent that the power of NATO was overwhelmingly superior to that of the Warsaw Pact countries. Since 1945, the preconditions for a major war have simply not materialized.

An interesting insight followed the dismemberment of the Soviet Union. Although the Republic of Russia still holds nuclear parity with the United States, its power relationship has suffered dramatically. The power transition perspective implies, as does common wisdom, that the probability of war between Russia and the United States declined markedly in the 1990s despite nuclear equality. The parity model supports the current evidence that essential structures for a confrontation are simply no longer in place and corroborates the expectation that peaceful coexistence should continue.

Figure 13.2 also sketches the path of China and Russia. This initial relationship dramatizes the disparity between a preponderance of power with and without the equality of nuclear capabilities. After World War II the stability among these counties was maintained not by parity, but by preponderance. However, conditions have changed. As Russia's power diminished by partition and economic decline, China correspondingly gained in relative strength. Indeed, indications are now apparent that China may soon achieve a level of mutual deterrence with the Russians. The necessary conditions for conflict caused by a conventional power overtaking between China and Russia seem to be emerging. Despite the current quiescence, the conditions for a confrontation between these two nuclear powers are expected to be achieved in the not-too-distant future.

A third pair, resulting from the Soviet partition, is worth noting. Although detailed statistics on the Ukraine are sketchy, mutual nuclear deterrence with Russia is likely, given the 750 delivery vehicles and 1,000-plus warheads based in the Ukraine. Nevertheless, conventional power parity is not achievable and Russia will remain preponderant, and therefore one can foresee no nuclear confrontation in this pair despite mutual deterrence implications.

Let us now examine differences drawn from two nuclear perspectives. Both the parity and preponderance models account for much of what has transpired in the international order during the nuclear era. Specifically, the models have competing explanations for the absence of nuclear war since 1945. There are serious differences. First consider assumptions. The parity model makes different assumptions about the actions of individuals during the stage of compellence and MAD. Recall that under Massive Retaliation, a nuclear power is not expected to act when it has an advantage. Under MAD, it is precisely the willingness of both actors to act that insures deterrence. By contrast, the preponderance model accounts for peace with equal precision, but holds assumptions constant across massive retaliation and MAD. Elites are expected to behave in the same manner under conventional conditions as they acted after the introduction of nuclear weapons. In sum, the power transition model challenges the notion that elites undergo a major behavioral transformation when faced with high costs of war. Prudence, maybe; behavioral transformation, never.

There are also serious differences dealing with effects of nuclear weapons deployment. Intriligator and Brito argue that once nations have achieved MAD, the buildup of an excess capacity of offensive weapons gives additional and relatively inexpensive security. In reality, increasing "overkill" capability insures that the opponent is less tempted to risk a confrontation. Power parity suggests, on the other hand, that among conventionally equally matched nations, deployment of nuclear weapons may lead to escalation and massive war. In fact, a war initiated under these conditions may end, like World War I and World War II, with demands for "unconditional" surrender despite the immense suffering on both sides.

Another fundamental difference between these perspectives concerns proliferation of nuclear weapons. Since Brodie, deterrence advocates have argued that the terror of nuclear devastation has dramatically changed the behavior of individuals in the international order. Thus, nuclear proliferation, while obviously not propitious, will lead to further stability by assuring a veto system for major nuclear powers (Waltz 1993; Intriligator and Brito 1987). From a parity perspective the danger of nuclear weapons is compounded by proliferation. Recall that nuclear deterrence is intrinsically unstable as parity is approached. Thus, the more nations who possess nuclear capabilities sufficient to destroy their chosen opponents, the greater the likelihood that a conventional confrontation will escalate to a nuclear encounter.

The strong aversion of practitioners to proliferation of this technology is consistent with this perspective. Indeed, if nuclear parity is a stabilizing factor, would it not be useful to allow limited nuclear arsenals to offset other nuclear capabilities and add stability to all regions? If parity in nuclear weapons leads to stability, why do the United States and Russia wish to retrieve nuclear weapons from the Ukraine or Belarus? At least from the American perspective, a nuclear Ukraine would have the ability to withstand potential pressures from Russia, ensuring stability in a region where disputes over territorial spoils are likely. On the other hand, if nuclear parity leads to the escalation of conventional conflicts to nuclear levels, then the restraints imposed seem valid; certainly they seem justified.

The position of nonproliferation is fully consistent with power transition logic and policy. Nuclear nations do not advocate the proliferation of nuclear weapons. Clearly, domestically induced nuclear proliferation has not stopped, nor is it likely to stop in the future, but efforts to curtail the international spread of such weapons is common to all nuclear nations. Furthermore, since such proliferation is occurring in areas far distant from the nations that wish to prevent the spread of nuclear weapons, their actions are consistent with the notion that deterrence is tenuous and not ultra stable.

Stable deterrence suggests that a "thick" defense system against nuclear attack would destabilize deterrence. This view is contested by the parity notion. Under tenuous deterrence, if the dominant nation gains a marginal advantage over the potential challenger, stability is reimposed. However, if the converse is true and the challenger gains the advantage, stability is threatened. Thus, consistent with the argument proposed by the Reagan administration in favor of the Strategic Defense Initiative, the acquisition of defensive capability by the United States, supporting the status quo, would enhance stability. However, if such a capability were developed by the challenger who seeks to gain on the dominant power, instability could be created. Moreover, stable deterrence fears defensive deployment, while tenuous deterrence encourages such activity. The willingness of nuclear nations to cooperate in the deployment of defensive systems suggests that they all fear the possibility that deterrence could be violated, even by an actor who can expect enormous destruction in response. Again, such actions are consistent with tenuous deterrence, but not with classical deterrence theory.

Conclusions

In the absence of a nuclear confrontation, the deductions about the stability of nuclear deterrence depend largely on the persuasiveness of the logic that supports each view. Nevertheless, indirect evidence supporting secondary deductions is compelling. Parity anticipates that a decline in power by one contender leads to conflict, while preponderance argues the opposite. From the balance perspective, the partition of the Soviet Empire should have led to major instability among the great powers; it has not. Limited confrontations among small powers and domestic upheavals within nations of the old Soviet Union have emerged, but the level of cooperation among Russia, the United States, and China has increased. Few can argue with conviction that the prospect of nuclear war among the main contenders has risen.

In summary, let us restate the original challenge posed by Einstein: "nuclear weapons have changed everything save man's way of thinking." Is Einstein right? Are we drifting into unparalleled catastrophe? Or is there sufficient evidence that our thinking has modified? Has the human race, faced with forbidding choice, opted for the renunciation of nuclear war over self-destruction? The jury is still out, but the evidence does not allow one to repose. In this period of relative tranquility perhaps we should reassess this question and formulate alternatives that may ensure that the specter of nuclear war recedes from humanity's horizon. Even though we are endangered, I contend we have not changed the potency of rational thought, and eventually we will manage a collaborative solution.

NOTES

1. Hart returned to his 1950s prediction in the 1967 revision of his work and concluded that "I ventured to predict that the new development (nuclear weapons) would not radically change the basis or practice of strategy and would not free us from dependence on what are called "conventional weapons," although it was likely to be an incentive to the development of more unconventional methods of applying them." (Hart 1967:xv). This view opens the door to deterrence as a strategy of "war fighting" rather than as a strategy of war prevention, particularly after tactical nuclear weapons are developed.

2. Policy debate at UCLA.

3. For a more academic presentation of this crucial aspect of deterrence, see Bundy et al. 1982.

4. Reports of considerations of a preemptive strike abound (Griffith 1975), but there is little evidence that the Soviet Union would be deterred from exercising the nuclear option for reasons other than because they chose to adhere to a retaliation policy.

5. To build their model, Intriligator and Brito postulate that the deterrer, nation i, and

the challenger, nation j, can anticipate the levels of destruction that would ensue from initiating nuclear war. They postulate further that each actor can determine a threshold below which an actor is willing to initiate conflict and a second one above which the same actor is no longer willing to do so. Thus, i will initiate war only if the prospective costs of retaliation fall within the cone of war and below its "acceptable" loss level i_1. Critical to deterrence, nation i will not initiate conflict, but will resist j's attack if the costs fall between i_1 and a higher i_2. Finally, i will be "deterred" from conflict and will yield to j's demands when anticipated costs exceed i_2. Similar constraints, but not necessarily identical levels, affect the actions of j.

6. The key to absolute deterrence is the assurance that costs are indeed high and the desired losses can be inflicted on the opponent, upon demand. The empirical record supports the contention that nuclear weapons are indeed more destructive than all previous tools of war. No conventional war has inflicted losses on total populations in excess of 20 percent of the total population affected. Maybe with such figures in view, Robert Mc-Namara argued that to achieve stability, the deterred had to be convinced that population losses would exceed 20 percent (McNamara 1969). For equivalent arguments that show that results can be achieved with smaller megatonnage, see von Hippel 1983. Technological developments have now far surpassed these levels. Currently, a nuclear exchange could lead to the destruction of over 60 percent of the population on either side. Other nations could suffer even more severe devastation if they become the target of a nuclear war. While we clearly do not know where the threshold of "unacceptable" damage is, we can, like Intriligator and Brito, presume that it exists at costs that exceed those inflicted upon participants in World Wars I and II.

7. By symmetry, similar conditions emerge when one traces movement along the vertical axis, except in this case the beneficiary is j rather than i.

CHAPTER 14

The Rites of Passage:
Parity, Nuclear Deterrence, and Power Transitions

Frank C. Zagare

If a myth is a notion based more on tradition or convenience than on fact or reality, then modern deterrence theory, and its intellectual forebear, balance-of-power theory, qualify as myths of absolutely the first rank (Vasquez 1991). My purpose in this chapter is to explain this contention by highlighting the logical and empirical inconsistencies of both theories, to suggest that these problems evaporate when interstate conflict is viewed through the lenses of power transition theory (Organski 1958; Organski and Kugler 1980), and to develop the implications of an incomplete-information model of conflict initiation for the transition perspective. By extending the framework provided by power transition, additional insights into the conditions associated with both crises and interstate wars are gained.

Modern Deterrence Theory

Because no single, authoritative exposition of its major premises exists, an outline of modern deterrence theory must be pieced together from a variety of sources. Nevertheless, there seems to be a wide consensus among theorists about both the provenance and the broad contours of deterrence theory.

It is generally agreed that the roots of modern deterrence theory lie in the classical model that has variously been labeled "political realism," "realpolitik," or "power politics." This conceptual framework—which some trace back to Thucydides—rests upon the supposition that international politics resembles Hobbes's "state of nature" where life is "nasty, brutish, and short." In the realists' paradigm, the international system, like Hobbes's anarchistic presocietal state, is seen as lacking an overarching authority or sovereign. Thus, each state in the system must "rely on [it]s own strength and art for caution against all others."

In a system where every state must provide for its own security, most realists hold that the most efficient mechanism for maintaining order is the balance of power. When power is equally distributed among actors in the system, or among the major partitions of actors (as the argument goes), peace is likely since no one state has an incentive to upset the status quo and challenge another. By contrast, balance-of-power theorists argue that an asymmetric distribution of power provides no check on stronger states driven, either by nature (Morgenthau 1948) or by systemic imperatives (Waltz 1979), to maximize their power. Or as Mearsheimer (1990:18) writes, "power inequalities invite war by increasing the potential for successful aggression; hence war is minimized when inequalities are least."

Modern deterrence theory builds upon this theoretical base and extends its domain by considering the consequences of war in the nuclear age. In this regard, two distinct yet compatible strands of the theory can be discerned: structural (or neorealist) deterrence theory (Kaplan 1957; Waltz 1979; Intriligator and Brito 1987) and a decision/game-theoretic thread that can be traced to what Young (1975) calls manipulative bargaining theory (Ellsberg 1959, 1961; Schelling 1960, 1966; Jervis 1972; Snyder 1972). As Allison (1971) has so clearly demonstrated, each of these two approaches to deterrence shares a common conceptual orientation with classical realism. In the strategic literature, these two strands converge to form the pastiche of modern deterrence theory.

Structural Deterrence Theory

Like classical balance-of-power theory, structural deterrence theory sees the key to international stability in the distribution of power within the system in general, and among the great powers in particular. Most structuralists hold that when a parity relationship is combined with the enormous absolute costs of nuclear war, a deliberate (i.e., rational) war is at once unthinkable and virtually impossible (e.g., Brodie 1946). Those who ascribe to this view see the nuclear balance as unusually robust and stable, crediting the absence of a major superpower conflict in the postwar period directly to the devastation that nuclear weapons can work.

The most astute structuralists, ruthlessly following the logic of their arguments, have advanced important counterintuitive policy implications of the underlying theory. Early in the nuclear age, for instance, Morgenstern (1959:74–77) suggested that it was in the interest of the United States to provide the Soviet Union with an invulnerable strategic retaliatory force. One reason for Morgenstern's prescription was the destabilizing impact of an asymmetry that, paradoxically, might cause the weaker state (the Soviet Union) to preempt the stronger (the United States).

Similarly, Waltz (1981) and other structuralists suggest that nuclear weapons be proliferated to certain nonnuclear states to transform potentially unstable dyadic relationships into ultrastable pairs. Here, too, the argument rests on the theoretical consequences of the confluence of parity and the high costs of nuclear war. When both conditions are present, outright conflict is extremely unlikely.

Fortunately, most policy makers have instinctively rejected proliferation policies and the superpowers, at least, have resisted transferring nuclear technology to third states except in the most unusual circumstances. The reluctance of policy makers to accept such prescriptions is but one indication that the underlying theoretical argument supporting them is suspect. Another clue is empirical: unlike power transition theory, classical deterrence theory is hard put to explain the absence of nuclear conflict when preponderance rather than parity reigns (Kugler and Zagare 1990). Though contemplated, the United States did nothing to exploit the huge advantage it enjoyed over the Soviet Union during the first half of the 1950s. The Soviet Union itself, despite a stormy and conflictual relationship, considered but ultimately chose not to impose its imperium over the Chinese in the late 1960s and early 1970s.

Decision-Theoretic Deterrence Theory

Unlike structural deterrence theorists, who find the key to interstate stability in the structure and distribution of nuclear power, decision-theoretic deterrence theorists focus upon the interplay of outcomes, preferences, and rational choices in determining interstate conflict behavior. Beginning where structural theorists leave off, the decision-making strand of classical deterrence theory almost uniformly posits an interactive situation in which nuclear war is so costly that only an irrational leader could consider it a means of conflict resolution. Thus, a critical deduction of structural deterrence is accepted and embedded as an axiom by this group of modern deterrence theorists.

The assumption that nuclear war is irrational explains the fixation of these theorists with the game of "Chicken" (see, inter alia, Schelling 1960, 1966; Hopkins and Mansbach 1973; Brams 1975, 1985; Jervis 1979; Powell 1987). In Chicken, conflict is the worst possible outcome. Since, by assumption, no rational decision maker could possibly prefer waging a nuclear war to any other outcome, accidental war came to be seen as the major threat to the stability of the postwar international system (Intriligator and Brito 1981). Most decision theorists, therefore, turned their attention to discovering useful tactics for managing crises and other lesser conflicts, although some critics (e.g., Rapoport 1964) warned of the inherent risks associated with the prescribed stratagems.

Schelling's (1960, 1966) recommendations to policy makers are the most well known. In coercive bargaining situations, statesmen are enjoined to win by seizing the initiative and irrevocably committing to a hard-line strategy: Schelling suggests that decision makers could do this by, for instance, "burning their bridges" behind them. In even more colorful language, Herman Kahn

(1962) once suggested that policy makers could gain the upper hand in a crisis by ripping the steering wheel from the automobile of state. Alternately, one could feign irrationality, make a public commitment to a noncooperative policy, or, by linking a present conflict to future conflicts, make capitulation less likely by making it more costly.

It is no mere coincidence that statesmen have been equally loath to follow the prescriptions of manipulative bargaining theorists like Schelling and Kahn, especially when facing a major adversary in an acute crisis. In crises, statesmen typically have eschewed commitment tactics. Rather, they have sought "to retain wide freedom of choice as long as possible and to avoid becoming boxed into an irrevocable position" (Young 1968:218). Or as Snyder and Diesing (1977:490) put it, "states seldom 'manipulate risk'—i.e., increase the risk of crisis getting out of control—as a coercive tactic . . . . They tend to avoid actions that create a chance of accidental violence and to avoid words that might arouse uncontrollable passions in the adversary's public and press."[1]

A second major empirical problem for this strand of modern deterrence theory arises in explaining actual stability in the superpower strategic relationship in light of theoretical instability implied by the underlying bargaining model. Specifically, in Chicken, the status quo is not a Nash equilibrium and, hence, should not survive rational play.[2] Thus, like the structuralists, decision theorists are hard put to explain the absence of both major war and chronic crisis during the postwar period (Zagare 1990).

Power Transition: An Alternative View of Deterrence

Unlike classical deterrence theory, which is riddled with logical inconsistencies (Zagare 1990) and empirical inaccuracies (Kugler 1984, Vasquez 1993:89), power transition theory offers a theoretically rich and empirically consistent perspective from which to view the dynamics of interstate conflict. Like balance of power, power transition is structurally based. But unlike balance of power, power transition argues that a parity relationship is a necessary—though not sufficient—condition for major-power war. Thus, power transition holds that major-power wars only occur when two great states are approximately equal in strength.

The power transition perspective explains, inter alia, the absence of war between the United States and the Soviet Union during the most heated days of the cold war, and between the Soviet Union and China during the most intense period of that rivalry; it also accounts for the instinctual rejection of proliferation policies by most statesmen and strategic analysts, and the deep concern of decision makers with marginal disadvantages and windows of vulnerability.

For instance, if it is true that the very existence of massive stockpiles of nuclear weapons virtually assures the stability of the status quo, as some modern deterrence theorists suggest (Jervis 1984), then marginal differences should be relatively unimportant in strategic circles. But if parity conditions are, at best, tenuously stable, then these distinctions become critical. Only power transition is consistent with the extremely close attention policy makers pay to seemingly minor distinctions in offensive and defensive nuclear weapons capabilities (Kugler and Zagare 1990).

This is not to say that power transition is a completely specified theory of conflict initiation. While it more thoroughly accounts for the outbreak of major-power wars during the last two centuries than does balance-of-power theory (Organski and Kugler 1980), the power transition perspective, itself, leaves some important questions unanswered. In particular, because it provides only a set of necessary conditions for the onset of major interstate wars, power transition cannot distinguish, a priori, those transitions that culminate in war from those that do not. Why, for example, did the Germans, the British, and their allies fight two global wars while, under similar parity conditions, the United States and Great Britain were able to avoid major catastrophe? Why didn't the United States and the Soviet Union wage nuclear war as the Soviets moved toward parity at the end of the 1960s? Although the empirical record associated with these and related riddles is consistent with the power transition perspective, they have not yet been answered adequately within its theoretical confines.

Game-Form

My purpose in this chapter is to move, tentatively, beyond the set of necessary conditions provided by the power transition, but not by refining the underlying structural model. Rather, my plan is to supplement and extend its boundaries by using game theory. I begin by offering a summary of an incomplete information model of asymmetric deterrence that builds upon power transition's conceptual base. Then I discuss the broader implications of this model for power transition.

The underlying game-form for this extension is given by figure 14.1. This model, as originally specified in Kugler and Zagare (1990), rests upon several critical assumptions drawn from the transition perspective. First, the status quo is taken to be a real and crucial element in the calculus of war and peace. Each of the two players in the game is cognizant of the other's relative evaluation of this outcome. *Challenger* must decide, initially, whether to accept the status quo or try to upset it. If Challenger upsets the status quo by initiating a crisis, *Defender* must decide whether or not to defend it. If Defender resists, Challenger

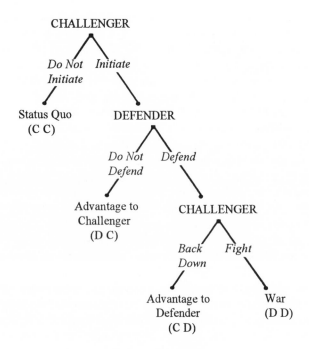

Fig. 14.1. Asymmetric deterrence game.

must then decide whether to back down, tacitly accepting the status quo ante, or fight to change the established order.

If Challenger chooses not to initiate, the status quo—outcome *CC*—persists. If Challenger initiates and Defender does not defend, Challenger gains the advantage—outcome *DC*. But if Defender resists, two other outcomes are possible. Defender will gain the advantage—outcome *CD*—when Challenger backs down; when Challenger fights, war—outcome *DD*—occurs.

For convenience, the four possible outcomes of this game, and the notation for the utility payoffs associated with them, are summarized in figure 14.2. Payoffs are given as (von Neumann-Morgenstern) cardinal utilities and are denoted by ordered pairs, the first entry being Challenger's utility, and the second Defender's. With little loss of generality, Challenger is assumed to receive 1 at *DC*, c_3 at *CC*, 0 at *CD*, and C_2 at *DD*. Similarly, Defender receives 1 at *CD*, d_3 at *CC*, 0 at *DC*, and D_2 at *DD*.

Defender

Status Quo (*CC*) (c_3, d_3)	Advantage to Defender (*CD*) (0,1)
Advantage to Challenger (*DC*) (1,0)	War (*DD*) (C_2, D_2)

Challenger (label at left, outside table)

Fig. 14.2. Outcomes and preference notation of asymmetric deterrence game.

In this model Challenger is assumed to strictly prefer an advantage to the status quo. Consistent with the power transition perspective, no such restriction is placed on Defender's preferences; Defender could most prefer either maintaining the status quo or facing down a challenge. Further, both players are assumed to prefer winning without a confrontation, either *CD* (for Defender) or *DC* (for Challenger), and the status quo, *CC*, to conflict, *DD*. This assumption is made to reflect the power relationship likely to exist when a power transition is in process. Namely, under parity conditions, neither player is so strong that it simply prefers to impose its will by subduing the other.[3] In other words, war costs are taken to be sufficiently high that both players prefer, ceteris paribus, to avoid outright conflict.

Finally, each player is presumed to know that the other's preferences satisfy these restrictions with the exception that neither knows whether the other prefers war, *DD*, to capitulation, either *CD* or *DC*. Thus, in this model, as in power transition, conflict is not, a priori, presumed unthinkable. Depending on the issues at stake, either Challenger or Defender, or both, may actually prefer war to some other outcome. Such a player, with a preference for conflict over capitulation, is called *hard*. *Soft* players are those with the opposite preference. Note that when Challenger is hard, it always fights when faced with a decision at the last node of the game tree of figure 14.1. When it is soft, it rationally backs down.

To model the players' uncertainty about each other's type (i.e., whether its opponent is hard or soft) the players' payoffs at *DD*, C_2 (Challenger), and D_2 (Defender) are taken as independent binary random variables (denoted by uppercase letters) with known distributions. More specifically, it is common knowledge that

$$C_2 = \begin{cases} c_2^+ \text{ with probability } p_{Ch} \\ c_2^- \text{ with probability } 1 - p_{Ch} \end{cases}$$

$$D_2 = \begin{cases} d_2^+ \text{ with probability } p_{Def} \\ d_2^- \text{ with probability } 1 - p_{Def} \end{cases}$$

where

$$c_2^- < 0 < c_2^+ \text{ and } 0 \leq p_{Ch} \leq 1$$

$$d_2^- < 0 < d_2^+ \text{ and } 0 \leq p_{Def} \leq 1.$$

Thus, if Challenger [respectively, Defender] prefers conflict to capitulation, (i.e., is hard), $C_2 = c_2^+$ [$D_2 = d_2^+$]; when this preference is reversed and Challenger is soft, $C_2 = c_2^-$ [$D_2 = d_2^-$]. Both players know the probability distributions on these values, both know they know, and so on. However, only Challenger knows the actual value of C_2 and only Defender knows the actual value of D_2.

The probability distributions p_{Ch} and p_{Def}, then, reflect the extent to which the players are seen to be hard or soft. Thus, they can also be interpreted as a measure of each player's threat *credibility*. Let me pause for a moment to explain why.

In both the strategic (Lebow 1981:15; Betts 1987:13) and the game-theoretic literature (Selten 1975), credibility is usually associated with rationality. Rationality, in turn, is defined, minimally, as behavior consistent with a player's preferences (Luce and Raiffa 1957). This suggests that credible—or rational—threats are precisely those threats that a player actually prefers to carry out. As a measure of this preference, then, p_{Ch} and p_{Def} also gauge threat credibility.[4] The higher these values, the more likely (that is, credible or believable) a player's threat, and conversely.

In sum, Challenger is assumed always to prefer an advantage to the status quo, and the status quo to conflict. Defender may prefer the status quo to all other outcomes, but prefers an advantage and the status quo to war. Both may prefer conflict to concession. More formally, the preferences of Challenger and Defender satisfy the following restrictions:

$$c_2^- < 0 < c_2^+ < c_3 < 1,$$

$$d_2^- < 0 < d_2^+ < \min \{d_3, 1\}.$$

Perfect Bayesian Equilibria

Given these assumptions, what will rational players do? Under what conditions is deterrence stable? When will Challenger try to upset the status quo? What distinguishes those crises that culminate in war from those that are resolved without resort to force? What factors are critical in separating peaceful from nonpeaceful transition periods?

To answer these and related questions, Zagare and Kilgour (1993) have determined all the *perfect Bayesian equilibria* of the game given in figure 14.1. A perfect Bayesian equilibrium is composed of a profile of rational strategies and a set of beliefs—one for each player. The strategies are plans of action for the players for every decision they may have to make during the play of the game, and also for every possible type the players may be (i.e., hard or soft). The beliefs are updated rationally (using Bayes's Rule) to reflect the other player's observed actions. Thus, a perfect Bayesian equilibrium completely describes rational strategic behavior under conditions of uncertainty. By knowing these equilibria one can determine the logical consequences of each set of beliefs and answer the questions posed above. To this end I next summarize the salient strategic characteristics of each of the four major types of perfect Bayesian Equilibria. Following that, I shall discuss the implications of these equilibria for power transition theory.

To understand the differences among the major categories of perfect Bayesian equilibria in the game of figure 14.1, it will be useful to refer to figure 14.3. Along the horizontal axis of this figure is graphed p_{Ch}, the a priori probability that Challenger is hard; along the vertical axis is graphed p_{Def}, the a priori probability that the Defender is hard. Also indicated on these two axes are several constants, such as d^*. These constants are convenient thresholds for categorizing and interpreting the equilibria of the asymmetric deterrence game of incomplete information.[5]

Deterrence Equilibria

Deterrence equilibria constitute the first major equilibrium category of the game of figure 14.1. All such equilibria are those in which Challenger never initiates a crisis. At a deterrence equilibrium, therefore, the status quo is the only rational outcome. Challenger's decision is independent of its type, although it may be contingent on Defender's strategy. Although the status quo may sometimes result when other equilibria are in play, all other equilibria also carry with them the possibility of other outcomes that depend on the players' types, their beliefs, and their choices. Thus, the status quo is fully robust under a deterrence equilibrium, but not under other kinds.

Significantly, a deterrence equilibrium can come into play under *any* set

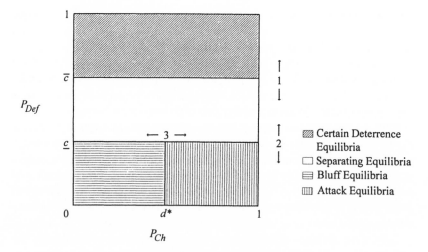

Fig. 14.3. Location of equilibria.

of player beliefs about threat credibility, save for the pure case (of complete information) in which Challenger's threat is perfectly credible, and Defender's is not. This means that deterrence could conceivably emerge under (almost) any conditions during a transition period. Nevertheless, while deterrence is (almost) always possible, it is not so often inevitable. The inevitability of deterrence depends upon which of two distinct types of deterrence equilibria, *certain* or *steadfast,* may exist. Next, I describe the distinguishing characteristics of each.

Certain deterrence. The key to certain deterrence is the credibility of Defender's retaliatory threat (p_{Def}). Certain deterrence equilibria exist only when Defender's credibility is so high that Challenger is completely dissuaded from initiating conflict.

The existence of a certain deterrence equilibrium does not depend on the a priori credibility of Challenger's threat. This equilibrium type, then, is invariant with respect to Challenger's credibility: when a certain deterrence equilibrium occurs, no rational Challenger—hard or soft—will choose to initiate.

Certain deterrence equilibria occupy the entire upper region of the unit square in figure 14.3. When a certain deterrence equilibrium exists, it uniquely exists and the status quo is the only rational outcome of the game; since there are no other rational behavior possibilities, peace is at hand.

Steadfast deterrence. A deterrence equilibrium may still exist even when Defender's credibility falls below the threshold required for certain deterrence. But then a deterrence equilibrium cannot occur alone—it coexists with equilibria of other types (see following), so its occurrence in actual play is less than certain. Moreover, for this type of deterrence equilibrium to occur, Defender must be steadfast in the sense of being committed to defend with at least a threshold probability, even when it is soft.

A steadfast deterrence equilibrium occurs only when Defender's threat is less credible than required for a certain deterrence equilibrium. The difference is that with steadfast deterrence Defender's threat to defend when hard is not in itself sufficient to sustain the status quo. Further commitment is necessary. Specifically, to offset the relative decline in Defender's credibility (i.e., the probability that it is hard), Challenger must believe that there is a high enough probability that even a soft Defender will resist. To rationally support this intention, Defender must believe that it is fairly likely that any Challenger who initiates is soft and will back down if Defender defends. Thus, the a posteriori credibility of Challenger's retaliatory threat is critical for the existence of a steadfast equilibrium.

Because steadfast deterrence is independent of any initial beliefs Defender might have of Challenger's type, equilibria of this category can be considered akin to what Bundy (1983) calls *existential deterrence*. Those who believe in this variant of modern deterrence theory hold that the vast destructive power of nuclear weapons, coupled with the belief that they would almost certainly be used in any overt superpower conflict, assures a stable international order (Jervis 1984) and a continuation of what Gaddis (1987) calls the Long Peace.

The irrelevance of Challenger's a priori credibility explains why a steadfast deterrence equilibrium can be found anywhere below the threshold required for certain deterrence. As can be seen from figure 14.3, other equilibria—all of which include the possibility of conflict—also occupy this area. Hence, the selection of a steadfast deterrence equilibrium cannot be considered certain.

Other Equilibria

In addition to the two types of deterrence equilibria, three other kinds of equilibria may occur in the game of figure 14.1: *separating, bluff,* and *attack.*

Separating equilibria. As figure 14.3 shows, separating equilibria lie between certain deterrence equilibria and attack and bluff equilibria. At a separating equilibrium a hard Challenger always initiates and a soft Challenger never does. Likewise, if challenged, a hard Defender always resists and a soft Defender always capitulates. Thus, the players' preferences are fully revealed by their strategy choices when a separating equilibrium is in play.

It is possible for the status quo to remain stable when a separating equilibrium is in play, but only when Challenger is soft. When Challenger is hard, it initiates for certain; Challenger then gains the advantage when Defender is soft, and precipitates overt conflict when Defender is hard. Three of the four possible outcomes of an asymmetric deterrence game (see fig. 14.2) can transpire under a separating equilibrium. But there are no circumstances under which Defender can gain the upper hand by facing down Challenger.

Bluff equilibria. When Defender's credibility is below the lower boundary for separating equilibria, two additional equilibrium possibilities arise. Which of these equilibria exist depends upon how credible Challenger's threat is. When Challenger's threat credibility is high, an attack equilibrium may come into play. When it is low, a bluff equilibrium may occur.

More specifically, bluff equilibria are possible when *both* Defender's and Challenger's credibilities are relatively low, that is, when both players believe the other probably prefers to capitulate rather than fight (see fig. 14.3). At a bluff equilibrium, the players' strategies depend upon their types. The Challenger initiates for certain in the unlikely event that it is hard. (After all, chances are that Defender is soft and likely to capitulate.) But if Challenger is soft, it adopts a mixed strategy, initiating with some positive probability. The more credible it is perceived to be, the greater this probability.

The equilibrium choice of a hard Defender is, as always, to defend. But at a bluff equilibrium, even a soft Defender defends with some positive probability. This conditional probability is a function of Defender's initial credibility, just as Challenger's conditional probability is a function of Challenger's credibility. But the lower Defender's credibility, the greater its tendency to bluff and resist a challenge when soft!

At any bluff equilibrium Defender's overall probability of capitulating or resisting if challenged is always the same, regardless of the value of p_{Def}. Challenger, therefore, always faces the same probability that Defender will defend. This (pooling) property serves to conceal Defender's type if it is soft—it is not possible to infer that a Defender who resists is likely or unlikely to be hard. As a consequence, Challenger is less and less willing to risk a challenge as its own credibility decreases—after all, the probability that it will have to back down if it challenges becomes greater and greater as its credibility drops. But the only way that Defender can achieve this constant level of defense readiness is to be prepared to defend when soft more frequently the lower its credibility.

More than at any other equilibrium, then, play under a bluff equilibrium is likely to be a "competition in risk-taking" (Schelling 1960, 1966). This should be no surprise, however, since the complete-information analogue of such games bears a structural resemblance to the game of Chicken. Recall that in Chicken neither player possesses a credible retaliatory threat. The difference is that games played in the region of bluff equilibria are characterized by players whose threats are likely, rather than certainly, low.

Putting this in a slightly different way, bluff equilibria describe optimal strategic behavior under the precise set of credibility conditions assumed in the decision-making strand of modern deterrence theory. By locating the bundle of assumptions that undergird this body of research in a more general credibility plane, the incomplete-information model is able to subsume both manipulative bargaining theory and the modeling efforts of its intellectual progeny (e.g., Powell 1990).

Attack Equilibria. Like bluff equilibria, attack equilibria occur only when the credibility of Defender is low. What distinguishes the two equilibria is Challenger's perceived credibility. When Challenger's credibility is relatively low—like Defender's—a bluff equilibrium arises; when it exceeds a certain threshold, the attack equilibrium choices predominate.

The attack equilibrium is the only equilibrium where there is no chance that Challenger will accept the status quo. At an attack equilibrium, Challenger—whatever its type—always initiates, and a soft Defender always capitulates. Thus, since only a hard Defender will resist, a war can occur if and only if a hard Challenger confronts a hard Defender. While this is unlikely under attack conditions, actual conflict remains possible. Typically Defender has few options and little defense. Like the Americans during the crises in Hungary in 1956 and Czechoslovakia in 1968, and the Soviets during the 1956 Suez crisis, Defender can only accept the inevitable; any other reaction would be contrary to its interests.

Attack equilibria occur precisely when the balance of credibility favors Challenger. Like power transition, then, the incomplete-information model finds that an asymmetric (credibility) relationship that favors Challenger is unstable (Kugler and Zagare 1990). By contrast, when Defender's credibility is high, and Challenger's is low, deterrence is likely to be certain. From the perspective of either power transition or the incomplete-information model of asymmetric deterrence, then, the explanation for the absence of a war between the United States and the Soviet Union, prior to the period of superpower parity, is evident.

Discussion

The incomplete-information asymmetric deterrence game described and summarized in the previous sections can be considered as a decision-making analogue of power transition theory. Like power transition, it posits two players with distinct and recognizable roles; it assumes that at least one player is dissatisfied with the status quo and is actively considering challenging it. Both players know their own preference between fighting and capitulating, but are unsure of the other's. Since the perfect Bayesian equilibria of this game describe rational strategies under specified credibility conditions, they provide additional information about the range of behavioral possibilities during a transition period.

Even when augmented with a decision-theoretic component, though, power transition may not be sufficient to completely distinguish between peaceful and nonpeaceful transitions. In the incomplete-information model, as in power transition, the conditions of war and peace may be present simultaneously. This occurs when a steadfast deterrence equilibrium coexists with a separating, bluff, or attack equilibrium, or even when a separating or a bluff equilibrium is about to come into play. For example, depending upon the players' types, the status quo could remain stable or outright conflict could evolve when the players select strategies consistent with a separating equilibrium.

The failure of both the aggregate structural model and its decision-theoretic extension to provide clear and unambiguous answers to critical questions, however, does not mean that either component is flawed. One possibility is that both theoretical constructs remain underspecified. Another, more likely, explanation is that this indeterminancy reflects both the essence of the real world and the limits of our ability to understand it.

This is not to say that some additional precision about the conditions necessary for a peaceful power transition have not been gained. For example, the incomplete-information model *does* specify the credibility conditions associated with certain deterrence. Moreover, using refinements of Nash equilibria (Fudenberg and Tirole 1991; van Damme 1991), some further insight is possible (Zagare and Kilgour 1993).

Specifically, when a separating equilibrium coexists with a steadfast deterrence equilibrium, the deterrence equilibrium is unlikely as long as Challenger is hard, since a hard Challenger strictly prefers the separating equilibrium, and a soft Challenger is indifferent. Given that a hard Challenger can choose to initiate (for certain), the status quo is unlikely to persist under rational play.

By contrast, it appears that there are no equilibrium refinements that rule out either steadfast deterrence or the bluff or attack equilibria, adding further support for the claim of power transition that the conditions for peace and major-power war are present simultaneously under parity. Again, from a theoretical point of view, the indeterminancy is disappointing. On the other hand, it is comforting. Decision makers are not compelled by their environment to choose war over peace, suggesting that structural constraints still leave room for human intervention, skillful diplomacy, and statesmanship. Nonetheless, since the possibility of war may loom in the background, the policy choices available to statesmen may be seriously constrained. Even a parity relationship between ostensively friendly states requires vigilance.

The incomplete-information model of asymmetric deterrence also corroborates the importance power transition places on the role of the status quo in peace and war decisions (Kugler and Werner 1993). Most balance-of-power and classical deterrence theorists, who (rightly) emphasize the strategic consequences associated with the high costs of nuclear war, (wrongly) ignore this other critical variable. In the incomplete-information model, both the value of the status quo and the players' evaluation of the conflict outcome are strategically significant. As will be seen, however, like power transition, the incomplete-information model reveals that past a certain point, there is no linear or other simple relationship between the absolute costs of war and deterrence stability. This conclusion stands in sharp contrast to that reached by many classical deterrence theorists (e.g., Intriligator and Brito 1984).

In the incomplete-information model, the threshold value for certain deterrence depends, in part, on Challenger's evaluation of the status quo, c_3. Specifically, as the value Challenger attaches to the status quo increases, the threshold for certain deterrence decreases, making deterrence more likely, and conversely. Or, put in a slightly different way, the more Challenger values the status quo, the more likely it is that deterrence will be the only possible outcome.

Challenger's evaluation of the status quo also is critical in determining the threshold value of the lower bound of the region of separating equilibria. As the value of c_3 increases, approaching that of Challenger's best outcome, the lower bound of the region of separating equilibria moves downward, shrinking the region of attack and bluff equilibria where conflict is more likely. Of course, the opposite is also true.

All of which suggest that a Defender can enhance the prospects of a peaceful power transition by making systemic adjustments that alleviate Challenger's dissatisfaction. Thus the incomplete-information model supports George and Smoke's (1975:531) observation that efforts to placate an opponent will reduce the need for overt deterrent threats, and will increase the likelihood that traditional deterrence tactics will succeed when and if they are practiced.

The Munich analogy notwithstanding, it is unfortunate that during the most heated periods of the cold war, decision makers in both the Soviet Union and the United States attempted to enhance deterrence by increasing the other's war costs. In the process they ignored the very real benefits attached to what Snyder and Diesing (1977:chap. 3) call an accommodative move. Note, however, that such political reconciliations do not necessarily imply appeasement or even self-abnegation (Wolfers 1951). Given the non-zero-sum nature of deterrence, an increase in Challenger's satisfaction with the status quo need not be at Defender's expense. It may be possible for Defender to simultaneously increase both its and Challenger's utility for maintaining the existing order. Trade concession or mutually acceptable arms control agreements constitute some likely examples.

Arrows 1 and 2 in figure 14.3 graphically indicate the boundary shifts in the various equilibrium regions implied by an increase or a decrease in a potential challenger's satisfaction with the status quo. Note that the boundaries of the region of certain deterrence equilibria and of the region of separating equilibria are similarly affected. If one rises, so does the other, and conversely, if one decreases, so does the other.

In this context it seems reasonable to suggest that recent dramatic events in Eastern Europe and the former Soviet Union have been accompanied by increased satisfaction with the prevailing system and rejection of the old world order. Most strategic analysts seem to believe that the probability of a nuclear war is lower today than at any time since the end of World War II. The downward shifts in the lower boundaries of both regions suggested by the descending component of arrows 1 and 2 explains why.

Conversely, the lower bounds of each of these equilibrium regions were likely higher, and their areas significantly smaller, during most of the cold war period. In other words, the regions of bluff and attack equilibria likely occupied a much larger area of the unit square when the Soviet Union was largely dissatisfied with the distribution of rewards in the international system. If so, then the incomplete-information model explains why this period of intense rivalry was punctuated by a series of acute crises: in Berlin, in Cuba, in Asia, and elsewhere.

One reason why the manipulation of the status quo is not generally thought of as a stratagem for practitioners of deterrence is the absence of this variable in most variants of classical deterrence. As a rule, and understandably so, the emphasis of the traditional strategic literature has been on the strategic consequences of the vast destructive power of nuclear weapons. Still, the fixation of classical deterrence theorists with the absolute costs of war—to the exclusion of other variables—explains the obvious oversight.

This is not to say that power transition and classical deterrence reach similar conclusions about the stabilizing impact of nuclear weapons. In fact, it is this issue more than any other that separates the two approaches. Power transition sees limits to the stabilizing impact of nuclear weapons. By taking into account not only the absolute costs associated with warfare, but also the marginal advantages of challenging the status quo, power transition admits the very real possibility of a major interstate conflict between nuclear powers. By contrast, classical deterrence sees a monotonic relationship between the absolute costs of war and deterrence stability. The greater these costs, ceteris paribus, the more stable deterrence.

The incomplete-information model of deterrence corroborates and refines the power transition theory's conclusions about the relationship between the costs of conflict and deterrence stability. To be sure, the prospects of certain deterrence *are* enhanced by decreasing the value of c_2^+, the value a hard Challenger associates with conflict. As this value approaches the value Challenger associates with capitulation, the lower bound of the region of certain deterrence equilibria moves downward, making certain deterrence more probable, and conversely (see arrow 1 in fig. 14.3). As one might expect, then, when conflict offers minimal advantages, deterrence becomes more robust.

Nonetheless, indiscriminate increases in the cost of conflict do not necessarily contribute to the likelihood of deterrence. When Challenger is soft and already prefers capitulation to conflict (i.e., when $C_2 = c_2^-$) further increases in the cost of conflict are redundant and unnecessary. Thus, like George and Smoke (1974:507), the incomplete-information model finds that deterrent "threats are often irrelevant or dysfunctional." Or, put in a slightly different way, there are distinct limits to the stabilizing impact of nuclear weapons. This suggests that prudent defenders have no reason not to pursue a policy of minimum deterrence, which relies "on the retention of only enough nuclear weapons to provide an assured destruction capability" (Kegley and Wittkopf 1989:351), rather than a maximum deterrent strategy rooted in an overkill capability. It is on this point, especially, that the normative implications of the incomplete-information model—rooted in beliefs and perceptions about the nature of deterrent threats—are at odds with classical deterrence theory, where subjective variables are generally ignored in favor of more objective factors and a fixed credibility assumption.

It is worth noting that the threshold that defines the region of certain deterrence is analogous to two thresholds that define an equilibrium region called *sure-thing* deterrence in a mutual deterrence model in which either player can challenge the other, initially (Kilgour and Zagare 1991). Thus, the conclusion that definite limits exist to the stabilizing impact of nuclear weapons does not depend upon the particular assumptions made here about the players' roles or the sequence of choices confronting them.

While the absolute costs of conflict do have implications for the probability of certain deterrence, they do *not* play a role in determining the boundary that distinguishes the region of separating equilibria from the more foreboding areas associated with bluff and attack equilibria. As noted above, the sole determinant of this threshold is Challenger's evaluation of the status quo, suggesting once again that, under certain conditions, increasing war costs may contribute little to deterrence stability. One of the hidden benefits of the incomplete-information model is that it clearly specifies the exact set of circumstances under which such conclusions apply. It is in this sense that this model augments power transition theory.

The costs of conflict to a *soft* Defender (relative to capitulation and the status quo) do, however, determine the boundary between bluff and attack equilibria. But here the relationship operates in an unexpected way. The greater the costs of confrontation to a soft Defender, the more likely an attack equilibrium; the lower, the more likely a bluff equilibrium. This suggests that under the shared conditions associated with the existence of these two equilibrium types (i.e., low Defender credibility), the high costs normally associated with nuclear warfare will actually make outright conflict more likely. So while it remains true that, under somewhat restricted conditions, nuclear weapons enhance central deterrence, they also have the opposite effect in areas peripheral to great-power rivalries.

This inverse relationship may account for a number of crises and brush-fire and proxy wars in the nuclear period. For instance, the Soviets might have been less willing to invade Afghanistan in 1979 if the world had not been nuclear, simply because the United States would have been more likely, ceteris paribus, to offer resistance had it not been facing a nuclear power. Much the same could be said about the American involvement in Vietnam. Thus the incomplete-information model explains why nuclear weapons may—up to a point—contribute simultaneously to the stability of basic, passive, or Type I deterrence, and to the instability of extended, active, or Type II and Type III deterrence (Kahn 1960, 1965; Betts 1987).

Arrows 1 and 3 in figure 14.3 indicate graphically the shifts in the various equilibria regions implied by a change in a hard Challenger's or a soft Defender's war costs. Notice that the lower bound of the region of certain deterrence equilibria either rises or falls (as the Challenger's cost vary), but that the lower bound of the region of separating equilibria remains unchanged. Also notice that the region of bluff equilibria and the region of attack equilibria are inversely related. As the region of attack equilibria expands (when the Defender's costs increase), the region of bluff equilibria contracts, and conversely.

Summary and Conclusions

This chapter begins with a discussion of modern deterrence theory. The roots of modern deterrence are traced to classical balance-of-power theory. Two distinct strands of the present formulation are identified: structural deterrence theory and decision-theoretic deterrence. I argue that each strand is riddled with logical inconsistencies and empirical inaccuracies.

Power transition theory is offered as an attractive alternative conceptual base upon which to construct a logically consistent and empirically accurate theory of interstate conflict. Unlike modern deterrence theory, which begins and ends with the assumption that war in the nuclear age is virtually impossible, the axioms of power transition allow for the possibility of nuclear war. In this sense, power transition is theoretically prior, incorporating the conditions of both balance of power and modern deterrence in a more general conceptual base.

To extend and refine power transition, an incomplete-information model of asymmetric deterrence is outlined. There are two players in this model: a Challenger who must decide whether or not to initiate a confrontation, and a Defender who must decide whether to protect the status quo. As in power transition, the preferences of players are permitted to vary: hard players are those who prefer conflict to capitulation; soft players are those with the opposite preferences. In the classical formulations of decision-theoretic deterrence, both players are assumed soft. Thus, by allowing preferences to vary, the incomplete-information model, like power transition, subsumes modern decision-making deterrence theory.

The incomplete-information model reveals that deterrence is possible under almost any condition, although some conditions are much more likely to support a stable status quo than others. In general, deterrence becomes more probable as the Challenger's evaluation of the status quo increases, as its perception of the Defender's credibility grows, and as the benefits of conflict decline. Like power transition, the incomplete-information model of asymmetric deterrence finds that the conditions of war and peace may be present simultaneously. It also finds, however, that certain conditions may exist in which deterrence is the only rational outcome. Specifically, when a Defender's initial credibility is sufficiently high, a certain deterrence equilibrium uniquely exists, and the survival of the status quo is assured.

At other times, under other conditions, stable deterrence is more problematic. At intermediate levels of Defender credibility, when a separating equilibrium exists, the critical variable is Challenger's preference between confrontation and capitulation. Hard Challengers will rationally initiate conflict; soft Challengers will refrain. If the status quo is upset, and Defender is hard, war is inevitable. Otherwise, the Challenger will be appeased and its demands satisfied.

Deterrence is even less likely to persist when Defender's credibility is low and either a bluff equilibrium or an attack equilibrium exists. A bluff equilibrium will exist when both Challenger and Defender are perceived soft. An attack equilibrium will exist when Challenger is perceived hard and Defender's credibility is low. A stable status quo is unlikely when either of these equilibrium types exist. Nevertheless, peace remains a remote theoretical possibility.

What, then, explains peaceful power transitions such as that between the United States and Great Britain, or the absence of a superpower war after the Soviet Union attained nuclear parity with the United States? The incomplete-information model suggests several possibilities. One is that the Defender in each of these cases was able to project unusually high credibility, thereby ensuring a certain deterrence equilibrium. Another possibility is that in these, and related cases, the Challenger was soft and that the actual transition games were played in the region of separating equilibria. Alternatively, it is possible that peaceful power transition games were played out in the region of a bluff or an attack equilibrium, and the players were lucky.

Like power transition, the incomplete-information model finds that increasing the costs of conflict does not necessarily lead to increases in strategic stability. Past a certain point, such increases are either unnecessary or counter-productive. Policies that promote an overkill capability, or those that are aimed at proliferating nuclear weapons, do not emerge as stabilizing choices within the confines of either theoretical framework.

Finally, as in power transition, the incomplete-information model finds that Challenger's evaluation of the status quo is a critical determinant of its decision to initiate conflict. When this value is very high, deterrence becomes virtually certain. And as it increases, the likelihood of conflict decreases. It is perhaps the value of this variable more than any other that accounts for successful transition periods. It is unfortunate indeed that most modern deterrence theorists have ignored its stabilizing possibilities, concentrating instead on the more dangerous—and more limited—tactic of manipulating the absolute costs of warfare.

NOTES

This material is based upon work supported by the National Science Foundation under Grant #SES-9123219. Any opinions, findings, and conclusions or recommendations expressed in this chapter are those of the author and do not necessarily reflect the views of the National Science Foundation.

1. One reason may be that Schelling's commitment strategy is an optimal choice "only under fairly restrictive conditions" (Morgan and Dawson 1990).

2. See, for example, Gamson and Modigliani 1971, for a description of the ebbs and flows of the superpower relationship between 1946 and 1963.

3. Deterrence is not possible when an unsatisfied player prefers conflict to the status quo (Zagare 1987:chap. 4).

4. For a detailed discussion of the connection between these values and threat credibility, see Kilgour and Zagare 1991.

5. For a detailed discussion, see Zagare and Kilgour 1993.

Part 5
Extensions beyond the Power Transition

CHAPTER 15

Beliefs about Power and the Risks of War: A Power Transition Game

Bruce Bueno de Mesquita

Of all of the possible consequences of international interactions, none inspires greater apprehension or attention than wars of cataclysmic proportions. These wars are the focus of theories concerned with power transitions or hegemonic decline (Organski 1958; Organski and Kugler 1980, 1989b; Gilpin 1981; Doran 1991) and are the concern here.

Theories about such conflicts contend that system-transforming wars are large, especially costly conflicts between about equally powerful adversaries who hold different views of how the international system should be run. The rival powers are presumed to be contending for a leadership role in that system and are hypothesized to be brought to the brink of war when differences in growth rates lead one rival to the point that it can challenge the dominant state. Considerable evidence suggests that the power transition perspective captures important features of international politics (Garnham 1976a; Weede 1976; Organski and Kugler 1980; Houweling and Siccama 1988a; Kim 1989). Yet, other evidence calls into question some core features of the theory (Bueno de Mesquita and Lalman 1992; Kim and Morrow 1992). Here I propose to link the power transition theory to a game-theoretic approach. I deduce a modified power transition theory that preserves key elements of the original formulation, while providing greater precision in the derivation of hypothesized relationships between power transitions and the risk of wars with the potential to be system transforming. In this way I hope to provide a somewhat new angle of vision regarding cataclysmic wars. I begin by specifying the game-theoretic model and its relationship to the power transition theory. Then I derive testable hypotheses and subject them to empirical scrutiny.

The International Interaction Game

Bueno de Mesquita and Lalman (1992) propose a game of international interactions. As their game is the basis for the analysis to follow, I begin with a brief summary of its structure and assumptions.

Although relations between nations are complex in their particulars, still there are essential features in the relationships between states. These features relate to sequences of decisions that lead to friendly or to hostile relations. In describing these essential features it proves helpful (but not essential) to assume that foreign policy choices can be treated as if they are made by a "weak" unitary actor. In particular, in the game I assume that there is a single decision maker in each country who is cast as being responsible for choosing the strategies and tactics for conducting foreign affairs. That leader, however, need not be responsible for selecting the foreign policy goals of the nation. These goals are determined by an unspecified domestic political process that is exogenous to the game.[1] It is exactly in the sense that the unitary actor chooses strategies but not goals that I mean to describe this assumption as a weak unitary actor assumption.

Any international relationship is assumed to arise in a context that provides one or another state with an opportunity to take the initiative in shaping relations, an initiative that may govern the future development of events (e.g., the rules of the system), much as is proposed in power transition theory. Each state shapes its relationship with the other through the selection of strategies. States may, for example, make demands or choose not to. Leaders may elect to acquiesce to those demands or they may choose to negotiate their differences. And, of course, leaders may choose to use force rather than give in to a rival's demands.

Different combinations of strategies result in different political outcomes. If, for instance, two states exchange demands and each state uses force in an effort to accomplish its goals, a war ensues. If the interaction between two states involves no demands, then each has elected to live with the status quo for the time being. The selection of strategies is a function of the value states attach to alternative outcomes and the beliefs they hold regarding how their adversary will respond to their strategic decisions.

I assume that decision makers respond to circumstances by making the choice that maximizes their expected utility from that stage of the game onward. They cannot precommit themselves to a future course of action, but they can act in anticipation of their opponent's choices. That is, I assume subgame perfection in which forward looking decision makers consider the consequences of their current strategic choices for an entire sequence of interactions (Selten 1975).

I begin with state A, which has an opportunity to initiate an interaction by making a demand (D^A) or not making a demand ($\sim D^A$) of another state. In power transition parlance, A is the *challenger*. The demand made by A may be about anything. I am less concerned here with the specific content of disputes than with the process by which international interactions evolve.

Once A makes its move, state B—the power transition's dominant state— has the opportunity to select a course of action by making a demand or not. Thus an initial sequence of $\sim D^A$, $\sim D^B$ yields the maintenance of the status quo. If the sequence is D^A, $\sim D^B$, B is said to have acquiesced to A's demand. Should A forego the opportunity to initiate a demand (so that the putative challenger does not challenge), thereby allowing B to initiate the course of events ($\sim D^A$, D^B), then A has the opportunity to acquiesce to B's demand ($\sim D^A$, D^B, $\sim d^A$) or to make a counter demand ($\sim D^A$, D^B, d^A). In this way, the game of international interactions allows for the possibility that the challenger or the dominant state makes the first *overt* move.

If the initial sequence of strategic choices were D^A, D^B, then by A's second move there is also a crisis. Failure by both parties to abide by the status quo and failure by either party to acquiesce to the other's demand results in a military crisis (Powell 1987; Lalman 1988). In a crisis, a state may choose to escalate the dispute further by using force or it can attempt to defuse the situation by offering to negotiate. If there is an offer to negotiate, the rival can reciprocate by not using force or it can exploit the opportunity to attack. Anytime a state escalates a dispute by using force it can expect one of two responses by its adversary. The adversary can capitulate to the attacker's demands, thereby cutting its battlefield losses, or it can retaliate, escalating the dispute to the state of war.

The extensive form of the international interaction game is depicted in figure 15.1. In the crisis subgame, A offers to negotiate by not initiating the use of force ($\sim F^A$) or A escalates the dispute by using force to back up its demand (F^A). If A chooses to escalate the dispute, then B must choose between capitulating to A's first strike (F^A, $\sim F^B$) or striking back by using force itself (F^A, F^B). In the latter case, the strategy sequence D^A, D^B, F^A, F^B results in a war initiated by state A.

Should A elect to offer to negotiate ($\sim F^A$) at the outset of the crisis, then B's choices can lead to negotiations ($\sim F^A$, $\sim F^B$) or to escalation ($\sim F^A$, F^B). If B chooses the escalatory path, then A must make a final strategic determination: to capitulate to B's enforcement of its demand ($\sim F^A$, F^B, $\sim f^A$) or to retaliate ($\sim F^A$, F^B, f^A), resulting in a war initiated by B. Thus the international interactions represented in figure 15.1 can culminate in the following outcomes displayed in table 15.1, with the crisis outcomes italicized.

The structure of the game establishes a simplified view of the foundations of all international relations, but it does not provide sufficient information to make predictive statements about behavior. The game requires further specification in the form of assumptions that establish the feasible range of preferences over the events at the terminal nodes. The value national leaders associate with each of the game's outcomes, and therefore the set of admissible preferences, is determined according to the following assumptions:

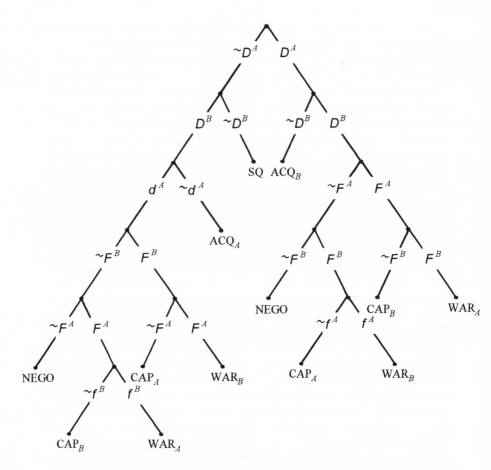

Fig. 15.1. The international interaction game. (From Bueno de Mesquita and Lalman 1994.)

TABLE 15.1.   Outcomes of the Crisis Subgame

Nonviolent	Violent
Status quo	*Capitulation by A*
Acquiescence to *A*'s demand	*Capitulation by B*
Acquiescence to *B*'s demand	*War initiated by A*
Negotiation	*War initiated by B*

Source: Bueno de Mesquita and Lalman 1994.

ASSUMPTION 1. *The players choose the strategy with the greatest expected utility given that they are playing sub-game-perfect strategies.*

ASSUMPTION 2. *The ultimate change in welfare resulting from a war or from negotiations is not known with certainty. Hence, arriving at a war node or at negotiations yields an expected value, assessed according to the subjective probabilities of gaining welfare and the subjective probabilities of losing welfare. We restrict the probabilities in such lotteries: $0 < p < 1.0$. All probabilities are treated as subjective unless stated otherwise.*

ASSUMPTION 3. *In contrast to assumption 2, capitulations result in changes in welfare that are certain rather than probabilistic. The probability that the capitulating state loses is 1.0, as is the probability that the challenging state wins its demand.*

ASSUMPTION 4. *All nations prefer to resolve their differences through negotiations rather than through war.*

ASSUMPTION 5. *Measured from the status quo (SQ) are $U^i(\Delta_i)$, the expected gain in utility by successfully obtaining one's demands, and $U^i(\Delta_j)$, the expected loss in utility by acceding to the adversary's demands. The value of these terms is restricted such that: $U^i(\Delta_j) < U^i(SQ) < U^i(\Delta_i)$.*

ASSUMPTION 6. *Each outcome has a set of potential benefits and/or costs appropriately associated with it. We make restrictions on the various costs such that: $\alpha, \tau, \gamma, \varphi > 0$ and $\tau > \alpha$.[2] $\alpha_i(1 - P^i)$ is the expected cost in lost life and property for nation i associated with fighting away from i's home territory; $\tau_i(1 - P^i)$ is the cost in lost life and property that i expects if it fights at home as the target of an attack; $\gamma_i(1 - P^i)$ is the cost in life, property, and lost face or credibility from absorbing a first strike to which the attacked party gives in; and $\varphi_i(P^i)$ is the domestic political cost (separate from life and property) associated with using force rather than diplomacy to try to resolve differences.*

ASSUMPTION 7. *The magnitude of actor i's demand ($U^i(\Delta_i)$), if any, is determined by the domestic political process in nation i.*

A broader discussion of these assumptions is contained in Bueno de Mesquita and Lalman 1992, along with an alternative perspective. Table 15.2 summarizes the restriction on preference orderings implied by these assumptions.

With the game structure now established, I turn to a comparative static analysis to evaluate the game's implications for power transition arguments. In doing so, I impose the assumptions of power transition theory on the relevant terms in the international interaction game to see how they influence behavior at the critical juncture during which decision makers may choose between war and peace.

TABLE 15.2. Possible Preference Rankings for Nation i on the Outcomes of the Game

Outcomes	Ordinal Restrictions on Orderings	Possible Positions in Preference Order
SQ	$> \text{Acq}_i, \text{Cap}_i$	7–3
Acq_j	$>$ all other outcomes	8
Acq_i	$> \text{Cap}_i$	5–2
Nego	$> \text{Acq}_i, \text{Cap}_i, \text{War}_i, \text{War}_j$	7–5
Cap_j	$> \text{War}_i, \text{War}_j$	7–3
War_i	$> \text{War}_j$	5–2
Cap_i	None	4–1
War_j	None	4–1

Source: Bueno de Mesquita and Lalman 1994.
Note: 8 is the highest ranking and 1 is the lowest.

The Role of Uncertainty about Power in System-Transforming Wars: A Theoretical Appraisal

According to many theorists who contend that power preponderance encourages peace and that power parity exacerbates the risk of war, a critical issue is whether the rivals are about equal in power or not. Yet, the game implies—and the evidence supports—the contention that wars expected to lead to many deaths and much destruction and also expected to engender little domestic opposition can occur only if each side *believes* its probability of success is greater than 0.5 (Bueno de Mesquita and Lalman 1992). That this is so is readily seen by a comparative static analysis that imposes key power transition conditions onto the game structure.

In the game, the difference in power—the probability of success—of the rivals turns out *not* to be particularly critical to decisions to use force or not. What is important is whether the expectation of success is large enough (or the relevant expected costs are small enough) that the rivals each believe the other will (or will not) retaliate effectively if attacked. In particular, if nation B does not think A has the wherewithal to mount a meaningful counterattack and if B prefers to force a capitulation by A rather than compromise with the challenger, then B would exploit A if A ceded the initiative to B. If A believes that B is the type who will exploit a first-strike advantage and if A believes its own first-strike advantage is large enough (either because it thinks B would capitulate if attacked or because it thinks it can do better by initiating a war than by giving in to its opponent), then A would start a war rather than risk exploitation or having to fight a defensive, retaliatory war. According to the game, the central source of uncertainty concerns whether the adversary is *believed* to prefer to retaliate if coerced.

Those known to bear high domestic political costs are less likely to retaliate and therefore more likely to initiate war than are those for whom a prospective war is relatively popular. The very popularity of a prospective war can facilitate reaching a negotiated settlement. One possible source of uncertainty in choosing between violence and peace, then, can be the level of domestic opposition to the use of force. Power transition theory is not concerned with this domestic political cost. Consequently, I set its value to zero in the comparative static analysis.

Another source of potential uncertainty, more in line with the concerns of power preponderance theorists, is the likelihood that one or the other side will succeed if a dispute turns violent. Indeed, this probability—which is assumed to be a function of the relative power of the challenger and the dominant state—is at the core of the power transition. In keeping the discussion within the power transition perspective, I treat this source of uncertainty as a function of the *difference* in the hegemon's (or the challenger's) beliefs about the chance for success.

Suppose, for instance, that actor A possesses private information that encourages its leadership to believe that it can wage a successful war even if B strikes first. Suppose that B, the dominant state, is uncertain of A's relative power or prospect for success in a violent dispute with B. This difficulty in calibrating accurately the challenger's power might arise, as hypothesized by Organski and Kugler (1980), Gilpin (1981), Thompson (1988), and others, during periods of rapid growth by the challenger. Alternatively, the difficulty in estimating one's chances for success might be the consequence of some combination of differences in rates of growth and of shifts in alliances, as argued by Morrow (1991), Kim and Morrow (1992), Iusi-Scarborough and Bueno de Mesquita (1988), and others. For current purposes it does not matter whether differential growth rates or alliance politics is the fundamental source of changing prospects of success; rather, what matters is that there is uncertainty regarding the relative power of the contending sides.

Assume that B believes with probability π^B that its chances to succeed in a war with A are large enough that A would prefer to capitulate to B rather than retaliate if attacked. Suppose, then, with probability $1 - \pi^B$, B thinks its prospects of success are not so great—A has closed the power gap—so that B anticipates a costly retaliatory strike if it attacks A. The problem for B is that it does not know whether A believes its chance for success is small enough that A would capitulate if attacked or whether A would strike back, yielding a war. A, of course, does know this about itself and so can form a policy over whether to initiate violence; propose a peaceful, negotiated resolution of its differences with B; risk exploitation; or prepare to fight a defensive war.

If A offers to negotiate with B, then B must choose between trying to exploit A, while risking a war, or negotiating with A, while possibly forgoing the

benefits of exploitation. If B decides to respond to A's offer to negotiate by using force, then with probability π^B, B gains $U^B(\Delta_B - \varphi_B(P^B))$—the value for
B of a capitulation by A—and with probability $1 - \pi^B$, B ends up starting a war
with A. That gamble is worthwhile for B so long as the expected utility for B is
superior to entering into negotiations with A. If the dominant state thinks it can
hold on to its preeminent position then it is unlikely to bargain away its control
over the international system. It would prefer to risk war. If, however, it believes
that it cannot hold on to its dominant status, then negotiation is a more attractive means of managing its decline.

The determination of the worth of using force versus negotiating depends
critically upon B's uncertainty regarding its own chances for success. Let P^{*B}
be B's subjectively estimated probability of success if A, the challenger, is believed to have grown sufficiently powerful that A would retaliate (effectively)
if attacked. That is, P^{*B} is B's probability of success with probability $1 - \pi^B$,
while P^B is B's estimate of its chance for success if A is the weaker type that
capitulates when attacked, a prospect believed to arise with probability $\pi^B \cdot P^B$
$\geq P^{*B}$. In the event that $P^B = P^{*B}$ then there is a power transition in the terms
of the international interaction game.

If B mistakenly tries to exploit A only to find itself at war with A, its expected value from the war is:

$$P^{*B} U^B(\Delta_B - \alpha_B(1 - P^{*B}) - \varphi_B(P^{*B}))$$

$$+ (1 - P^{*B})U^B(\Delta_A - \alpha_B(1 - P^{*B}) - \varphi_B(P^{*B}));$$

a value that arises with probability $1 - \pi^B$. If B chooses to negotiate with A
rather than trying to coerce its adversary, B can expect:

$$\pi^B [P^B U^B(\Delta_B) + (1 - P^B)U^B(\Delta_A)]$$

$$+ (1 - \pi^B) [P^{*B} U^B(\Delta_B) + (1 - P^{*B})U^B(\Delta_A)].$$

The value associated with negotiating is also dependent on the uncertainty. If B has the larger threat implied by a likelihood of success as large as P^B
when $P^B > P^{*B}$, then it can expect to cut a better bargain at the negotiating table
than can be anticipated if its chances for success at war are only as large as P^{*B}.
That is, as B's true chance for success approaches P^B, negotiation and war become more attractive for B.

What difference does the gap make between B's possibly higher prospect
for success (P^B) and B's possibly lower prospect for success (P^{*B})? To answer
this question—which is truly at the heart of power transition theory—I restate
the above expectations by solving for π^B. If:

$$\pi^B > \frac{(1 - P^{B*})\alpha_B + \varphi_B P^{*B}}{(1 - P^B)U^B(\Delta_B - \Delta_A) + (1 - P^{*B})\alpha_B - \varphi_B(P^B - P^{*B})} \tag{1}$$

then B attacks A in the hope of coercing it into a capitulation. If π^B were smaller than the right-hand side of equation (1), then B would pursue negotiations, leading to a peaceful transition.

Expression (1) reveals several important features of prospective power transition wars and illuminates some aspects of the success enjoyed by Organski and Kugler and others (Garnham 1976a; Weede 1976; Houweling and Siccama 1988a; Kim 1989) in their empirical investigations of this theory.

First, a peaceful transition occurs through negotiation, provided that π^B is less than the right-hand side of expression (1). If we assume power transition conditions (including that $P^B = P^{*B}$), then π^B is likely to be less than the right-hand side of expression (1) as A and B become more satisfied with each other's world view. In the limit, if B's utility for A's set of demands approaches in value its utility for its own demands about the rules of the international system (so that A and B are both members of the satisfied coalition), then the right-hand side of expression (1) approaches infinity, insuring that π^B will be smaller. This is consistent with Organski and Kugler's notion of a peaceful transition. The challenger is satisfied with the rules of the dominant state and so does not have a reason to fight. Presumably, this was true between the United States and the United Kingdom when the latter was passed in power by the former.

Second, satisfaction is sufficient for a peaceful transition, but it is not necessary. Such a transition can also arise if the expected costs in lost life and property are large relative to the confidence B has that A is the type who can be exploited. So, growth in A's power can itself be sufficient to provoke a peaceful transition, provided the costs are expected to be large enough. This comparative static result seems to contradict the classical power transition view with respect to circumstances that can precipitate a peaceful transition. After all, power transition wars are expected to be extremely costly and the challenger is presumed to have about equaled or overtaken the dominant state in power.

Third, notice that there *is* a critical power transition but that it is not necessarily related to whether A's power equals and is overtaking B's power, as contended by Organski and Kugler. Rather, the critical transition in the game structure occurs when B believes that A has achieved a chance for success large enough that A is expected to retaliate effectively if attacked.

Just before that transition, A is likely to launch a preemptive attack against B to secure whatever first-strike advantage there is. Beyond that threshold, B will negotiate with A, setting the stage for a peaceful transition. Overall, A might pursue a purely preemptive strategy, a purely negotiating strategy, or a mixed strategy, depending upon the particular values A attaches to the lottery over fighting a retaliatory war or negotiating versus the value of initiating the war (presuming that B is known to be the retaliatory type). The key is that B's belief threshold is *not* critically dependent on the difference between A's power and B's power, at least not in the international interaction game. Rather, the belief threshold is a complex mix of B's estimate of relevant costs, benefits, and risks.

Where, then, does the power transition theory's belief that the two rivals must be about equal in power come from? To answer this question, I continue with a comparative static analysis in which power transition conditions are imposed on the international interaction game. In particular, suppose $P^B = P^{*B}$. Of course, this can be true when P^B equals any value in the interval $(0,1)$, including, but not limited to $P^B = P^{*B} = P^A = 0.5$, which is the specific value assumed by Organski and Kugler. Suppose for the moment that large expected losses in life and property are a necessary condition for a system-transforming war, in keeping with the assumptions of Organski and Kugler (1980) and Gilpin (1981). In particular, let $\alpha_B = U^B(\Delta_B - \Delta_A)$ in expression (1) and let each of these be very large compared to anticipated domestic costs (so that φ_B approximately equals zero). Then, for a large, apparently popular war (in that there are few domestic costs connected to waging the war), in which B is fighting to retain control over the international system and A is challenging that control, $\pi^B \geq 0.5$ in order for B to try to exploit A. If A is the type who would retaliate if attacked and A has gambled that B will negotiate rather than try to exploit nation A, then A, the challenger, will wind up fighting a war begun by the hegemon. If A does not gamble on B's willingness to negotiate, then A will launch the war. This latter circumstance can arise, of course, whether the challenger is the type who would capitulate or retaliate if attacked. So, apparently, under conditions assumed by Organski and Kugler (1980), the critical *belief* threshold is at 0.5, although the power or probability-of-success threshold need not be. In fact, the probability of success, per se, is not germane.

To show that the power or probability level at which the transition takes place does not matter, let $P^B = P^{*B}$ and let the domestic cost terms in expression (1) equal zero. Then, P^B is exactly at the point of transition in expectations that could foment a power transition war. How does the belief threshold that leads to a choice between war or negotiations vary with respect to P^B under these conditions? The answer is that:

$$\frac{\partial \pi^B}{\partial P^B} = 0 \tag{2}$$

This means, of course, that the belief threshold that implies peace or war does not vary with respect to the dominant state's probability of victory. The same is true, by extension, to the relationship between the belief threshold and the challenger's chance of success. The magnitude of P^B does not systematically alter the chance of war, but the magnitude of the belief about the rival's ability to retaliate effectively does, under power transition conditions imbedded within the international interaction game.

Propositions and Tests

A crude test of the comparative static implications of the international interaction game for the power transition is possible. The conditions of expression (1) under power transition restrictions can be approximated and evaluated in light of a broad data set on international disputes—including power transition wars—for the years from 1816 to 1974.

To test expression (1), I define the following variables:

$$\text{EQUAL} = 1 - |\text{Power of } A - \text{Power of } B|$$
$$\text{EQUAL}_A = 1 - |0.5 - P_A|$$
$$\text{EQUAL}_B = 1 - |0.5 - P_B|$$
$$\text{DISSAT}_A = U^A(\Delta_A - \Delta_B)$$
$$\text{DISSAT}_B = U^B(\Delta_B - \Delta_A)$$
$$\text{GAME} = 1/(\text{DISSAT}_B + 1)$$
PTWAR = A and B are both major powers and each used force against the other in a dispute.

The power of A (or of B) is defined in accordance with the composite capabilities index developed by the Correlates of War Project, so that EQUAL captures the Organski and Kugler notion of equality in power between rivals. In power transition theory this condition is viewed as a necessary (but not sufficient) condition for a power transition war.

P_A and P_B are A's and B's subjective estimates of their probability of success vis-à-vis the other, taking into account the individual capabilities of A and B and also taking into account their expectations about support or opposition from third parties. These two terms are estimated as in Bueno de Mesquita and Lalman 1992. EQUAL_A and EQUAL_B are, then, measures of the perceived degree of equality in available or expected power in a prospective dispute between nations A and B. As such, these are alternative, subjective estimates of the core power-transition concept of equality of power. As with the variable EQUAL, the larger these terms are, the more likely a power transition war, according to Organski and Kugler (1980).

DISSAT_A and DISSAT_B are designed to capture the perceived level of dissatisfaction of the rival camps represented by nations A and B. It is based on Bueno de Mesquita and Lalman's (1992) estimation of the utility each side attaches to its own set of policy demands and the demands it has received from its opponent. The larger the dissatisfaction variable is, the more dissatisfied the relevant actor is with the relationship it has with its prospective opponent. Organski and Kugler hypothesize that dissatisfaction is a necessary ingredient for a power transition war. As will be evident in the logit analysis reported below, $EQUAL_A$ is multiplied by $DISSAT_A$ (and likewise for the comparable terms for nation B) to capture the interaction between dissatisfaction and power equality. The expectation from power transition theory is that when rivals are about equal in power and the challenger is dissatisfied with the dominant state's policies, then the risk of a power transition war is heightened.

GAME is a crude approximation of the conditions stipulated on the right-hand side of expression (1) from the international interaction game. Because I do not have an *ex ante* measure of *expected* costs (as denoted in expression 1), I construct the game variable under the assumption that the specific impact of costs is constant and that the impact of the stakes in the dispute ($U_B(\Delta_B - \Delta_A)$) is as indicated by expression (1). The game variable is constructed under the comparative static assumption that $P^B = P^{*B}$, so that it is evaluated exactly at the point of transition. The larger this variable is, the greater the belief threshold must be that must be satisfied in order for war to be waged. Consequently, the larger this variable is, the less likely it is that there will be a war. The main hypothesis I wish to test, then, is that this variable is inversely related to the likelihood of a power transition war under power transition constraints.

The dependent variable, PTWAR, is operationalized strictly on *ex ante* criteria. Unlike Organski and Kugler, I do not make any assumptions about the actual costs in a war in designating it as a power transition conflict or not. The reason for avoiding their battle death criterion (Organski and Kugler 1980:46) is that this can only be known *ex post* and so risks sampling on the dependent variable. I do, however, implement their *ex ante* criteria; namely, that the rivals are major powers and that the stakes are so high as to "insure that the contestants were really trying to win" (1980:46). Thus, in the test I present, I look only at disputes in which the *ex ante* demands (measured as the stakes implied by the demands—the dissatisfaction variables) exceeded the median for demands among the 707 dyadic observations I have for relations among European states between 1816 and 1974.

The hypotheses, then, are:

H1. EQUAL is positively associated with PTWAR.
H2. $EQUAL_A * DISSAT_A$ and/or $EQUAL_B * DISSAT_B$ are positively associated with PTWAR.
H3. $EQUAL_A$ and $EQUAL_B$ are positively associated with PTWAR.
H4. The coefficient for $EQUAL_i * DISSAT_i$ is greater than the coefficient for $EQUAL_i$ by itself.
GAME: GAME is negatively associated with PTWAR.

Before proceeding with the test, I list, in table 15.3, the conflicts that have been categorized, *ex ante,* as power transition wars. It will be evident that I have included a broader set of events than Organski and Kugler would deem appropriate. I include all cases of the reciprocal use of force—even low levels of force, as in the Berlin crises of 1948 and 1961—between great powers to reflect my own evaluation that system transformations can occur (1) after cataclysmic wars like World War II; (2) peacefully, as with the U.S.-U.K. transition; or (3) through intermediate-sized wars, like the Seven Weeks' War (Bueno de Mesquita 1990a). Since Organski and Kugler appear to be theoretically motivated by a desire to explain system transformation, I include in the operationalization of the dependent variable disputes with the *ex ante* potential to be system transforming by their criteria, whether, in fact, the event turned out, *ex post,* to be of that magnitude or not.

The dependent variable focuses on violent disputes. Power transition the-

TABLE 15.3.   Power Transition Wars

Year	Challenger	Target	Dispute
1850	AUH	GMY	Dispute over Hesse
1853	RUS	FRN	Crimean War
1853	RUS	UK	Crimean War
1859	AUH	FRN	War of Italian Unification
1866	ITA	AUH	Seven Weeks' War
1866	GMY	AUH	Seven Weeks' War
1870	FRN	GMY	Franco-Prussian War
1888	RUS	FRN	
1897	FRN	UK	Clash over Fashoda
1914	GMY	FRN	World War I
1914	GMY	UK	World War I
1914	AUH	RUS	World War I
1914	GMY	RUS	World War I
1914	AUH	FRN	World War I
1914	AUH	UK	World War I
1937	ITA	UK	Abyssinian War
1939	UK	GMY	World War II
1939	FRN	GMY	World War II
1940	UK	ITA	World War II
1940	FRN	ITA	World War II
1948	RUS	UK	Berlin Airlift
1948	RUS	USA	Berlin Airlift
1948	RUS	FRN	Berlin Airlift
1961	UK	RUS	Berlin Crisis
1961	USA	RUS	Berlin Crisis
1961	FRN	RUS	Berlin Crisis

orists are interested in whether the above set of variables can sort out those events from other disputes when it is known in advance that the conflict involves high stakes. Consequently, I test the hypotheses against the subset of all European disputes expected to involve *high* stakes across the period from 1816 to 1974. The results of the logit analysis are reported in table 15.4.

The logit analysis focuses on circumstances that, *ex ante,* were clear candidates for power transition wars. In these 168 disputes, the challenger held an above average level of manifest dissatisfaction as estimated using the methods developed by Bueno de Mesquita and Lalman (1992). The results strongly support the version of the power transition theory that is imbedded within the game-theoretic structure. The coefficient for the GAME variable is extremely significant and in the predicted direction. The other variables, which do not involve the game structure as an added, comparative static feature to the power transition theory, do not fare nearly as well.

EQUAL has the right sign, but it is not significant, calling into question the importance attached to the relative power of the principal belligerents. The two terms that represent the interaction between perceived equality in power and dissatisfaction have the *wrong* sign, although their net effect is in the predicted direction (i.e., $EQUAL_A - EQUAL_A * DISSAT_A$ *is* positive). That is, perceived power equality combined with dissatisfaction raises the risk of a power transition war (as predicted by hypothesis 2), but not as much as perceived equality in the absence of dissatisfaction (in contradiction of hypothesis 4).

TABLE 15.4. **Conditions Preceding Power Transition Wars High Stakes Disputes**

	Coef.	Std. Err.	Prob.
Constant	48.62	24.14	0.05
EQUAL	2.58	5.54	0.64
$EQUAL_A$	21.44	19.90	0.27
$EQUAL_A * DISSAT$	10.45	8.30	0.21
$EQUAL_B$	55.68	19.80	0.01
$EQUAL_B * DISSAT$	−21.21	7.74	0.01
$DISSAT_A$	6.94	6.08	0.26
GAME	−238.21	79.14	0.00
$N =$		168	
$\chi^2 =$		20.38	
Probability $<$		0.01	

Note: The correlation between $DISSAT_A$ and $DISSAT_B$ is .97, so I have dropped $DISSAT_B$ from the analysis.

The impact of perceived power equality is much greater on the part of the defender than it is on the part of the challenger. Although the direction of the relationship is the same in both cases, only the coefficients for the defender are statistically significant. This is consistent with other expectations derived from the structure of the international interaction game. Since a war is defined as reciprocated violence, it is clear that if the defender, B, does not fight back, then there cannot be a war, although there can be a capitulation. This result represents mixed support for hypothesis 3.

Conclusion

Power transition comparative static conditions were imposed on the game of international interactions. The game-theoretic analysis implied modifications of standard power transition hypotheses. The modified hypotheses were tested, as well as more standard power transition expectations, against a large data base. The results strongly support the game-theoretic modifications and less strongly reinforce standard power transition views.

NOTES

1. Bueno de Mesquita and Lalman (1992) propose two variants of the unitary actor assumption. The one used here corresponds to the domestic variant.

2. It may facilitate remembering what each cost term, α, τ, γ, φ refers to by using the following mnemonic device: α is the cost borne by the Attacker for fighting Away from home in a war; τ is the cost borne by the Target in a war; γ is the cost borne by a state that Gives in after being attacked; and φ is the domestic political cost associated with the use of Force.

CHAPTER 16

The Conditions and Consequences
of Dyadic Power Transitions:
Deductions from a Dynamic Model

Kelly M. Kadera

Power transitions are intriguing phenomena that occur between two competing nations. In this paper, I propose that they are the result of a complex interactive process in which national power growth and dyadic conflict behavior are interdependent. Previously, this process has been explored by two logics that explain how the power[1] distribution between two competing nations affects their propensity to go to war. Scholars favoring the balance-of-power theory argue that approximate parity brings peace. Another group, favoring the power transition theory, believes power parity is a war-prone condition. Empirical analysis testing these two explanations has produced mixed results (noted below). While my goal is not to settle the balance of power–power transition debate, I do use it as a means of suggesting key variables in the power–conflict story and the ways in which these variables might be related. The interactive process can then be expressed in a formal, dynamic model. Once the model is complete, I use it to answer two important questions about power transitions. What types of patterns in national power growth lead to and result from power transitions? Can we expect power transitions to be highly conflictual in nature?

The Balance-of-Power Explanation

The balance-of-power explanation in explicitly developed form evolves in bits and pieces, claiming roots in the classical balance-of-power theory. In the modern era, writers such as Morgenthau (1985) and Dougherty and Pfaltzgraff (1981) have provided synthesized, complete versions of the balance-of-power theory.[2]

Scholars of this persuasion (e.g., Morgenthau 1985; Claude 1962; Kissinger 1979) argue that an approximately equal distribution of power across a system of nations tends to produce a peaceful equilibrium. Extension of this argument from a system of nations to a dyad has been achieved in the deterrence literature (see, for example, Snyder 1961; George and Smoke 1974). Here, the view is that two approximately equal rivals will deter each other from initiating an attack, making war unlikely. Justification for moving from systemic to dyadic analysis is presented by Siverson and Sullivan (1983:474), who claim that "the underlying rationale for the hypotheses at both levels is similar enough—at least in terms of general theory building—that it would be advantageous to compare them." Additionally, they concur with Singer's (1980:359) proposition that most system properties come from characteristics of its parts. Lastly, it may simply be the case, as it is here, that we are interested in the behavior of a dyad as opposed to the behavior of a system of nations.

The balance-of-power scholars reason that when the power distribution is roughly equal, each alliance's or nation's power serves as a "check and balance" against aggression by the opponent(s). But if the balancing mechanism breaks down and one alliance or nation becomes preponderant, war is more likely. Thus, preventing an opponent from gaining an advantage is crucial to survival, because a preponderant nation will attack its competitor (Waltz 1959:232). We might liken this to the *para bellum* adage: if you want peace, prepare for war.

According to this logic, war may occur in one of two ways. First, the reason balance is essential for peace is that by nature, nations are aggressive and seek to maximize power. Nations with unchecked power are likely to exercise these aggressive tendencies, potentially by initiation of war with weaker nations. The logic of the balancing mechanisms certainly does not exclude, however, the possibility that war is initiated by the weaker state(s) in an effort to return the power distribution to equilibrium. If other balancing efforts fail to prevent domination or are insufficient, lower-ranked nations may possess no alternative to war.

Support for the balance-of-power explanation is found in Ferris's (1973:117) work. This investigation led to the conclusion that "more wars were found to be characterized by change in the direction of a greater inequality in the power capabilities relationship for three of the five time lags examined."

The key propositions of the balance-of-power explanation that I have isolated for this analysis are: (1) under relatively equal distributions of power war is less likely, (2) under unequal distributions of power war is more likely, and (3) the stronger, or preponderant, nation is most likely to initiate war.

The Power Transition Explanation

According to power transition theory, a typical nation experiences three stages of development: (1) potential power, (2) transitional growth in power, and (3) power maturity (Organski 1958:340). We can think of these stages as descriptive of portions of an S-shaped curve that measures a nation's power across time. Initially, a nation's power grows gradually while its economy and political system are still underdeveloped. In the transitional stage, a nation experiences rapid growth as its economy industrializes and as its government bureaucracy expands. Finally, the growth levels off and possibly declines vis-à-vis other nations experiencing rapid accumulation of power in their second stage of development.

Because all nations do not develop simultaneously, at any given time there will be a mix of nations at various points in their stages of development. The power transition explanation is principally concerned with nations that overtake each other during their more developed stages.[3] Some of these nations will not be "satisfied with the way the international order functions and the leadership of the dominant nation" (Kugler and Organski 1989b:173), and as their power levels approach that of the dominant nation, they will be increasingly able to act on this dissatisfaction. It is at this time, when a dominant nation is being overtaken by a dissatisfied powerful nation, that war is most likely. The overtaking nation initiates such a war because it "anticipates greater benefits and privileges if a conflict is successfully waged than if the current status quo is preserved" (Kugler and Organski 1989b:175). War is least likely when the dominant nation's power far exceeds that of the others because the dominant nation has no reason to initiate a war, and because the potential challengers do not possess enough power to initiate war, assuring their probability of success is low.

The War Ledger supports the overtaking aspects of Organski's original notions (1958). However, concerned with the limitations of Organski and Kugler's (1980) empirical investigation, Houweling and Siccama expanded the analysis to thirty-six dyads between the three (or four) strongest nations for eight test periods of twenty years length. Claiming more confidence in their own work than in the few-case analysis of Organski and Kugler, Houweling and Siccama (1988a:101) likewise conclude that "differential growth rates and specifically power transitions among great powers are indeed a potent predictor of consecutive outbreak of war."

In sum, the key propositions of the power transition to be explored here dynamically are: (1) war is less likely when one nation's power is clearly dominant, (2) war is more likely when another nation's power threatens to overtake that of the previously dominant nation, and (3) the weaker nation is the likely war initiator.

A Dynamic Model of Power and Conflict

Both the balance-of-power and the power transition explanations of war are reasonable claims and enjoy some empirical support. I will attempt to clarify the relationship between power distributions and conflict, not by settling the balance of power–power transition debate, but by drawing on the debate in order to formulate an alternative dynamic explanation. The alternative proposed in this section draws both from the balance-of-power and the power transition explanations in several important aspects.

First, my alternative explanation is expressed as a formal model. Both the balance-of-power and power transition explanations are "logics-in-use" (Kaplan 1964:8–11). They are rich and meaningful stories about why nations go to war, but neither demonstrates formal rigor. In order to design a formal model, I extract key variables and the relationships that might exist between these variables from the balance of power and power transition literatures. The arguments made by Morgenthau (1985), Organski (1958) and Organski and Kugler (1980) are then mapped into a system of mathematical statements. In other words, I draw on the "logics-in-use" found in the balance-of-power and power transition literatures in order to design the "reconstructed logic" of a formal model (Kaplan 1964:8–11). The advantage of using a reconstructed logic rests on the explicit statement of the original logic's assumptions and structure. The symbolic expression of arguments makes obvious any inconsistencies. This allows the modeler to insure a logically sound deductive system that eliminates inconsistencies in a manner most complimentary to the original logic-in-use.

Second, because I use arguments from both explanations of war, the formal model is a hybrid that incorporates arguments from both of the explanations of war. One advantage of building a hybrid reconstructed logic lies in the modeler's ability to choose which arguments from each logic-in-use make sense, and under which conditions, or at which times. Some evidence of the need for a hybrid explanation already exists. Mansfield came to such a conclusion when investigating the power distribution–war behavior relationship at the international system level (1992:21). Under the condition of relatively low capability concentration, the balance-of-power explanation was supported, while under the condition of relatively high capability concentration, the power transition explanation was supported. When incorporating economic and political constraints into a dyadic rivalry model of conflict, Wolfson, Puri, and Martelli (1992:147) suggested how it was "possible for both approaches to apply in different periods of time." Where logical gaps appear or when inconsistencies arise, the modeler can make choices that adjust the original structure.

Another advantage of constructing a hybrid logic is found in the modeler's opportunity to expand the original logics-in-use to their useful extremes. As they are currently offered, the balance-of-power and power transition explanations address the outbreak of war, not the levels of conflict sent between rivals. In addition, comparisons of these explanations, especially those of an empirical nature, tend to focus on the conditions for the outbreak of war, that is, unequal power levels or equal power levels coupled with unequal growth rates. Of course, more logical possibilities exist. In the model-building section, I draw logical extensions from arguments about the outbreak of war to form arguments about increasing and decreasing amounts of dyadic conflict. The hybrid model is therefore more general than either the balance-of-power or power transition explanations.

Furthermore, the modeler may alternatively distill from elaborate logics-in-use the least complex, most parsimonious arguments possible. While the power transition explanation of war is clearly a treatment of dyadic war between a dominant nation and a single challenger, the balance-of-power explanation is sometimes a systemic treatment and sometimes a dyadic one. The hybrid model presented here relies on strictly dyadic comparisons.

Last, the hybrid model provides a dynamic picture of dyadic interactions. By focusing on the conditions for the outbreak of war, comparisons of the two explanations have presented a largely static version of each.[4] The stories told by Morgenthau (1985) and by Organski (1958) and Organski and Kugler (1980) offer a much richer view of national power growth, dyadic conflict, and the interactions between the two. If we ignore these stories, the world of rivalries seems static and boring. If we capture them in a general, dynamic model, the world of rivalries becomes the interesting place it is. In this chapter, I argue that it is therefore more useful to begin with a generalized picture of power growth and conflict behavior. We need to speak of levels of power and conflict, of what causes them to change over time, and of how these two variables might interact.

Power Growth and Decay Equations

I begin with the understanding that power represents the ability of one nation to control another nation's behavior, consistent with views held by both balance-of-power and power transition scholars. In the tradition of the balance-of-power school, Morgenthau defines it as "man's control over the minds and actions of other men" (1985:32). Organski and Kugler, proponents of the power transition school, see national power as the "ability of one nation to control the behavior of another for its own end" (1980:5). These definitions share the idea of *control over behavior.*

Next, I propose that this ability to control others can be stored up and depleted. Again, some support for this proposition can be found in both of the bodies of literature. Morgenthau argues that power is based on "physical" threats, both realized and anticipated (1985:33). The term *physical* implies certain aspects of power are concrete. The idea that power can be both *realized* and *anticipated* implies that the ability to make threats may either be acted upon or set aside for possible future use. Troops, for instance, may be deployed or kept at home. As examples of national power, the power transition authors mention the use of persuasion, fear, and force (Organski and Kugler 1980:5–7). The use of persuasion and fear are instances in which abilities are stored. The nation attempting to control another does not directly call upon its tangible resources. The rival nation, however, is painfully aware of the resources available to its opponent and is fearful nonetheless. The use of force, on the other hand, is an instance in which the nation attempting to control another actually uses its available resources.

Hence, power can be thought of as an accumulated ability to use threats, persuasion, fear, and force in order to control others. In this sense, power is a national capacity. Now, I ask what factors affect a nation's propensity to accumulate this ability and then consider what factors cause a nation to use up or spend this ability. Members of the power transition school treat the compelling forces of power growth as largely *internal*. Here, national power is a reflection of economic and technological development and the concomitant social and political progress of nations (Organski 1958:5). In other words, power accumulation is internally driven by natural tendencies toward growth and modernization. Based on this view, power accumulation can be seen as the result of a natural growth pattern in which power begets power. Nation X's power growth, dp_x/dt, is a positive function of its own power level. One way to simply express this relationship is:

$$dp_x/dt = \alpha p_x,$$ (1)

where α is a positive constant.

An understanding of how a nation spends its ability to control others can be found in the balance-of-power literature. According to Morgenthau, rival nations constantly struggle to check and balance each other's power levels (1985:192). One way to do this is to decrease the opponent's power level.[5] What methods are available to a rival attempting to decrease its opponent's power level? I assume that conflict is the means by which nations act.[6] The use of conflict is successful in depleting the opponent's power supply because it either directly destroys some of the opponent's stockpiled abilities or it requires the opponent to use some of its stockpile in order to respond in some fashion. Perhaps the targeted nation will suffer a decline in its gross national product after the erection of trade barriers or perhaps it will need to pour resources into the design and deployment of an elaborate antimissile network. This view depicts a nation's power level as *externally* affected. Thus, nation X's power decay is governed by nation Y's conflictual behavior toward nation X, c_{yx}, which is aimed at decreasing nation X's power level, p_x, and to which nation X must respond. This relationship might be formally written as:

$$dp_x/dt = -\beta c_{yx},$$ (2)

where β is a positive constant.

I would like to modify the $-\beta c_{yx}$ term so that the impact of nation Y's conflict, c_{yx}, on nation X's power decay is tempered by nation Y's level of power, p_y. This modification means that nation Y's conflict is more effective, or requires more of a response, when nation Y is strong. For example, Japanese trade restrictions on American goods would have more impact on the United States economy than would Argentinean trade restrictions. Two competing nations' growth in power can therefore be represented by two parallel equations:

$$dp_x/dt = \alpha_x p_x - \beta_x p_y c_{yx} \tag{3}$$

$$dp_y/dt = \alpha_y p_y - \beta_y p_x c_{xy} \tag{4}$$

where:

p_x is nation X's power level,
p_y is nation Y's power level,
c_{yx} is the conflict nation Y directs at nation X,
c_{xy} is the conflict nation X directs at nation Y, and
α_x , α_y ,β_x , and β_y are positive constants.

The first term in each equation indicates that the more powerful a nation is, the more capable it is of accumulating additional power. This is consistent with Morgenthau's rule that nations constantly strive to increase power.

The second term in each equation represents the competition that rival nations experience. This term can be thought of as the cost associated with responding to an opponent's conflict, or the cost of competition. One rival exercises its power, that is, directs conflict, with the intention of decreasing the other rival's power. A similar type of analysis of the original balance-of-power and power transition logics is now used in order to develop equations explaining each rival's conflictual behavior.

Base Conflict Equations

By conflict, I mean any hostile behavior initiated by a specific nation, the actor, and intended for a specific nation, the target. This need not be violent. We can think of conflict as ranging anywhere from verbal insults to initiation of war.[7] War, then, is indicated by a very high critical value of conflict, which I have not yet determined. In this sense, the general model is not directly a model of war initiation.

I propose that conflict may be additionally thought of as the flexing of a nation's power. The more power a nation has at its disposal, the more it is able to actually use. In a competitive dyadic rivalry, the use of national power is manifested in conflictual behavior. It can therefore be said that nation X is able to direct increasing amounts of conflict toward nation Y as it is increasingly powerful.[8] Because a nation's ability to act conflictually is dependent on its level of power, that is, $c_{xy}(t) = f(p_x)$, the two concepts of power and conflict are intimately linked. Thus, I begin building the conflict equations with the following:

$$dc_{xy}/dt = \gamma_x p_x \tag{5}$$

$$dc_{yx}/dt = \gamma_y p_y \tag{6}$$

where γ_x and γ_y are positive constants.

If national power enables rivals to act conflictually, what makes them motivated to act more or less conflictually? I start by discussing a fundamental force that motivates a nation to direct increasing amounts of conflict toward its opponent. According to Morgenthau (1985:192–94), in a balance of power setting, rival nations act to counter each other using a "pattern of direct opposition." If conflict is the means by which nations oppose one another, then we can take this to mean that nation X is increasingly conflictual toward nation Y because nation Y is increasingly conflictual with nation X.

Incorporating the motivating force into equations 5 and 6 produces this pair of equations:

$$dc_{xy}/dt = \gamma_x p_x c_{yx} \tag{7}$$

$$dc_{yx}/dt = \gamma_y p_y c_{xy} \tag{8}$$

These equations might be compared to the action-reaction terms in Richardson's (1960b) arms race model, where the increase in one nation's armaments causes an increase in the opponent's armaments. Equations 7 and 8, however, modify the action-reaction component of conflict with the effect of national power. Furthermore, note that when either nation X's power level is zero ($p_x = 0$ because nation X no longer exists) or Y ceases hostilities ($c_{yx} = 0$), then X's conflict toward Y will no longer increase ($dc_{xy}/dt = 0$). This captures the idea that both the ability to act and the motivation to act are necessary in order for a nation to become increasingly conflictual.

The conflict equations expressed in (7) and (8) capture the primary forces governing conflict behavior. The ability and motivation to act conflictually endure regardless of the particular condition experienced by a rivalry. I previously argued, however, that one of the advantages of my model is its capacity to specify certain conditions under which each of the two original logics makes sense. Another way to put this is that the model accounts for different forces coming into play under different conditions. I now move to a specification of those specialized conditions and the forces that play a role under each condition.

Conditional Conflict Equations

I propose that there are three distinct conditions of conflict behavior that can be defined in terms of the power relationship between nation X and nation Y. While all three conditions share the ability and motivation forces as specified in equations 7 and 8, each requires the addition of another term. I first distinguish these three regions, which are depicted in the $p_x - p_y$ plane of figure 16.1, and then move to a specification of the equations that apply under each individual condition.

The three different regions of conflict behavior, based on the relationship between p_x and p_y, are suggested by Morgenthau in the following passage:

> This balancing of opposing forces will go on, the increase in the power of one nation calling forth an at least proportionate increase in the power of the other, [until the nations concerned change the objectives of their imperialistic policies—if they do not give them up altogether—or][9] until one nation gains or believes it has gained a decisive advantage over the other. Then either the weaker yields to the stronger or war decides the issue. (1985:193)

The first condition, then, is defined by the boundary denoting a decisive advantage, $d^* \geq |p_x(t) - p_y(t)|$. I call this first condition the "balance of power" condition because it is here that the rivals' power levels are essentially equal and a balance is maintained.

Beyond this balancing condition, why is it that weak nations sometimes yield but at other times force the strong nation to become more conflictual? We might reason that at a certain point, the difference in power is so great that the weaker sees the absolute futility of competition and yields. Let's call this difference $\hat{d}$, and distinguish it from d^*, the difference representing a "decisive advantage." The second condition, defined by $d^* < |p_x(t) - p_y(t)| \leq \hat{d}$, will be called the prevention and consolidation condition because the weaker nation will continue to battle against the dominant nation in hope of avoiding the third condition while the stronger nation will try to ensure its advantage. The third condition, defined by $|p_x(t) - p_y(t)| \geq \hat{d}$, will be called the submission and dominance condition because the weaker power will have given up all hope of competing with the vastly stronger rival. Thus $|p_x - p_y|$ [10] is in the region defined by $0 < d^* < \hat{d} < +\infty$.

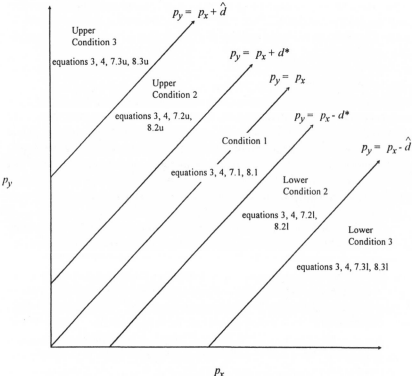

Fig. 16.1. Conditions of conflict behavior.

In sum, the behavior of the power-conflict system of equations should be different in each of the three areas defined by a graph in the $p_y - p_x$ plane, as in figure 16.1. Under condition one, balancing occurs. Under condition two, the weaker nation will fight in an attempt to prevent movement into condition three. If these efforts fail, condition three is reached, and the weaker will have to concede. I will present the conflict equations for each condition individually.

Condition 1. $|p_x - p_y| \leq d^*$: Balance of Power

The first condition is represented by the area between the two lines $p_y = p_x + d^*$ and $p_y = p_x - d^*$ in figure 16.1. The argument that applies to this condition begins with the proposition that when the difference between two rival nations' power levels is small, that is, $|p_x - p_y| \leq d^*$, the balancing mechanism established by equations 3, 4, 7, and 8 is appropriate. Morgenthau

(1985:198–201), for example, argues that when small shifts from equality arise from, say, armaments buildups, there is a tendency for the mechanism to restore parity. In the short term, this process does not produce exploitable advantages for either nation.

When the difference between p_x and p_y grows larger, however, new forces come into play. As the system moves closer to imbalance or a decisive advantage, that is, as the power relationship approaches a line described by either $p_y = p_x + d*$ or $p_y = p_x - d*$ in figure 16.1, the nation hoping to be stronger will want to ensure movement beyond $d*$, and the nation dreading the weaker position will want to avoid this. In this model, both nations act to affect the distribution of power by the only available means, directed conflict. Although the stronger and weaker nations have different motives, both behave similarly. Thus, the following parallel conflict equations will apply under condition 1:

$$dc_{xy}/dt = \gamma_x p_x c_{yx} + \kappa_x / (\epsilon + (d* - |p_x - p_y|)) \tag{7.1}$$

$$dc_{yx}/dt = \gamma_y p_y c_{xy} + \kappa_y / (\epsilon + (d* - |p_x - p_y|)) \tag{8.1}$$

where κ_x and κ_y are positive constants and ϵ is a very small positive constant.[11]

The denominator in the second term indicates how close the discrepancy in power levels, $|p_x - p_y|$, is to a decisive advantage, $d*$. I use the absolute value of the difference between p_x and p_y because the identity of the stronger nation is irrelevant in terms of conflict behavior. As either one of the two nations moves closer to a decisive advantage, the $(d* - |p_x - p_y|)$ term in the denominator grows smaller. As the denominator grows smaller, the entire fraction grows larger and directed conflict increases at a greater rate. When the decisive advantage has been reached, the second term will be κ/ϵ. Since ϵ is a very small constant, κ/ϵ will be very large and conflict will increase at the greatest rate. When the power levels of nations X and Y are perfectly balanced, that is, $|p_x - p_y| = 0$, the denominator will be as large as is possible and the entire fraction will be as small as possible. Even though directed conflict will still increase, it will do so at the smallest rate possible. In other words, the second term, representing the approach to a decisive advantage, plays a negligible role when the difference between power levels is small.

Condition 2. $d* < |p_x - p_y| \le \hat{d}$: Prevention and Consolidation

What kind of arguments apply when the nations' power levels fall in the upper region bound by the lines $p_y = p_x + d*$ and $p_y = p_x + \hat{d}$, or the lower region bound by the lines $p_y = p_x - d^*$ and $p_y = p_x - \hat{d}$ in figure 16.1? In contrast to the first condition, two lines of reasoning are relevant here—one for the weaker nation and one for the stronger. The weaker nation tries to prevent movement beyond the point of futility, $\hat{d}$. It anticipates this when it sees that the opponent's *growth rate is larger than its own*. If nation X is the weaker, this anticipation can be expressed as a function of $(d(p_y - p_x)/dt)$. The weaker nation is most capable of prevention while its own power is large vis-à-vis its rival's power. In other words, conflict for the purpose of prevention has an inverse relationship with the quantity $|p_x - p_y|$. Morgenthau explains:

> Since in a balance-of-power system all nations live in constant fear lest their rivals deprive them, at the first opportune moment, of their power position, all nations have a vital interest in *anticipating* such a development and doing unto others what they do not want the others to do unto them. (Morgenthau 1985:228)

The power transition scholars are not silent on this subject. Recall that their explanation predicts a higher likelihood of war initiation by the weaker, or dissatisfied, rival when two conditions are met: (1) the rivals are in rough parity, that is, $|p_x - p_y|$ is small, and (2) the weaker's growth rate is larger than the stronger's, that is, $p_x < p_y$ and $dp_x/dt > dp_y/dt$. However, it seems unnecessary for the weaker to further increase the conflict directed toward the stronger when its own power is promising to overtake the stronger's. The equations for the weaker nation, then, will be designed so that a rival is increasingly conflictual when its own growth rate is slower than the opponent's and decreasingly conflictual when its own growth rate is greater than the opponent's.

The stronger nation tries to consolidate its advantage through conflictual actions ranging from diplomatic pressure to initiation of war:

> Hence all nations who have gained an apparent edge over their competitors tend to consolidate that advantage and use it for changing the distribution of power permanently in their favor. This can be done through diplomatic pressure by bringing the full weight of that advantage to bear upon the other nations, compelling them to make the concessions that will consolidate the temporary advantage into a permanent superiority. It can also be done by war. (Morgenthau 1985: 228)

As the weaker nation concedes or relinquishes power, the difference between the two rivals' power levels increases, and the system gets closer to $\hat{d}$. The closer the system is to the point of futility, the less necessary the stronger nation will find it to consolidate its power.

The second condition must therefore be further refined to distinguish the case where nation X holds the advantage from the cases where nation Y holds the advantage. The equations for the condition where $p_x > p_y$ are as follows:

$$dc_{xy}/dt = \gamma_x p_x c_{yx} + \sigma_x(\hat{d} - (p_x - p_y)) \tag{7.2l}$$

$$dc_{yx}/dt = \gamma_y p_y c_{xy} + (\omega_y/(p_x - p_y))*(d(p_x - p_y)/dt) \tag{8.2l}$$

where σ_x and ω_y are positive constants. The equations for the condition where $p_x < p_y$ are as follows:

$$dc_{yx}/dt = \gamma_{xy} p_{xy} c_{xy} + (\omega_{yx}/(p_{xy} - p_{yx}))*(d(p_{xy} - p_{yx})/dt) \tag{7.2u}$$

$$dc_{yx}/dt = \gamma_y p_y c_{xy} + \sigma_y(\hat{d} - (p_y - p_x)) \tag{8.2u}$$

where ω_x and σ_y are positive constants. Note that the feeling of anticipation is modified by the ability to act on the feeling in equations 8.2l and 7.2u, which represent the argument about what makes a weaker nation act to prevent movement past$_d$. The placement of the term $(p_y - p_x)$ in the denominator of equation 7.2u captures the inverse relationship between the difference in power and the ability to act conflictually.

It should also be noted that if the weaker nation's growth rate is larger than the stronger's, that is, $p_x < p_y$ and $d(p_y - p_x)$ is negative in equation 7.2u, then the weaker nation becomes less conflictual toward the stronger. When the weaker nation's power grows more rapidly than the stronger's, the weaker has no reason to increase its conflict level and can afford to decrease it instead.

Condition 3. $|p_x - p_y| > \hat{d}$: Submission and Dominance

The third condition is shown in figure 16.1 by the positive region above the line $p_y = p_x + \hat{d}$ and by the positive region below the line $p_y = p_x - \hat{d}$. Under this condition, the weaker nation chooses to yield, thereby avoiding a more conflictual situation. Its decision to resign is based on two elements— the futility of the situation coupled with the cost of conflict in terms of power loss. Not only is conflict useless at this point, it is expensive as well. According to the power transition literature, the weaker nation is dissatisfied, but is no longer able to act on this dissatisfaction. If nation X is the weaker, as its position worsens, or as the quantity $(\hat{d} - (p_y - p_x))$ becomes increasingly negative, it is less and less capable of sustaining conflict. And as long as nation X directs the slightest amount of conflict toward nation Y, that is, as long as $c_{xy} \neq 0$, it is motivated to decrease this conflict in order to avoid further cost to its power capacity. The more conflictual nation X is, the more effort it will exert to lessen its conflict.

The stronger nation, on the other hand, has achieved a "permanent" advantage and no longer needs to press for consolidation. In this sense, nation Y is quite satisfied. As its position improves, or as the quantity $(\hat{d} - (p_x - p_y))$ becomes increasingly negative, nation Y finds it less and less necessary to further its cause. Nation Y will also find conflict to be expensive. As long as the stronger nation is directing conflict toward the weaker, that is, as long as $c_{yx} \neq 0$, it will be motivated to decrease this expensive behavior.

Once again, we need two sets of equations, one for the case in which nation X is the stronger, and one for the case in which nation Y is the stronger. The equations for the condition where $p_x > p_y$ are as follows:

$$dc_{xy}/dt = \gamma_x p_x c_{yx} + \lambda_x(\hat{d} - (p_x - p_y))c_{xy} \tag{7.31}$$

$$dc_{yx}/dt = \gamma_y p_y c_{xy} + \lambda_y(\hat{d} - (p_x - p_y))c_{yx} \tag{8.31}$$

where λ_y and λ_x are positive constants.
The parallel equations for the condition where $p_y > p_x$ are as follows:

$$dc_{xy}/dt = \gamma_x p_x c_{yx} + \upsilon_x(\hat{d} - (p_y - p_x))c_{xy} \tag{7.3u}$$

$$dc_{yx}/dt = \gamma_y p_y c_{xy} + \upsilon_y(\hat{d} - (p_y - p_x))c_{yx} \tag{8.3u}$$

where υ_x and υ_y are positive constants.

The formal model is hence complete. It should be noted that while the power equations apply universally under all three conditions, the conflict equations change depending on the particular power relationship being experienced. Under the first condition, the conflict equations 7.1 and 8.1 are combined with the power equations 3 and 4 to form a system of differential equations. Under the second condition, when $p_x > p_y$, conflict equations 7.2l and 8.2l are combined with equations 3 and 4 to form a similar system. Under the second condition, when $p_x < p_y$, equations 7.2u and 8.2u are used instead of equations 7.2l and 8.2l. If the power relationship between two nations places them in the third condition, and $p_x > p_y$, then equations 7.3l and 8.3l form a system of equations along with equations 3 and 4. If the two nations are in the third condition and $p_x < p_y$, equations 7.3u and 8.3u replace equations 7.3l and 8.3l in the system of equations. The appropriate system of equations is placed in the region representing each condition in figure 16.1.

Simulating Dyadic Behavior

Now that the model-building exercise is complete, I turn to an analysis of the model itself. The purpose of this section is to use the model in order to deduce certain conclusions about power transitions. In particular, I am interested in whether transitions are peaceful or conflictual and in the type of power relationships that produce and result from transitions. The issue of whether transitions are peaceful or conflictual is an interesting one that rises from the balance of power–power transition debate. Proponents of the balance-of-power explanation of war would expect transitions to be relatively peaceful (i.e., free from war) since they are thought to occur when rivals are approximately equal in power and stably balanced. Proponents of the transition explanation, on the other hand, would expect transitions to be relatively conflictual (i.e., war-prone) since they signal the weaker nation's challenge of the previously dominant nation. The hybrid model should provide expectations of its own concerning conflict behavior near power transitions. It should additionally provide expectations concerning the patterns of power growth and decay before and after a transition.

Expectations or conclusions can be determined from a system of differential equations in one of two fashions. If the system can be analytically solved, the equations are rewritten so that they explicitly describe the behavior of p_x, p_y, c_{xy}, and c_{yx} as time passes. In other words, p_x, p_y, c_{xy}, and c_{yx} are the left-hand side of the equations, as opposed to dp_x/dt, dp_y/dt, dc_{xy}/dt, and dc_{yx}/dt. If, however, the system cannot be solved analytically, it must be solved numerically. The principle underlying numerical solution is to provide just enough information to get the system of equations started. The equations can then generate all subsequent and previous values of p_x, p_y, c_{xy}, and c_{yx} through time. As misfortune would have it, the model proposed here requires a numerical solution.

Several steps are involved in the numerical solution. First, the parameter values are determined. Next, random sets of initial values, or conditions, are chosen. Multiple sets must be used because rigor demands that conclusions not be drawn from special or unique cases. Since the general analytic solution cannot be determined, many instances must be used in order to make broad-based claims about the model's expectations. Third, the equations are seeded with a set of initial conditions from which they determine subsequent and previous values of p_x, p_y, c_{xy}, and c_{yx}.[12] The equations are then reseeded with each additional set of initial conditions, which provides several more runs or simulations. Last, the many sets of resulting paths for p_x, p_y, c_{xy}, and c_{yx} through time are examined in order to distill the commonalities held across all initial conditions.

The selection of particular parameter values is done after several trial runs are made using varying values for each parameter. This is similar to the process of varying the initial conditions across several runs. Here, I vary the parameter values across several runs. I select those values that produce the most typical paths for p_x, p_y, c_{xy}, and c_{yx} through time. These paths are often referred to as trajectories. Although this selection process is a somewhat subjective one, I do employ some criteria. First, I avoid choosing parameter values that produce trajectory behavior that is vastly different from that resulting in other runs. The idea is to produce prototype simulations from which general statements can be made. Second, I want the values for all of the parameters to be similar in size, or magnitude.[13] This prevents any single parameter from having an overwhelming effect. At this point in time, I have no reason to believe that any single term in either the power or conflict equations has an effect that is substantially stronger than any other terms. Third, all parameters are, by definition, positive. If a theoretical or analytic issue arises that suggests the necessity to adjust any particular parameter value after the initial investigation, this will be done. Throughout all simulation runs, parameters are held constant at the following values:[14]

$$
\begin{array}{lll}
\alpha_x = .3 & \alpha_y = .3 \\
\beta_x = .5 & \beta_y = .1 \\
\gamma_x = .2 & \gamma_y = .1 \\
\kappa_x = .1 & \kappa_y = .15 & \epsilon = .001 \\
\omega_x = .6 & \omega_y = .5 \\
\sigma_x = .2 & \sigma_y = .15 \\
\lambda_x = 1.5 & \lambda_y = .9 \\
\nu_x = .9 & \nu_y = 1.5
\end{array}
$$

The only special argument currently used to set parameter values is reflected in the stipulation that $\alpha_x = \alpha_y$. This is done because I believe the two competing rivals should be approximately equal in their natural abilities. Such an assumption parallels Organski's (1958) presentation of competing dyads existing among only the most advanced nations. All nations in this category must possess high specific growth rates, that is, high α values, as they are the nations that have progressed the furthest in terms of national power. As a simplified example, I offer figure 16.2, in which two nations begin at time $t = 0$ with ten units of power apiece. If each nation's power grows without the negative effect of competition, as in equation 1, the nation with the higher specific growth rate is well ahead of the other by the fifth time period. A situation such as this, in which one nation has a high α value while the other has a low α value, would not be meaningful in the context of a competitive rivalry or in Organski's story of a dominant nation being overtaken by a challenger. Setting $\alpha_x = \alpha_y$ is a simplified way to assume the two nations have competitive specific growth rates.

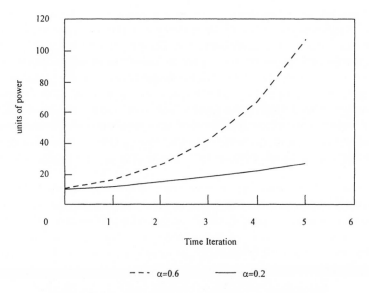

Fig. 16.2. Two nations' noncompetitive growth patterns.

Having settled on parameter values for the model, I now turn to the simulations themselves. As explained previously, I must decide on several sets of initial values for the four variables p_x, p_y, c_{xy}, and c_{yx}. In other words, I must assign values to $p_x(0)$, $p_y(0)$, $c_{xy}(0)$, and $c_{yx}(0)$. These particular values provide the model with starting points in time. The sets of initial conditions satisfy two simple requirements. First, all must be positive and less than or equal to fifty. That is, $0 < p_x(0)$, $p_y(0)$, $c_{xy}(0)$, $c_{yx}(0) \leq 50$. The purpose of this requirement is simply to limit the space under consideration. If such a limitation is not imposed, the initial values could take on a larger, unmanageable variety of values. This is a beginning step. If analysis suggests that interesting or unusual behavior is produced outside these boundaries, then they will be expanded. Second, all initial conditions will begin at a transition. Since transitions occur when two nations' power levels intersect, this requirement can be expressed as $p_x(0) = p_y(0)$. This will guarantee that the power and conflict behaviors that are produced are concomitant with dyadic power transitions, the phenomenon I wish to investigate.

Now, the equations are given a set of initial conditions from which both subsequent and previous values for each of the four variables can be determined.[15] In other words, each variable will have associated with it a time series of values. These time series can be plotted as trajectories, as in figures 16.3, 16.4, and 16.5. Each of these figures contains three graphs. The horizontal axes are all indications of the progress through time, while the vertical axes represent power levels, the difference in power levels,[16] and conflict levels, in that order from top to bottom. For each set of initial conditions that the equations are given, a unique set of trajectories is generated. Hence, figures 16.3, 16.4, and 16.5 are different from each other. The initial conditions that produced each set of trajectories can be read in the first and third graphs of each figure. The point at which a variable's trajectory intersects the vertical axis indicates the value of that variable when $t = 0$, or at the initial point in time. In the first graph of figure 16.4, for example, nation X's power level and nation Y's power level cross the vertical axis at $p_x = 10$ and $p_y = 10$.[17] In the third graph of figure 16.4, nation X directs ten units of conflict toward nation Y at $t = 0$. Nation Y reciprocates with six units of directed conflict at $t = 0$. The set of initial conditions that produced the unique trajectories in figure 16.4 are $p_x(0) = 10$, $p_y(0) = 10$, $c_{xy}(0) = 10$, and $c_{yx}(0) = 6$.

For each variable, it is generally possible to generate values for an infinite number of points in both future and past time unless a discontinuity is reached. In figures 16.3, 16.4, and 16.5, the system of equations generates all variable values to the left and to the right of the vertical axis at $t = 0$.[18] How, then, does one decide when to end a simulation? I employ the following rule. When either $p_x(t)$ or $p_y(t)$ becomes zero, the simulation ends. I assume that a nation must possess some positive level of power in order to exist. Hence, the endpoints of the trajectories in figures 16.3, 16.4, and 16.5 correspond to the points in time—future or past—at which either nation X or nation Y did not possess any power whatsoever. In other words, the model is not meaningful when $p_x(t)$ or $p_y(t)$ is nonpositive. I allowed conflict levels to become negative, having reasoned that negative conflict could be interpreted as some type of concessions or cooperative behavior.

Simulation Results: Conclusions from the Model

Approximately fifty different sets of initial conditions satisfying the above criteria were used to seed the dynamic model. Although each set of initial conditions produced a unique set of trajectories for p_x, p_y, c_{xy}, and c_{yx}, three distinct patterns of behavior emerged. Each distinct pattern is represented by a single simulation result in figures 16.3, 16.4, and 16.5. So, while the graphs in figure 16.3, for example, demonstrate the results of only a single simulation, many other simulation results have similar trajectory behavior. The trajectories in figure 16.3 can be thought of as a single, representative member of an entire family or class of like trajectories. The three distinct classes of behavior are distinguished primarily by the differences in how national power grows and decays before and after a dyadic power transi-

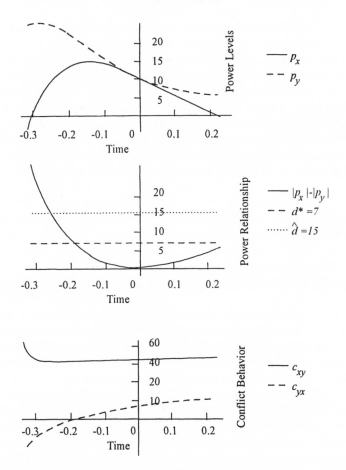

Fig. 16.3. An example of deflection: $\beta_y\, c_{xy}(0) = \beta_x c_{yx}(0)$, $p_x(0) = p_y(0)$ = 10, $c_{xy}(0) = 50$, $c_{yx}(0) = 10$.

tion. The first is characterized by the challenger's obvious inability to successfully complete a transition. All trajectories in the second class show the challenger rising to compete with the dominant nation for a period of time before returning to the weaker position. This second group might also be characterized by saying that the formerly dominant nation finds a way to recover its lost position. The third class is characterized by a single, distinct transition.

The first group of trajectories might be referred to as the *deflection* cases. They.are represented by the graphs in figure 16.3, where $p_x(0) = p_y(0) = 10$, $c_{xy}(0) = 50$ and $c_{yx}(0) = 10$. In the first graph, the weaker nation's power level,

p_x, approaches that of the stronger, but is deflected away at $t = 0$. Nation X appears to threaten nation Y, but never actually succeeds. An interesting feature of all the cases falling into the deflection class is that the difference between the values for conflict behavior, $c_{xy}(t) - c_{yx}(t)$, seems large compared to those for the other types of simulations. This disparity appears to diminish slowly as the nations move forward in time, or from left to right on the third graph of figure 16.3. The disparities between conflict levels in the other two simulations, as presented in the third graphs of figures 16.4 and 16.5, are smaller and diminish more quickly as the nations move forward in time.

Deflection occurs when a trajectory in the $p_x - p_y$ plane reaches the line $p_y = p_x$ (the line along which transitions occur) and the slope of the line tangent to the curve at that time is unity; that is when $dp_x(0)/dt = dp_y(0)/dt$. It can therefore be shown that all deflection cases are characterized by the following stipulation on the initial conflict values:[19]

$$\beta_y c_{xy}(0) = \beta_x c_{yx}(0) \tag{9}$$

Since $\beta_x = .5$ and $\beta_y = .1$, $c_{xy}(0)$ must be exactly five times larger than $c_{yx}(0)$ in order for deflection to take place. Equation 9 may be understood to mean that when nation X's power level is equivalent to nation Y's power level, nation Y must direct only one-fifth as much conflict at nation X as nation X directs at it in order to prevent nation X from achieving the advantage.

A different pattern emerges in the second group of trajectories. In figure 16.4, a single simulation result demonstrates the typical behavior in this category. Here, $p_x(0) = p_y(0) = 10$, $c_{xy}(0) = 10$, and $c_{yx}(0) = 6$. One nation's power level rapidly rises to its peak and then declines. Its competitor, meanwhile, takes the more methodical path, maintaining a relatively even level of power and achieving the advantage in the end. An appropriate name for this group might be the *tortoise and hare* trajectory class. This scenario might remind us of the British-Russian rivalry. In the late nineteenth to early twentieth century, the United Kingdom began to experience a rapid decline from her previous dominance while Russia slowly crept forward. Shortly after World War I, the United Kingdom was surpassed by the Soviet Union, whose perseverance paid off. The tortoise and hare story might also remind us of the Soviet-American military rivalry during the Cold War. In the mid 1960s, the USSR began to experience a rapid rise in its capabilities vis-à-vis the US. American capabilities were maintained, and perhaps even declined slightly, at about this time. In 1992, however, the hare's pace was broken by the fall of the USSR, leaving the tortoise as the victor. Unlike the first class of trajectories, I am unable to describe this class with a simple stipulation on the initial conditions, as in equation 9.

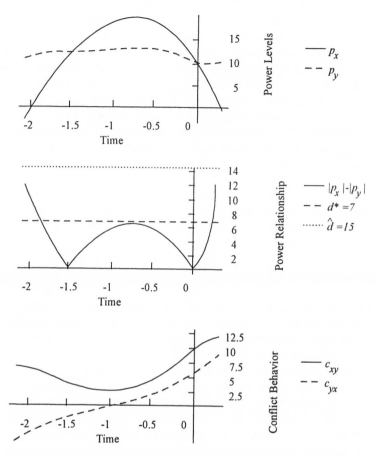

Fig. 16.4. An example of a tortoise and hare transition: $p_x(0) = p_y(0) = 10$, $c_{xy}(0) = 10$, $c_{yx}(0) = 6$.

The conflict behavior in this group of trajectories is interesting in two respects. First, the conflict values appear to be converging. In the third graph of figure 16.4, nation X is directing a great deal more conflict toward nation Y than Y is directing toward X when $t = -2$. In fact, nation Y is directing a negative amount of conflict, or cooperation toward nation X. As time passes, nation Y abandons its cooperative efforts and becomes increasingly conflictual. Nation X decreases its level of conflictual behavior until nation Y ceases cooperation, at $t \approx -1.0$. Second, the conflict levels are highest following a permanent transition. Two transitions appear in each simulation in this family. The first, when nation X surpasses nation Y, does not result in a permanent advantage. Rather, the power position of the two rivals switches again when the second transition occurs. The second transition, does, however, result in a permanent advantage, as nation X's power level plummets while nation Y's power level is maintained. Hence, I will call the second type of transition a permanent one. Both before and after the first transition, conflict levels are low. Nation Y's behavior might even be considered to be cooperative, and nation X's conflict is decreasing in response to the cooperative efforts made by nation Y. The permanent transition, on the other hand, produces the highest conflict levels of the entire simulation.

In the third group of trajectories, only a single, permanent transition occurs. This family of trajectories is represented in figure 16.5 by the behavior resulting when $p_x(0) = p_y(0) = 10$, $c_{xy}(0) = 10$ and $c_{yx}(0) = 50$. In the first graph, the decline of nation X is concurrent with the rise of nation Y. This class can be thought of as the growth and decay class. One nation's power level grows while the other decays. This group fits nicely with Organski and Kugler's (1980) version of a single, permanent power transition. There is a distinct and rapid movement from lower condition three (where nation X enjoys an overwhelming advantage) to upper condition three (where nation Y enjoys an overwhelming advantage).[20] This is distinct from the second group of trajectories, in which the dyad spends considerable time in the first condition, or in lower condition two or three. Again, I cannot determine a precise stipulation on the initial conditions that produce this type of behavior.

The general statements that can be made about conflict behavior in this group of trajectories are very similar to those made for the permanent transitions in the second group. In the third graph of figure 16.5, the conflict levels appear to converge as they move from the left to right. Beginning at $t = -.15$, nation X is highly cooperative while nation Y is highly conflictual. As time progresses, each nation's behavior becomes less extreme until nation X finally becomes conflictual at $t \approx -.06$. From then until $t = 0$, nation X becomes increasingly conflictual while nation Y's conflict behavior levels off. Although the small scale of the conflict axis in the third graph makes it difficult to see, both nations are slowly increasing their conflictual behavior after the transition at $t = 0$ until nation X's power level becomes zero at $t \approx .04$. By the end of the simulation, the conflict behaviors appear to be coupled. Furthermore, they seem to be highest following the permanent transition. Although nation Y's conflictual behavior is highest early on, it is rapidly decreasing at that time. It is increasing following the transition. Nation X's corresponding behavior is cooperative when Y's conflict is at its peak and is most conflictual following the transition. Taken together, and accounting for the direction of change, nation X and nation Y can be considered to be most conflictual following the transition.

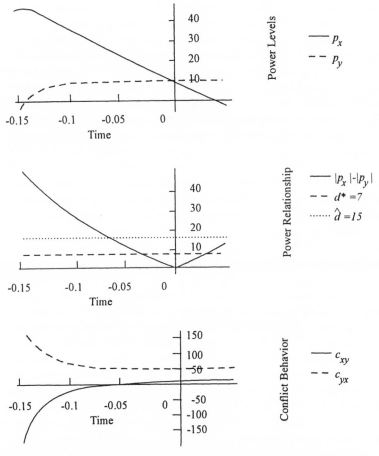

Fig. 16.5. An example of a single transition: $p_x(0) = p_y(0) = 10$, $c_{xy}(0) = 10$, $c_{yx}(0) = 50$.

A Summary

I began this work with inquiries about the patterns of growth in national power and in dyadic conflict behavior that produced and resulted from power transitions. Believing that the power transition and balance-of-power explanations of war initiation could provide some insight, I drew on them in order to design a formal, dynamic model. The model was a hybrid of the two original explanations, and it demonstrated how conflict behavior changes with varying conditions in the dyadic power relationship. The model, a system of first-order differential equations, was solved numerically using multiple simulations. The results of these simulations provide some answers to my initial questions.

The most obvious finding is that there are three different patterns in national power growth that are related to power transition. One comes close to producing a transition, but ultimately does not. The next produces two transitions, only one of which is decisive in the end. The last produces a single permanent transition. These three cases allow for a much more diverse world than either the balance-of-power or the power transition explanations do separately. The balance-of-power explanation focuses on two types of transitions, neither of which emerges as a distinctive category in the simulations. The first type includes nondecisive transitions, or those that result in a power difference of no more than d^*. They are seen as inconsequential results of a balancing mechanism and therefore do not produce wars. Decisive transitions transpire when the balancing mechanism breaks down, and they do indeed give rise to wars. Conversely, power transitionists focus only on single permanent transitions, highly similar to the type in the third class of simulation trajectories. By itself, neither of the traditional explanations accounts for deflection or tortoise and hare transitions.

Another type of result relates to the tendency of nation X's conflict behavior and nation Y's conflict behavior to become coupled as time progresses. This was seen, although to varying degrees, across all three classes of trajectories. The fact that c_{xy} and c_{yx} tend to converge is important for two reasons. First, it suggests that the model might be refined to include only one conflict variable, c, which measures the total level of conflict experienced by the dyad at any given time. Second, it implies that neither nation can be clearly considered as the aggressor. In contrast, the power transition explanation does portray one nation as the most aggressive by naming the challenger as the likely initiator of war. The balance-of-power explanation, while naming the stronger as the likely initiator of war, does not exclude the possibility that the weaker may be highly conflictual in an attempt to restore the balance.

A third conclusion shows that a pair of rival nations is most conflictual immediately following a permanent power transition. Short-lived transitions, as seen in the tortoise and hare class of trajectories, do not produce high levels of conflict. Moreover, periods between the two transitions are relatively peaceful. If, during this period, the $|p_x - p_y|$ trajectory remains below d^*, as it happens to in figure 16.4, then it may be considered to be a fairly *peaceful balance*. Peaceful balances are predicted by the balance-of-power explanation. Hence, the conclusions concerning conflict behavior appear to be consistent with both the power transition explanation's predictions about conflictual transitions and with the balance-of-power explanation's predictions about peaceful balances.

Finally, it is worth noting that the prescriptions for "winning a transition" may be somewhat counterintuitive. In at least two ways, the ultimate winner is not a fierce fighter in the typical sense. In the deflection case, nation *Y* succeeds in fending off nation *X*'s advances in power by directing only a fraction of the conflict that nation *X* does. If nation *Y* already holds the advantage, it may make little sense to be more conflictual than is absolutely necessary to maintain that advantage. Conflict is costly in terms of national power, so using it sparingly is wise. In the tortoise and hare case, the ultimate winner has a slower initial growth rate than the ultimate loser. A slow but persistent rate of growth in national power is more effective than a fast but tiring one.

These conclusions serve to inform both the process of modeling national power growth and dyadic conflict and substantive issues raised by the power transition–balance of power debate.

NOTES

1. The term *capabilities* is often used as well, referring to the typical indicator of power. In this chapter, "power" will be used unless "capabilities" is the term used by the original authors being discussed.

2. Yet one of the largest difficulties with balance-of-power theory is that its central concept has too many meanings. Haas (1953), for example, found at least eight different common uses of the phrase "balance of power." The explanation presented here uses just one of those meanings, namely "stability and peace in a concert of power" (Dougherty and Pfaltzgraff 1981:24).

3. Less powerful nations are only of interest later on in their development.

4. There has been a slight effort to capture dynamics by including a test for whether growth rates are equal or unequal (see, for example, Organski and Kugler 1980; Houweling and Siccama 1988a). Nonetheless, this approach is still limited to a test of certain variable values at a single point in time.

5. Of course, another way to do this is for a rival to increase its own power level. Here, however, I am interested only in the forces that cause a nation's power level to decay.

6. This is not to say that conflict is the only means by which nations act. It is simply the means that is important in this model.

7. Similarly, Richardson (1960a) considered "deadly quarrels" to include any violent action resulting in death, from homicide to riots to interstate war.

8. The idea that conflictual behavior depletes national power is captured indirectly in the second term of the power equations. The expenditure is actually made in order to respond to the opponent's conflict. The response, an action in itself, is conflictual in nature and requires the expenditure of power reserves.

9. I intentionally leave this possibility out of my discussion because it essentially refers to the case where the basic assumptions about what drives nations no longer hold. I am concerned here only with the scenario where these assumptions do hold.

10. We might be alternatively interested in *d* as a certain critical proportion between p_x and p_y.

11. This prevents the denominator from becoming zero.

12. This procedure was performed using the *Mathematica* software developed by Wolfram.

13. This excludes ϵ, which is very small. It is used solely as a method to prevent discontinuities in equations 7.1 and 8.1. I do not wish its effect to be important otherwise, although one could certainly make arguments about the potential for its contributions to be meaningful for the analyses.

14. The constants d^* and $\hat{d}$ are arbitrarily set at 7 and 15, respectively. The choice of their particular values does not change the findings of this work as long as $d^* < \hat{d}$.

15. Note, starting at $t = 0$ and generating four subsequent values and three previous values, for example, yields the same results as beginning at $t = -3$ and generating seven subsequent values.

16. The second graph shows the system's movement through the three conditions defined by the magnitude of the difference between p_x and p_y.

17. $p_x(0) = p_y(0)$ is a requirement for all initial conditions, as specified previously.

18. A system of differential equations can generate previous values of variables if $-t$ is substituted for t in all equations.

19. Because $p_x(0) = p_y(0)$ in all simulations, and because $\alpha_x = \alpha_y$, the $\alpha_x p_x$ and $\alpha_y p_y$ terms are subtracted from each side of the equation.

20. The system, of course, goes through lower condition two, condition one, and upper condition two before reaching upper condition three.

CHAPTER 17

The Logic of Overtaking

James D. Morrow

Power transition theory (Organski 1968; Organski and Kugler 1980) is the first and foremost of a set of theories that strive to explain the origins of big wars (for applications to a larger range of wars, see Kim herein; Houweling and Siccama herein). According to power transition theory, big wars occur when a *challenger* rises to the point when it surpasses the *dominant state* in capabilities. The point in time when the two are equal is the *transition point.*

The challenger's growth relative to the dominant state could be powered by many factors. In power transition theory, the *power transition*, the growth in power produced by industrialization and the resulting demographic growth and increases in political capacity, drives the faster growth of the challenger. In other theories of big wars, the financial burdens of international hegemony (Gilpin 1981; Kennedy 1987), the need to service debts from prior wars, and the failure to dominate new leading economic sectors (Thompson 1988) account for the relatively slow growth rate of the dominant state. Regardless of the source of differential growth rates, all these theories assume that the challenger is growing more rapidly than the dominant state and that its growth provides the motivation for war.

Underlying all these arguments is the idea that shifts in power provide a motivation to fight. I refer to this idea as the logic of overtaking. Shifts in capabilities could provide a dynamic motivation to fight that is not present in most international relationships. The challenger must consider when it should strike to change the international order. The dominant state must decide if it will resist the challenge when it comes. But shifts in power occur among dyads other than just the dominant state and challenger. The logic of overtaking should apply to any dyad where one member's capabilities are growing relative to the other's. I call the former the *rising state* and the latter the *declining state*. I use dominant state and challenger to describe the declining and rising states in power transition theory, where the dominant state is the most powerful in the international system.

I aim to provide an answer to a simple question here—how do rational actors respond to long-term, predictable shifts in capabilities? The answer I provide is based on a formal model presented and tested in Kim and Morrow 1992. My informal presentation here strives to provide the reader with an intuitive appreciation of how shifts in power might cause war. My argument examines the logic of overtaking in general. It has important implications for power transition theory and cyclic theories of war driven by long-term changes in capabilities. Power transitions are a special case of such shifts in capabilities. My hypotheses about when shifts in capabilities make war more likely should hold for power transitions between the dominant state and its challenger.

The argument proceeds in a straightforward fashion. I describe the decisions both sides face and discuss the assumptions underlying the analysis. The decision to resist by the declining state is analyzed first, followed by the rising state's choice of when to challenge the declining state. I discuss the implication of the argument for power transition theory and other theories of war. I conclude with an assessment of the weaknesses of the argument and with questions for further research.

The Basis of the Logic of Overtaking

The logic of overtaking is an argument about the timing of war. Shifts in capabilities introduce a dynamic element to decisions for war. The status quo, whether it be the structure of the international system or just a simple border dispute, favors the declining state. It has used its greater initial power to arrange the status quo as it wishes. The rising state disagrees with the declining state on some issues in the status quo. It has grievances against the status quo. But it lacks the capabilities to force the changes it would like at the beginning of its power transition. As the rising state's capabilities grow, it considers the possibility of challenging the status quo. But it must consider when to challenge. If it waits, it will be more powerful and so its challenge will be more likely to succeed. If it challenges now, it will change the status quo sooner. The declining state must decide whether to resist a challenge if one is made. Early in the transition, it is willing to fight to preserve the status quo. But its willingness to fight declines with its relative capabilities. At some point, the declining state simply accepts the now-greater power of the rising state and agrees to its demands. These questions of timing create the logic of overtaking.

My argument rests on five basic assumptions. First, I focus on the rising power's decision when to use its growing capabilities to force change in the status quo. I ignore the parallel question of why declining states do not eliminate challengers well before those challengers approach them in capabilities. It is an important question these theories must face, but the literature is silent on it. Second, explanations of wars must account for why both sides chose war rather than a peaceful resolution. Here I must account for both why the rising state tries to use force to change the status quo and why the declining power resists rather than submitting. War is costly; both sides might be better off agreeing on some settlement short of war. Why do they fight, then? Third, the sides make dynamic comparisons when considering war. Shifts in capabilities add the question of when to fight to the question of whether to fight. The actors compare war now to war later. Otherwise, shifts in capabilities cannot be said to cause the ensuing war. Fourth, I assume war ends the conflict by resolving all issues between the two states. I adopt this assumption to parallel power transition and cyclic theories of big wars.

Fifth, I assume that misperceptions of the shifting balance of capabilities are independent of position. This is the assumption of rational expectations— perceptions of future changes in capabilities are accurate, on the average, even though the perceptions of particular states may be inaccurate. This point deserves further explanation. Misperceptions of future changes in capabilities can cause wars. If rising states overrate their future growth and declining states underrate their own decline, then these misperceptions will make war more likely. Rising states will ask for greater concessions and declining states will be unwilling to grant them. But it seems equally plausible to believe that rising states underestimate their future growth and that declining states exaggerate their decline. In that case, the change in capabilities makes war less likely, not more likely. Misperceptions of future shifts could make war more or less likely. In the absence of an argument that explains why rising states exaggerate their future growth and why declining states discount their own decline, it seems premature to assume that they do. It may be that misperceptions cause wars, but there is no relationship between long-term shifts in capabilities and misperceptions of those shifts. It is difficult to argue, then, that shifts cause war. Some shifts end in war, others do not, and there is no systematic way to differentiate them.[1] I avoid this question by assuming that there are no systematic misperceptions of the changes in capabilities.

Characterizing Power Transitions

Long-term shifts in capabilities can be driven by many factors. Less-developed nations can gain capabilities rapidly through industrialization, demographic growth, increases in political capacity, and dominance in emerging new technologies. The growth of the absolute capabilities of existing powers is slowed by the financial burdens of foreign and military commitments and debt from prior wars. For whatever reason, the long process of modernization has led to differential growth rates in absolute capabilities. Different growth rates create shifts in relative capabilities. These shifts typically favor states with fewer capabilities but large populations. They can add capabilities rapidly as they develop. Industrialized states find adding capabilities more difficult.

In the typical pattern, powers that are initially weak increase relative to

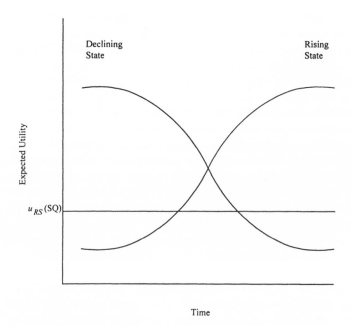

Fig. 17.1. Shifts in expected utility during a power transition.

stronger powers. As time passes, the capabilities of the rising state reach and then surpass those of the declining state. The *transition point* occurs when the capabilities of the two sides are equal.

These shifts in absolute capabilities produce a relative decline for the declining state and a relative gain for the rising state. These shifts in relative capabilities change what each side expects to gain from war. As the rising state's capabilities grow, it expects to gain more from war. As the declining state's relative capabilities wane, it expects to gain less from war. Each side's expected utility for war specifies what it expects to gain from war. Expected utilities are determined by the dyadic correlation of forces, each side's willingness to take risks, and expected aid from third parties. Figure 17.1 graphs how each side's expected utility for war changes over the course of a power transition.

Figure 17.1 also gives the rising state's utility for the status quo, $u_{RS}(SQ)$. The status quo reflects the greater capabilities of the declining state before the transition begins. The declining state uses its superior initial position to establish an international order in its interest. This order need not be as broadly conceived as the international order is in power transition theory. It could simply reflect the resolution of outstanding border disputes and other dyadic issues between the two states. The rising state does not completely accept this order. Otherwise, there would be no conflict between the two. The lower $u_{RS}(SQ)$, the more dissatisfied the rising state is with the order of the declining state.

The status quo persists in the absence of a challenge by the rising state. If the rising state challenges the status quo, it makes demands for change that force a response from the declining state. The declining state could make concessions in the status quo to the rising state or it could resist those demands, triggering a war. War, as assumed earlier, resolves all outstanding issues between the sides. Each side's expected utility at a particular time gives the value it attaches to the range of outcomes possible after a war at that time weighted by their probability of occurring. If the declining state grants concessions to the rising state, I assume that each side values those concessions the same as their expected utility for war at that point in time.

This sequence of decisions is captured in the game tree in figure 17.2. The rising state must first choose whether to challenge the status quo. If it does not, the status quo persists and the rising state can challenge later in the transition. If it challenges the status quo, the declining state must decide whether to resist that challenge. Resistance provokes war; acceptance leads to concessions.

The rising state's decision to challenge depends upon the response it anticipates from the declining state. If the declining state will acquiesce to a challenge, then the rising state always wants to make a demand. If the declining state will resist a challenge, the rising state only challenges when it prefers war now to accepting the status quo for now and challenging it later. The analysis

Fig. 17.2. Game tree of a challenge. Boxes are decision nodes; dots are terminal nodes.

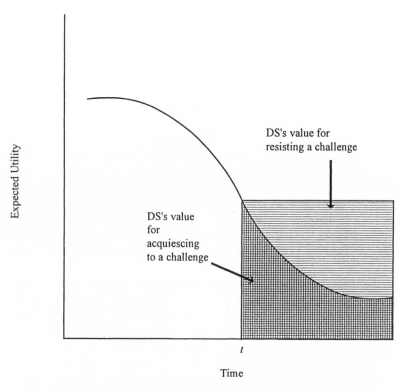

Fig. 17.3. Declining state's decision during a power transition. Horizontal area gives DS's value for resisting a challenge. The vertical hatched area gives DS's value for acquiescing. DS resists when its cost for war is smaller than the difference between the two regions. T_{crit} is the time when the cost of war equals the difference between the two regions.

begins with the declining state's choice because the rising state's choice hinges on it.

The Declining State's Choice

The declining state's choice between resisting and acquiescing to a challenge depends on the value of each alternative. Figure 17.3 gives a graphical representation of the declining state's value for resisting and acquiescing at time t. Resisting a challenge triggers war. War settles the issues at dispute between the sides but also imposes the costs of war on both. Because war "freezes" the outcome, the declining state receives the outcome of the war for the remainder of the transition period. The declining state's expected utility gives its value for the likely outcomes of a war. The rectangle shaded with horizontal lines gives

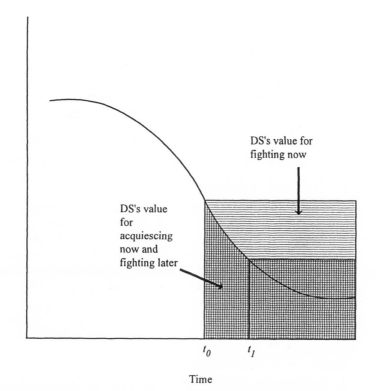

Fig. 17.4. Why declining state does not acquiesce now and fight later. All shaded areas give DS's value for resisting a challenge at time t_0. The cross-hatched area gives DS's value for acquiescing now and resisting later. Because DS pays the cost for war in either case, resisting now is preferable to acquiescing now and fighting later.

the declining state's value for the outcome of the war over the remainder of the transition period. The net value of war is this rectangle minus the costs of war.

The declining state's value for acquiescing to a challenge depends upon whether it will resist or acquiesce to future challenges. Acquiescing to all future challenges places a floor on what it can secure for itself in the future. This minimum value it can secure is given by the area under the curve of its expected utility for war for the remainder of the transition. This area is hatched with vertical lines in figure 17.3. Can the declining state do better than this if it acquiesces at time t? No. If it is better off acquiescing now and fighting against a later challenge than acquiescing to all future challenges, then it is even better off resisting now than acquiescing now and fighting later. Figure 17.4 illustrates this point. If the declining state acquiesces at time t_0 and resists at time t_1, it reduces the value it receives for resisting now by the region shaded with horizontal lines. But it pays the costs of war in either case.[2] Consequently, fighting now dominates acquiescing now and fighting later for the declining state. If it acquiesces at any time, it will always acquiesce at all future times.

Acquiescing for the remainder of the transition can be better than resisting for the declining state. Return to figure 17.3. The declining state suffers the costs of war if it fights but not if it acquiesces from time t on. The difference in its value between resisting and acquiescing is the diagonally shaded area above the curve. At time t, if this area is greater than the costs of war as measured in utiles, the declining state resists a challenge. If the costs of war are greater, it acquiesces to a challenge and will acquiesce to all future challenges. I call the point in time where this area equals the costs of war the *critical time*. The declining state is indifferent between war and peace at the critical time. Before the critical time, it resists challenges; after the critical time, it always acquiesces to challenges. The critical time completely specifies the declining state's strategy.

The critical time occurs later in the transition period as the difference between resisting and acquiescing to challenges grows. The declining state is willing to fight later into the transition as this difference grows. What makes this difference increase? First, the smaller the costs of war, the later the critical time occurs. Lower costs of war require fewer net policy benefits for resisting over acquiescing for the declining state to prefer war. The intuition here should be obvious. Second, the steeper the descent of the declining state's expected utility, the greater the net policy benefits of resisting. Figure 17.5 shows this point graphically. The descent of the declining state's expected utility is greater in the picture on the right. The shaded area above the curve is greater in the diagram on the right. The intuition here is simple. The faster the descent of the declining state, the greater the benefit that fighting provides by freezing the outcome where it is now. The future is worse in the diagram on the right, so fighting now is more attractive there.

What causes the declining state's expected utility to fall off more rapidly as the transition progresses? First, the faster the rising state's growth rate relative to the declining state, the faster the fall of the declining state. The quicker the rising state is growing, the more willing the declining state is to fight before the situation deteriorates further. Second, the more risk-averse the declining state, the more rapidly its expected utility falls off over the transition. Figure 17.6 shows representative expected utility curves for three different utility functions to illustrate this point. The risk-neutral utility function is linear, the risk-acceptant function concave, and the risk-averse function convex. The expected utility of all three for war drops as the transition proceeds. For a given time t in figure 17.6, the shaded area above the curve gives the net policy benefit from resisting a challenge. It is greatest for the risk-averse actor (upper horizontal shading), then the risk-neutral actor (vertical shading), and smallest for the risk-

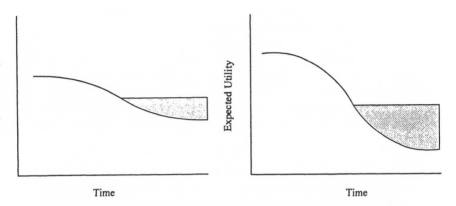

Fig. 17.5. Steeper decline makes war more attractive. The descent of DS is faster in the picture on the right. The shaded area, which gives the policy value of resisting a challenge, is greater with a faster decline. The faster the decline, the more likely war.

acceptant actor (diagonal shading). The more risk averse the declining state, the later in the transition the critical time occurs.

The Rising State's Decision

The rising state's decision depends upon the response it anticipates from the declining state. After the critical time, the rising state always challenges the status quo because the declining state always acquiesces to a challenge. Before then, a challenge will be resisted, leading to war. The rising state makes a challenge only when it prefers war now to accepting the status quo for now and challenging later. Figure 17.7 represents this comparison graphically. War now fixes the status quo at the outcome of the war for the remainder of the transition. The rising state's utility for the policy outcome of a war at time t is its expected utility for war at time t times the remaining time in the transition. The area of the rectangle shaded vertically in figure 17.7 gives this value.

If the rising state waits to challenge, the status quo continues unchanged. The rising state knows that the declining state will acquiesce to any challenge after the critical time. If it waits until the critical time to challenge, it can change the status quo without war. The horizontally shaded area in figure 17.7 gives the value of waiting until the critical time to challenge. This value consists of two parts. The first part is its value for the status quo that persists from the current time to the critical time. It is given by the rectangle with base from t to t_{crit}

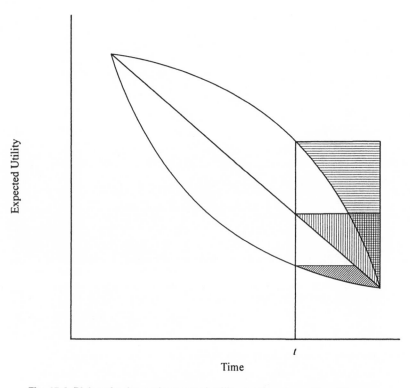

Fig. 17.6. Risk attitudes and expected utility during a transition. The area above the three curves gives the net policy benefit from resisting for risk-averse, risk-neutral, and risk-acceptant actors. The benefits are greater the more risk-averse the declining state is.

in figure 17.7. The second part is the rising state's value for the stream of concessions over time that the declining state makes after the critical time. The area under the expected utility curve between t_{crit} and the end of the transition gives this value. The sum of these two parts gives the value to the rising state for waiting until the critical time to challenge the status quo.

I do not yet consider the question of when the rising state should challenge the status quo if it prefers war to waiting. It may be that if the rising state prefers war to waiting at time t, it may be better off waiting a little bit and then making a challenge that triggers a war. But the question of interest here is whether the rising state prefers war to waiting at any time before the critical time. If not, the power transition does not lead to war. If so, it does. After completing the analysis of what factors make the rising state more willing to go to war, I return to the question of when it provokes war through its challenge.

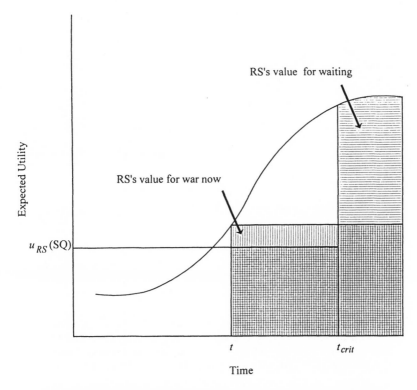

RS's value for waiting

RS's value for war now

Expected Utility

u_{RS}(SQ)

t t_{crit}

Time

Fig. 17.7. Rising state's decision during a power transition. Horizontal-hatched area gives RS's value for waiting until after t_{crit} to challenge DS. The vertical-hatched area gives the value of a challenge now (i.e., war). RS will challenge at time t if the difference between the two regions exceeds its cost of war.

The rising state prefers war to waiting when the difference between the two is greater than the costs of war. In figure 17.7, this difference is the area shaded with only vertical lines minus the area shaded with only horizontal lines. This difference is negative in figure 17.7, and so the rising state prefers waiting to war at time t, regardless of the costs of war. When this difference is positive and exceeds the costs of war, the rising state prefers war to waiting.

What causes this difference to grow, and so make war more likely? First, the smaller the costs of war, the more likely the difference exceeds them. War is more attractive when its costs are low. Second, the lower the value of the status quo to the rising state, the greater the benefits from fighting early to overturn it. Lowering the status quo increases the vertically shaded rectangle in figure 17.7. The intuition here is simple; the greater the dissatisfaction of the rising

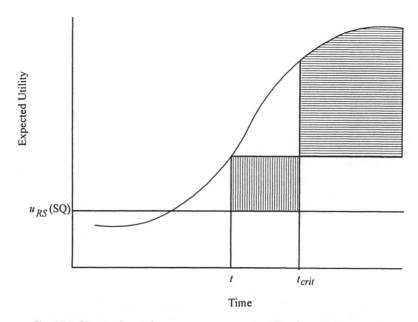

Fig. 17.8. Slower rise makes war more attractive. The rise of RS is slower in the picture on the right. The vertically shaded area minus the horizontally shaded area gives the value of war now over waiting. A slower rise increases this difference and makes RS more willing to go to war.

state, the greater its motivation to change the status quo. It is more willing to bear the costs of war as its opposition to the existing order grows.

The slower the rise of the expected utility of the rising state over time, the greater its willingness to go to war. Figure 17.8 illustrates this point. The difference between war now and waiting is greater when the growth rate is slower. Why wait and suffer with a status quo opposed to your interests when the benefits of waiting are small? The rising state's evaluation of its chances of overturning the status quo in its favor grow slowly in the picture on the right of figure 17.8. This observation leads to the third and fourth hypotheses about when rising states are more willing to fight. Third, the lower the growth rate of the rising state's capabilities, the more willing it is to consider war. The slower its relative growth rate, the slower its expected utility for war grows over time, and the less advantage there is to waiting.

Fourth, the more risk-acceptant the rising state, the more willing it is to go to war. Risk attitudes have two effects on the rising state's decision. First, they change the growth of its expected utility over time. For risk-averse actors, their expected utility for war rises rapidly at first and then slows down as the transi-

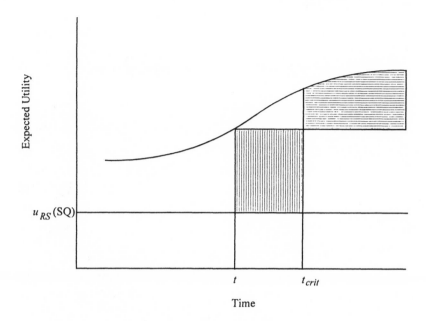

Time

tion progresses. The expected utility of risk-acceptant actors rises slowly at first and then more rapidly. At first glance, one might think that risk-averse rising states would be more likely to go to war. But the second effect of risk attitudes changes the rising state's value for the status quo, $u_{RS}(SQ)$. The greater the risk acceptance of the rising state, the lower $u_{RS}(SQ)$ is. Raising the value of the status quo to the rising state lowers the benefits of attacking now over waiting, and this effect is large enough to overwhelm the first effect of risk attitudes. The insight behind this point is the basic intuition behind risk attitudes. Rising states are comparing in part the risk of war against the certainty of the status quo. Risk-acceptant actors are more likely to accept that risk than other actors; risk-averse actors are more likely to choose the certainty of the status quo than other actors.

Return now to the question of when the rising state strikes if it wants war. It chooses the time before the critical time that maximizes its net benefits. Figure 17.9 shows this calculation graphically. Two times are important here, the time when its expected utility for war matches its value for the status quo, labeled t_D for time of dissatisfaction, and the critical time. Before t_D, the rising state prefers the status quo to the outcome it expects from war. It has absolutely no incentive to fight then; the status quo is better than the outcome of a war. The rising state faces two considerations in choosing the time to strike between

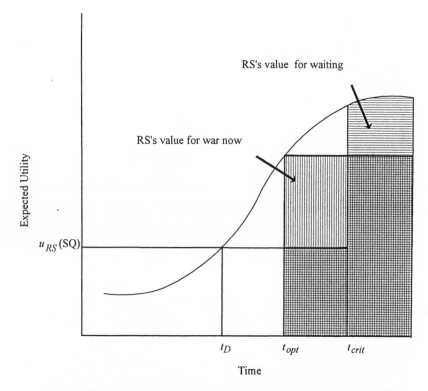

Fig. 17.9. When should rising state trigger war? The optimal time for RS to trigger war is roughly halfway between t_D and t_{crit}. RS maximizes the difference between the horizontal and vertical hatched areas to select the optimal time to challenge.

these two times. One, it wishes to increase the size of the vertically hatched rectangle in figure 17.9. If the growth of its expected utility for war was linear over time, the midpoint between the two times would do so. When the growth of its expected utility is not linear, as in figure 17.9, the optimal time to strike is not exactly the midpoint but should be close to it. Two, it wishes to reduce the horizontally shaded area in figure 17.9. Striking closer to the critical time shrinks this area. Combining these two considerations, the rising state's optimal time to attack should be shortly after the midpoint of the point where its expected utility equals its utility for the status quo and the critical time.

This time is generally close to the transition point, when the two sides are equal in capabilities. The time when the rising state's expected utility equals its utility for the status quo occurs early in the transition. For war to occur, the critical time must fall late in the transition period. If it occurs earlier in the transition, the horizontally shaded area in figure 17.9 is large and likely to be greater than the vertically shaded area. The rising state prefers waiting to war in such a case. The halfway point between these two times should be near the transition point then. The optimal time for the rising state to trigger war is after the midpoint. If war occurs, it begins after the transition point but while the sides are still roughly equal in capabilities. Transition points themselves are not conducive for war, but rough equality of capabilities is. If shifts in capabilities produce a dynamic motivation for war, then rough equality, not transition points, describe the relation of forces when war starts.

What Makes War More Likely?

Pulling together the decisions of the declining and rising states, the following factors make war more likely. First, the lower the expected costs of the war, the more likely war is. Second, the greater the dissatisfaction of the rising state, the more likely war is. Third, the more risk-averse the declining state, the more likely war is. Fourth, the more risk-acceptant the rising state, the more likely war is. Fifth, war is most likely when the sides are roughly equal in capabilities; the weaker has no less than two-thirds of the capabilities of the stronger.

Several factors should not be related to war. First, the relative growth rate of the rising state should not matter. A faster growth rate makes the declining state more willing to fight, but the rising state less willing to fight. Second, the transition point itself is no more dangerous than any other time, when the sides are roughly equal.

Low costs are the single most important cause of wars triggered by the logic of overtaking. The other contributing factors affect only one side's decision, while low expected costs increase both sides' willingness to go to war. Additionally, the critical time occurs later in the period when the declining state expects the war to have low costs. This expectation delays the critical time, which increases the motivation of the rising state to trigger a war. In figure 17.9, moving t_{crit} to the right—later in the transition—increases the vertically shaded area and decreases the horizontally shaded area. These changes make war more likely. If shifts in capabilities lead to wars, the sides are unlikely to go to war if they expect war will be costly.

Rethinking the Role of Power Transitions in Big Wars

Power transition theory focuses on a particular shift in capabilities, the surpassing of the most powerful nation by another. The hypotheses presented here provide a way to distinguish peaceful transitions from those that end in war. As Organski (1968) argues, the rise of a satisfied challenger is less likely to end in war. But risk attitudes, notably that of the declining state, also affect the likelihood of war during a transition. Although power transitions may provide an additional motivation toward war, there is no inexorable logic of overtaking that drives the sides into cataclysmic war. War occurs because both sides would rather fight than settle their differences peacefully.

But the argument here also presents a deeper challenge to power transition theory. Lower expected costs make war more likely; war is very unlikely when the sides expect high costs. This conclusion directly contradicts power transition theory. Power transition wars, by definition, are large, costly wars. Both sides have sizable capabilities and make any effort possible to win the war. The intervention of allies is not seen as an important factor in these wars. Allies are weak relative to the contenders and unable to shift the correlation of forces between them (Organski and Kugler 1980:26). Presumably, the sides can anticipate that a power transition war will be a life-and-death struggle over the future of the international system. But why fight then?

The argument here suggests that the logic of overtaking plays a different role in the origins of big wars. Big wars are always general wars, and there are four good reasons why general wars should be long and bloody (Blainey 1973:196–97). General wars: (1) have an equal distribution of capabilities across the sides, which makes decisive victory less likely; (2) involve fighting on multiple fronts, which makes it unlikely that one side will win everywhere; (3) require intracoalitional bargains as well as an intercoalition bargain to end; and (4) leave few uninvolved powers that can decisively swing the outcome by intervening. The explanation of big wars must center on why some wars expand to general wars and others do not. The logic of power transition theory deems the question of allies irrelevant and so does not even try to explain why big wars are general wars. General wars typically begin small and then expand to encompass all the major powers. This expansion is why big wars occur. Big wars, like big oaks, grow from small seeds. The logic of overtaking may explain the origin of the acorns of big wars.

The logic of overtaking presented here concludes that war is unlikely when expected costs are high. But reality does not always match expectations. Big wars begin as small wars, often between a major and a minor power and then expand into general wars. The logic of overtaking may explain why these small wars that trigger big wars occur or may explain the initial spread of a major–minor war to other major powers. There are other wars triggered by the logic of overtaking that did not expand into big wars. I suggest that the Seven Weeks' War and the Franco-Prussian War are two such wars.[3] They were good candidates to expand into big, general wars. The 1860s may provide a good case of the big war that did not happen for study. Understanding the causes of big wars requires us to look at the cases where they did not occur as well as those cases where they did occur.

Some Self-Criticism

The logic of overtaking presented here leads us to some new questions for research. The model makes assumptions that we may wish to relax in future research. I make these assumptions because they characterize how others have thought about power transitions. This argument attempts to provide a logic of overtaking that is broadly consistent with arguments that others, particularly Organski and Gilpin, have made about the consequences of long-term shifts in power. In this concluding section, I criticize my own argument. My objective here is to raise some issues for discussion that power transition theories have not considered.

First, the argument does not allow the declining state to crush the rising state before it emerges. The changes in capabilities that power transition theory examines can be foreseen far in advance, even by decades or more. We should wonder why declining states do not view the future rise of a challenger with alarm and take decisive action to make sure that it does not rise. For instance, why did Great Britain allow Germany to unify and then grow in power until it outstripped Britain? The British could have formed a coalition with the French and Austrians and guaranteed that Germany remained divided. One outstanding question facing power transition theory is why dominant states do not use their dominance to crush nascent challengers.

A second question is why war resolves the issues in dispute between the two sides. This assumption simplifies the analysis greatly and reflects the common argument that big wars resolve the issues in dispute. If we relax it, a transition could lead to several wars, as did Germany's ascent before the world wars. Oddly enough, relaxing this assumption might make war impossible in the model. After a war, both sides compare the outcome of that war to their expected utilities for another war when deciding whether to fight. Unless the outcome of the initial war exactly matches their expected utilities at the time of the second decision for war, at least one side expects to benefit from another war. Another war is likely then, and the outcome of the original war lasts only until the second war. The value of fighting the first war is reduced because its outcome does not persist after the second war. In other words, the value of winning the first war is lower when the loser can fight another war to overturn its outcome. Hence, the declining state's willingness to fight is reduced. Allowing multiple wars diminishes the value of fighting, making war less likely.

The final assumption I wish to discuss here is the lack of alternative responses to decline. The argument here focuses on the question of fighting versus adjusting to the demands of the rising state. But there are other strategies available to the declining state. Alliance and armament strategies can provide a way to compensate for a decline in relative capabilities. Dominant states do use these strategies to try to cover their deteriorating position in the system. As I argued earlier, alliances and the spread of small wars are the key to understanding big wars. It would be nice to incorporate those strategies directly into a model of decline. However, several problems immediately crop up. *N*-actor models are far more difficult to analyze than two-actor models. The trade-off between arms and allies is not understood well yet. The motivations of the lesser powers must be included in such a model. Nevertheless, response to decline is a critical question that power transition theory must consider in the future.

NOTES

1. One could argue that any shift in capabilities makes misperception more likely and so raises the likelihood of war. But we will be hard pressed to tell which shifts lead to war and which do not, except by observing what happened after the fact.

2. This argument assumes that the costs of war are constant across the transition period. A plausible alternative is that the costs of war grow as the transition progresses. The greater the rising state's capabilities, the more likely fighting will be protracted and bloody. This alternative assumption strengthens the argument. Higher costs of fighting in the future increase the attraction of fighting now instead of waiting.

3. See Bueno de Mesquita 1990a and Morrow 1993 for analyses of these two wars and the possibility that they could have spread into general wars.

CHAPTER 18

The New Open Door Policy:
U.S. Strategy in the Post–Cold War Era

A. F. K. Organski and Ronald Tammen

The power transition theory established the connection between development, growth rates, and the beginnings of major wars. Since it was first presented, the theory has also been used to predict the outcome and evaluate the consequences of major military conflicts. The theory has made of military conflict a successful laboratory for understanding how the international system works (Organski 1958; Organski and Kugler 1980).

In this paper, however, we look away from war, to the strategy of a dominant power in an environment of minimum danger, one where there is no real or imagined challenge to its rule. Power transition theory tells little about the dominant power's choices when it enjoys massive power advantages over all other major powers and potential challengers are not yet a problem. What will the dominant power's strategy be under these conditions?

Unbelievably, since the collapse of the Soviet Union there has been a continuing stream of allegations that the American leadership has lost its way. In the "bipolar" world that recently collapsed, each superpower had the task of guarding against the other. But with one superpower left, what is the survivor to do? The absence of an answer to this question serves as an allegation that American elites have lost their international bearings.

We strongly disagree. Critics have missed reality by a country mile in their images of each superpower balancing the other in the cold war period (Kugler and Organski 1989a; Organski forthcoming). They have been equally incorrect in their view of an incoherent U.S. strategy in the post–cold war world. It is not the United States but rather its critics who appear to have lost their way.

The United States clearly has a strategy in place. There are fundamental political guidelines we expect would be abided by whoever is in power in Washington, Democrats or Republicans, in much the same way that Democrats or Republicans abided by the containment policy over the decades. One can account for why the strategy has been "chosen," and trace its origins. The components of the strategy fit well together. There can be legitimate disagreement about its merits but not about its existence.[1]

The Shift in Coalitions

Analysis of strategy leads inevitably to talk of coalitions. Strategies reflect the identity, the aims, and the power of the members of the coalitions that sustain them. Governments similarly respond to pressures of the powerful who support them.

First, a brief mention of some basic points. We assume rationality. People who represent interests in the coalition running foreign affairs shape and direct foreign dealings to fit their interests. We assume the substance of national strategy to be a set of policies shaped by hard-won compromises in tugs-of-war among the representatives of the economic, security, and political interests who exercise control over resources and the flow of information. The international environment provides opportunities and sets constraints on their choices. It is the relative success of those representing these interests in capturing attention and resources for their cause that tells us what the overall strategy is all about.

We judge that the identity and the pecking order of interests represented in the foreign affairs coalition has changed substantially since the cold war days. We shall talk here only of what appears to us by far the most important of such changes. The change in question is a shift from an arrangement where political and military interests were preeminent to a fundamentally different one where political business interests have top priority in foreign affairs.

In the cold war period, politico-military interests had priority in the use of human and economic resources to shape the direction of foreign affairs. Politico-military interests defined the threat, determined what resources were required to meet it, and decided how such resources were to be allocated. Although other interests (i.e., business or information interests) were obviously represented in the coalitions, the interests of the politico-military actors dominated. President Eisenhower erred in his famous assessment that the military-*industrial* complex was behind security policy.

Here is an example of what we mean: containment policy was the central foreign affairs thrust of the United States for over forty years. In containment policy the political mandate was to guard against Soviet expansion, regardless of cost. Containment had two elements. One was the direct confrontation with the Soviet Union; the other was the containment of real or imagined Soviet expansion through the interdiction of Marxist forces, or their allies, from winning power in their own countries or in struggles with their neighbors. Much of this activity was only of very peripheral importance either for the security of the United States or for the functioning of the international order, but it dominated the American foreign policy agenda. It was critical in the choice of allies, the granting of assistance, and involvement in military operations. This, in the last analysis, is why the United States went into Vietnam and did not get out in time. This is also why the United States intervened in Africa and Latin America. Very often what was done had little to do with security but a lot with shielding presidents, their administrations, and their parties from the politically fatal charge that they were responsible for having lost an ally or neutral country to the Communist camp. Domestic political dynamics were the driving force in all such decisions and security concerns were a very distant second.

The politico-military coalition that had dominated American foreign affairs for so long was brought to its knees by the spectacular disintegration of the Soviet empire. With that fall the two political mainsprings that drove American strategy disappeared. One could no longer argue that the United States was in mortal danger and that maximum military readiness was required. Dark hints that the world was still a dangerous place did not carry much weight. It was also no longer possible to "lose" countries to the Communist camp. Political leaders were free of that incubus and political and military interests began to part ways. Their union finally came to an end with the economic recession in the midst of the Bush administration. The wrenching change in foreign policy priorities could be seen in President Bush's trip to the Far East. It had been planned as a trip to show the flag, but was hurriedly recast as a mission to win trade concessions. Hosts and guests would rather soon forget that trip. For President Bush, who was the last true representative of the politico-military coalition that had run United States foreign policy for forty-five years, it was a demeaning moment.

The Clinton administration established a politico-business coalition to coordinate foreign policy. The changeover occurred with remarkable smoothness, so much so that it has virtually gone unrecognized. Economic elites are now at the table with political leaders when foreign policy decisions are made. This new foreign affairs coalition has set about creating a political climate wherein mass publics, who might be adversely affected by the "new world economy," could be persuaded to mute their opposition. The traditional "foreign policy team," so important when security concerns held central stage, appears to have been downgraded. Critics' musings that the mediocrity of the team was a signal that the president himself would be his own secretary of state, secretary of defense, and national security adviser, or that the president was signaling that in his administration foreign policy was of secondary importance, miss the point completely. Clearly no president can entirely take over the making and execution of foreign policy in his own hands, nor is foreign policy less important. If anything, it is more important today. But the content of foreign policy has changed because the foreign affairs interests that the United States respond to are profoundly different. The old foreign policy team has found itself peripheral to the new set of national interests. Political military affairs are a smaller part of the action now and for the foreseeable future. The new coalition of interests has revolutionized U.S. foreign affairs, and there is a new strategy in place.

Strategy in the Present Period

Presently, the United States holds massive power advantages over other nations, leaving no room for ambiguity about who and what guarantees the peace in the post–cold war era (Organski 1993; Organski and Arbetman 1993; Organski and Kugler n.d.). It is the very condition power transition theory presents as conducive for international peace and stability (Organski 1958, 1968; Organski and Kugler 1980). The international and domestic power realities are a sine qua non for the strategy the United States has adopted and is executing. If and when these power distributions change the strategy will need to change as well.

Creating the Open Door

Foreign economic policy is a critical part of the new U.S. strategy. The cold war habit of thinking of foreign policy or strategy purely in politico-military terms has been hard to break. But it is quite clear that while some political and military questions are still very important, such matters, on average, are likely to have much lower priority. The reason why is common knowledge: the security of the United States and that of its principal allies is not even remotely at risk.

The shift in strategy of the United States is far more fundamental than it appears on the surface. One can advance a hypothesis that, after World War II, the internationalist contingent in American leadership circles seized upon the Soviet threat as a tool to achieve certain less obvious goals. They may have used the new set of circumstances to insure that the United States did not slip back into isolationism and to maintain the international leadership role the United States had achieved in the war. They took advantage of the Soviet threat to maneuver the United States into its cold war role of guardian against Soviet expansion. They transformed the Soviet refusal to allow the West any access to Eastern Europe, and their constant ruthless stamping out of independence and freedom there, into evidence that the Soviet Union was a militarily expansionist power threatening the West. On the domestic side, they found it possible to put together a coalition of interests that permitted the raising and channeling of mammoth funds to cover the costs attendant to any attempt to deal with this image of the Soviet Union.

At the end of World War II, there was no world economy. The United States could have withdrawn into isolation. Over the cold war years, however, a world economy was created with the United States at its center. The leaders of the United States economy and military forces, as well as important factions of the political elite, opposed any retreat into isolationism; American internationalism was secure. However, by the time the cold war ended, American elites were no longer willing to pay the high costs of military expenditure that had been the price to affirm the American international position.

The new international strategy of the United States has profoundly reorganized international priorities. Top priority has been withdrawn from security and extended to economic issues. The business of the government is again business: removing trade barriers to American goods and services. Opening new markets to American exports has replaced stopping Communism as the first priority of foreign affairs. Similarly, the fear of loss of jobs, or the hopes of creating new ones through expanding exports, have replaced the fear of Soviet expansion, or efforts to reduce the Soviet grip over the people it controlled in the cold war era. Jobs, the new political all-purpose justification for much that is done in the field of foreign affairs, is the political motive for American activities abroad.

Opening one's markets to economic interpenetration should prove destructive of sovereignty, particularly in the case of nations with weaker economies. How else is one to view government abandonment of its right to interpose itself between external pressures and its people's economic lives? A political field equivalent of this claim against economic sovereignty is the demand that nations give up their right to arm themselves with nuclear weapons and submit to international monitoring. The steps have been slow and often halting, but if the movement should continue it would revolutionize the way we think of international relations.

In a sense, of course, much of the present strategy is not new. The United States appears to be returning to at least some aspects of foreign policy of the "isolationist" days before World War II. It was then the accepted view that the United States ought to expand, but that the expansion should be economic. Political commitments and military expenditures were to be kept to a minimum. That would only be possible if political obstacles to trade were swept aside. That task was the proper role of government. The reader will recall that in that period the call was for "the open door." Doors were opened by Latin American and East Asian countries where governments were very weak. But, for the most part, governments that were powerful enough defended their markets against competition they could not win (Organski 1990, 1993, forthcoming). That central thrust of avoiding political commitments and military investments, as we are all aware, was reversed in the cold war era and has now been reversed once again in the post–cold war period.

At the end of World War II there were major concerns with the construction of a new international economic order and free trade. Even in the cold war period, when the focus in foreign affairs was on political and military issues, there were major initiatives sweeping away political obstacles to economic expansion. The most notable, even though it is seldom seen in this light, was the United States and the Soviet Union joining forces to help colonials demolish the West European empires. We are not arguing that the United States was the sole or even the most important force in bringing about decolonization. Colonialism contained its own seeds of self-destruction. Change in Europe, even more than in the colonies, pushed forward decolonization. Europe had modernized its political systems, and the political costs of keeping colonials in check had become too high. At the margins, however, the American role was critical.

Removing the iron curtain (the political obstacle par excellence of any kind of penetration from the outside) and permitting the United States and the rest of the West to enter the Communist domain was the major aim of American policy for the whole period of the cold war. The Soviet Union that joined in the decolonization effort of West European colonies could not imagine that thirty years later its turn would come.

There were also other initiatives to expand free trade before the cold war wound down: the longstanding struggle with the Japanese over opening their markets; the initiatives to insure the United States would expand access to the European Common Market; the political effort to "pool" the North American economies under the North American Free Trade Agreement (NAFTA); and American leadership and prodding on the General Agreements on Tariffs and Trade (GATT).

Clearly, concern with removal of trade barriers has a long history. The difference between previous efforts and present policies is that earlier (since World War II) other matters—usually security matters—always took precedence. Now they no longer stand in the way.

Privatization

Another aspect of the strategies of removing political barriers to economies requires comment: the United States government advocacy of free markets. Free markets are a prerequisite to free trade. Advocacy of them has been a powerful refrain since the Reagan administration and has won incredible acceptance from the elite of most nations. Indeed, it has become almost a new elite religion worldwide. Privatization, the creation of and reliance on private markets for almost any purpose, is the policy of choice for elites in countries with different governmental forms and at all levels of development. It is simply extraordinary that, while democracies still contend with Communist political systems, both are now accompanied by capitalist economies. The triumph of the private over the public sector, at least for the time being, appears assured.

This emphasis on the private sector, so strongly identified with U.S. policy preferences for itself and others, has made the United States once again a model for other nations to emulate. This is a clear signal of the nation's influence, almost a return to the period after World War II when the United States was considered at the apogee of its power and a model to the rest of the world. When one looks back, the 1960s were a period when many of the elite of the world looked to the Soviet system for a solution to their problems. The 1960s were the high point of Soviet influence. There followed a long period of indecision when neither superpower enjoyed much prestige. Then, in the second half of the 1980s, the United States regained its role as model for the international community.

Again, we do not suggest that the widespread acceptance of free markets in the last decade is all in response to American policy, only that, at the margins, the American role was clearly important. Other forces were at work. The move away from control economies and expansion of public sectors in democratic societies had become visible by the end of the 1970s. The change in China, shifting agriculture to private control and the initial steps of privatization in the secondary and tertiary sectors, and Thatcher's victory in Britain had come at the end of the 1970s. Both took place before the Reagan administration. China's action, particularly, was an indication that the tide had turned. Of course, most of all, the collapse of the Soviet world in the late 1980s propelled forward the belief that private markets were the solution. Perhaps the United States sailed the issue as the tide was rising.

Nevertheless, American elites played a central role in creating the present free market craze. This began with President Reagan's cajoling, demanding, negotiating, and wherever possible pressuring governments to deregulate and privatize. During the cold war, American elites called on the elites of other nations to support democratic systems. Democracy was all that was required. Today the standard has been changed to democracy *and* free markets. The importance of the change cannot be overestimated.

The Strategy's Politico-Military Components

Political and military commitments appear no longer tightly connected. Political commitments seem to be expanding while military resources are being kept in check or decreasing. For example, in central Europe the talk is of a substantial extension of political commitments unaccompanied, for the time being at least, by an increase of military resources. Of course military commitments have immediate and often large economic and political costs while the payments for political commitments can often be postponed and, with luck, need not be paid at all.

In the cold war era, political and military commitments were determined by the elite's perception of threat, and were then communicated to mass publics through a vast and complicated process. The estimation (and manipulation of perceptions) of threat was big political business in the cold war years. Even now with every new disquieting occurrence—the Russian election of December 1993, for example—observers today begin to hint darkly that dreadful danger may lay ahead, and call for the United States to stop drawing down its military forces and begin to talk tough.

In the new security environment the question that is posed by those responsible for the financial health of the nation as well as foreign affairs is the inverse of the question asked in the cold war. Then the issue faced by leadership was, in effect, how large an investment in the military one could get away with. For the new coalition leading foreign affairs the question appears to be, how little an investment in the military establishment is possible?

It would be foolish to expect that limits will be established by the exercise of self-restraint, the common good, good sense, or morality plays about how international politics should work. It is more realistic to expect that those positioned advantageously in the decision-making process of allocating resources will likely press their advantage all they can. The American experience in the cold war is a case in point. No one knows how much was spent in excess of any reasonable requirement of security. Perhaps it is better not to know. In the case of the Soviet Union, those in control took their advantage until they nearly bankrupted the country. Clearly the most effective way to establish limits is through competition with other actors pushing and pulling for their share.

Political Commitments

Presently, international political commitments have not been reduced, and there are proposals to extend them. The most important of these is the present struggle about extending NATO to central European countries searching for stability and protection. The concern with their vulnerability to Russian aggression is highly questionable in view of a dispirited and exhausted Russia with no interest and stomach for "foreign" adventures. They are on firmer ground in their concern about relations with one another and other Eastern European states, a resurgent Germany, and domestic challenges. The whole discussion about the expansion of NATO is very reminiscent of the political dynamics of the creation of the organization decades ago. The process is one where European political leaders rush to Washington begging it to commit itself to their protection. They are resisted by some elite factions worried about trade-offs with other American interests in Eastern Europe, or about the potential economic and political costs of such moves, but they are befriended by other factions of the elite anxious to have the United States commit itself to institutional expansion of its political influence. The arguments of those beating the drums for political expansion are many: expansion of NATO membership to selected countries in Eastern Europe is a "logical" European denouement of the victory in the cold war. We may not see again East European countries begging for security, economic, institutional, and cultural ties with the West. It is a historic opportunity to recast the European scene for the next thirty years and for the United States to establish influence over a major pool of human and material resources. This will prove critical if a new challenge to American leadership materializes. From the point of view of the new strategy two points should be kept in mind. "Partnerships," rather than full membership for the Eastern European nations lowers immediate economic and political costs. And lowering costs is a major preoccupation of the new coalition. Admittedly, President Clinton's Partnership for Peace amounts to backing into an expansion of American political influence in Europe. Not a very admirable maneuver, but the expansion cannot be missed.

Military Investments

On the other hand, the current economic and political coalition of interests has been under very strong pressure to reduce military and paramilitary commitments. As one would expect, a major part of the military investments being drawn down are the ones connected with providing a nuclear and nonnuclear deterrent targeted against the former Soviet Union. A large portion of that investment was, of course, in Europe. It was a huge investment and represented a very large share, some argue about half, of the total defense expenditure (Organski 1990: appendix C2). The politico-military leaders in charge of these resources resisted as much as they could, and the intelligence community fought hard to postpone the inevitable. Yet, the change in the international environment removed any reason, and more importantly any political justification, to continue very large investments; domestic pressures finally won.

The process, though complex, could be summarized as follows. A major source of pressure came from the enormous debts and deficits piled up in the last decade of the cold war. The pool of national resources earmarked for the military was extremely large, and for years was a tempting target for other elites. Reductions in other major expenditure categories, particularly entitlements, were politically impossible until the military investments, now suddenly indefensible on security grounds, were also cut. A more general source of pressure came from the prodding by the demands of economic and other elites who wanted their share of the national pool of resources. For forty years their immediate interests had been sacrificed repeatedly on the altar of security considerations. American business had seen American markets open to competitors in order to keep them on America's side in the struggle with the Soviet Union. They had seen enormous monetary and human resources poured into security and away from productive enterprise. They had previously little influence in the direction of foreign affairs, but now wanted resources freed for investment or for programs that had gone without for a long time.

The New Military Needs

The reduction of investments in military resources, connected with the disappearance of the Russian threat, give a rough indication of where the support levels for U.S. investment in military resources are to be found.[2] The issue is not likely to be settled for some time. Investments can be expected to go up again as a new challenger threatens to erase the power advantage the United States enjoys at present. But that is in the future.

Powerful countercurrents fight decreases much beyond the new levels, at least for a time. The new American interests in foreign affairs require military power for three contingencies, two of which deal with the operation of the international order and its stability. First is the protection of resources and transportation routes of commodities critical for the operation of the international system. The case of the Iraqi invasion of Kuwait and the American reaction to it in all its ramifications is a case study of a dominant nation protecting such sources. One still recalls in disbelief the accusatory statement of so many observers questioning the purpose of the United States in the Gulf War. Was it oil? Would the United States become the world's policeman? And the speaker would look knowingly at his or her interlocutor. One wonders why American leaders always find it difficult to tell the truth about such matters. Of course it was oil and aggression. The United States needed to guarantee availability and prices for itself and allies and guarantee the safety of nations that were the source of both. One should not forget that an important part of that operation was also the successful international legitimation of a military operation.

The other pressure for not cutting military expenditures beyond present levels is the threat of proliferation of nuclear weapons and the destructive effect on the stability of the system. Military power is widely recognized as essential in dealing with the problem, and deeper cuts in military muscle are seen as weakening the ability of the United States to develop solutions when confronting new nations wishing to acquire nuclear weapons. We have already noted that American (and allied) insistence that new aspirants to the nuclear club desist and submit to monitoring of their behavior is a body blow to the sovereignty of states, for sovereignty is rooted in the untrammeled right of governments to arm themselves. Monitoring the behavior of countries in the nuclear field requires interference in their internal affairs. Targets of such interference often resist international monitors and controls. Such resistance, of course, can be overcome, as has been the case with Iraq. However, international consent to press on with such controls is difficult to maintain.

The third factor pressing against further decreases in military resources is less obvious, yet very important: large military resources still provide important quid pro quos to any dominant nation in its dealings with members in the international order. Military resources are evidence that the United States has the capability to protect its friends, providing additional leverage in its negotiations with other nations. The widespread interest in being under the protection of the United States military might is not limited to any one region or very small powers, but includes nations in the medium and even great-power rank. Japan and some of the West European powers serve as examples. Europeans have demonstrated security questions to be beyond their abilities for the time being. For the present, credible protection can only be found with the United States.

We should note again that military resources are not enough. International legitimacy and hence consent are required. A dominant power's persuasiveness and negotiating skills are as important as military resources.

On Not Using U.S. Military Forces

We assume that the coalition of interests backing American foreign affairs will try, as we have already seen them do, to maximize their utility by cutting outlays but keeping gains at least at the same level. Ignoring the weak has clearly been one way to go. In choosing tactics and tools to discipline the "refractory members" of the international community tactics will be selected with an eye to minimizing costs.

Opponents of the intervention in the Gulf War often asked, is the United States now the world's policeman? What are the limits to intervention? A clear answer has been given. In the vast majority of cases where struggles over local issues and resources erupt, the United States will not intervene to restrain the warring faction battling to impose a settlement on others. With the exception of cases where nuclear proliferation and commodities critical to the functioning of the advanced economies are involved, United States military might will not be used to reverse outcomes or change distributions of power on the site. The balance of forces of the contestants will impose the settlement of such disputes in the end.

The tactics used are not difficult to understand. They involve ignoring the strife until the parties are ready to agree, helping them negotiate if they do, and continuing to ignore them, at least publicly, if they do not. The United States will not be a peacemaker. The parties must be ready to accept the outcome their own struggle produces, whether they are persuaded, defeated, or simply exhausted. Somalia, Bosnia, Haiti, and Palestine are illustrations of the pattern we hypothesize. The United States will serve primarily as negotiator and legitimizer, and once peace has been agreed, it will assist in peacekeeping. It is a strategy permitting the minimization of economic and, far more important, of political costs associated with interventions. Interventions are always expensive because military resources provide very dull tools at best and changing the behavior of other countries' leaders and populations through their use is very uncertain business. It seems plausible to argue that economic elites strongly oppose policies threatening the economic environment, while political leaders are particularly sensitive to political costs of foreign interventions, for it is they who have to pay them. That is why American leaders have said that they will consider peacekeeping but not peacemaking.

Many react with bitterness and contempt to this portion of the strategy. How could the United States allow Bosnia or Somalia or Haiti to occur? Is *this* the New World Order? Although one understands all these feelings, they show a misunderstanding about the meaning of maintaining international order. It does not mean protection of populations everywhere from the savagery of their own or neighboring leaders; it does not mean protection of such populations from mass punishments to bring their leaders back on the straight and narrow. Weak nations, regardless of whether they are satisfied or dissatisfied, cannot impact the functioning of the international order and can be ignored (Organski 1958). That is precisely what the United States is doing. The strategy has resulted primarily from internal pressures to minimize political costs that follow when military operations cost lives, material resources, or reputation. There is no reason to believe that this calculus will differ for the foreseeable future.

It is a nasty business, of course, and the coalition running foreign affairs wants no part of the responsibility. Inevitably, there is a lot of finger-pointing. American government leaders, pundits, and the media have blamed the American mass public for its inconstancy and contradictory ways. American government leaders have blamed the leaders of other nations. Such arguments are without merit. Remember the American leaders explaining they could not move in Bosnia because they had asked European leaders for and been refused cooperation? Critic after critic weighed in against the secretary of state for his "incompetent" performance. Why, the critics said, one should know how to deal with Europeans; one should not have asked but have told them! In our view, the explanation for inaction was an excuse, and a very transparent one at that. Had the United States wanted to intervene the Europeans would have followed. It was all too handy to ask and be turned down. The issue of the skills of the secretary of state in his dealings with Western Europe is a red herring. He did what he was expected to do.

Similarly, it is said that this policy of ignoring local fights, even where action comes close to genocide, is due to the fact that mass publics want it both ways. They want the United States to intervene in regional or civil wars when it is brought to their attention that bullies savage their neighbors or their own people. Yet as soon as there are American casualties the mass public recoils and turns on the government for sending American forces in harm's way. Of course, mass publics are being maligned; they do not destroy these policies unless they are mobilized and led by elites scavenging for attention and political advantage by offering themselves as defenders and protectors of the lives of soldiers. The responsibility rests with elites, not masses. Recall how patient and controlled mass publics were in accepting casualties when elites were rendered impotent to exploit their grief. Remember the drawn faces and the muffled crying of relatives receiving remains of soldiers killed in Lebanon, or Grenada, in peacekeeping operations during the Reagan administration.

There is also another reason for the high political costs of using military forces. The military establishment contributed to raising the political costs of any use of American military power in peacekeeping operations by establishing unwritten, but politically powerful, rules governing the use of such forces. In situations where the United States and its allies' security is not in danger, American forces should not be used. Were they to be used, however, military leaders could give their consent (politically essential to all but the strongest and most popular political leaders) if they were permitted to build up forces to levels guaranteeing the overwhelming of the opponents, quick victory, no casualties, and total support for the armed forces on the part of elites and mass publics back home. But such operations cost too much. This doctrine is said to have been imprinted on the military mind by the Vietnam experience and is a legacy of Secretary of Defense Caspar Weinberger.[3] This doctrine appears less concerned with issues of security than protection of the military as an institution in its constant competition with its sometime allies and sometime competitors, the political and economic elites.

The result of the interplay of all of these elements is that the foreign policy leadership of the United States has chosen to finesse the use of force by eliminating peacekeeping, with the exceptions mentioned earlier, from the foreign affairs agenda. It is wrong, however, to view the elimination of this peacekeeping function, so important to those concerned with issues of morality and justice, as a sign that there is no strategy and that the leaders do not know what they are doing. They do, and have chosen to eliminate costs they think they need not afford.

Sanctions

Sanctions and embargoes are used because they permit the leadership and major sectors of the public of the dominant power (and principal allies) to feel they are potent in the face of recalcitrant foreign leadership, yet minimize the use of force. Sanctions have become a disciplinary action of choice because those who impose them pay minor economic and even smaller political costs. In a sense, in the short term at least, the imposition of sanctions has more to do with self-images and feelings of important sectors of elite and mass publics in the dominant nation (and principal allies) than any immediate thought of changing the realities they were instituted to address. In the long run, of course, embargoes are supposed to change the distribution of power within the targeted nations by building opposition and putting political resources in their hands. Some question whether they really work that way. Be that as it may, it is clear that sanctions and embargoes are dull instruments, affecting their targets usually only after a substantial delay, and then often only hurting the masses and sparing the leaders. If the leaders cared for the suffering of their people, one supposes, international sanctions might be a more effective tool for disciplining international miscreants. The problem with such tactics is that the leaders of the target nation do not care. Indeed, they use the opportunity to turn the tables on their tormentors. They fight back by holding their people hostage. In struggles for power the leaders of weak nations have one major resource to hit back at the international community or the leaders of adversaries: their people's lives. It is their strategy to let their people agonize until their adversaries "can no longer stand" the sight of suffering and retreat.

Conclusion: The New Open Door

What are we to say in conclusion? Clearly, it is incorrect to argue that the United States has no post–cold war strategy and that it has lost its way. Not surprisingly, its new strategy is a function of the pulls and pushes that arise out of the new coalition of interests guiding the nation's foreign dealings. Priorities have changed. The United States views the world in terms of economic gains for its people. Economic penetration and new markets are the critical preoccupation of foreign affairs; political commitments follow at some distance and military investments trail the other two. The structure of priorities maximizes gains and pares costs. It is the very opposite ranking that governed foreign policy during the cold war. The nature of the international strategy of the United States in the post–cold war era, where American power is unchallenged, is coming into view.

There has been a clear cycle to American foreign dealings since it first passed every other major power in economic terms over a century ago. The cycle has moved from open door to containment and back to the open door. When the policy of forcing open new markets was first pursued in the decades before World War II, the United States towered over all other nations economically, but not politically. It held no real advantage over other nations in terms of power. It has a huge advantage now. We should expect, however, a new challenge down the road (Organski and Kugler 1980; Organski 1993; Organski and Arbetman 1993).

The strategy we have reviewed will not endure forever. We view this post–cold war period as a long one between two challenges to American dominance. One challenge has just been turned back. Another challenge is still far off, but now appears a probability (Maddison 1989; Organski and Arbetman 1993; Organski and Kugler n.d.). Toward the end of the present period the major powers will again need to deal with the consequence of one in their midst going through its power transition. We end as we began. It takes the strangest vision to argue, as so many observers have, that the leadership of the United States has lost its way. No, it is the critics who have lost their way, and not for the first time.

NOTES

The authors are very grateful to Ellen Lust-Okar of the Center for Political Studies for her sharp analytic comments and superb editorial skills. Professors James Morrow of the Hoover Institution at Stanford University and Randy Siverson of the University of California at Davis deserve our thanks for having convinced us that this research effort needed doing.

1. The Clinton administration's definition of its foreign policy views it as containing four principles: continuation of the U.S. international engagement; primacy of economic concerns; readiness to use multilateral and/or unilateral action in dealing with international security problems; and an attempt to enlarge the set of societies with free markets and democratic political systems. Administration officials have named this group of policies "The Policy of Enlargement," a particularly unfortunate name. For a good review see Lowenthal 1993.

2. The issue of decreases in military expenditures can be seen from a number of perspectives. There have been substantial decreases in military outlays from the peak year of 1987. (All data are in 1994 dollars.) From $347.8 billion in 1987, outlays decreased in 1994 to $268.9 billion. Estimates for subsequent years are $259.2 billion for 1995; $248.5 billion for 1996; $227.7 billion for 1997; $232.6 billion for 1998. Calculating decreases from the peak years in some sense distorts the reality of changes in military outlays. In the current administration outlays have decreased less than 10 percent. Highest expenditures for defense were substantially higher. Expenditures at the end of World War II were $466.3 billion in 1946; and $362.6 billion in 1968, at the height of the war in Vietnam. A second perspective comes from comparing U.S. expenditures with those of the other great powers. The comparison here is for the year 1989 and is in billions of 1989 U.S. dollars: U.S. $304.1; U.K. $34.6; Japan $28.4; Germany $33.6; China $22.3 (U.S. Arms Control and Disarmament Agency 1991; Kosiak 1993).

3. The Weinberger tests are:

(1) The United States should not commit forces to combat overseas unless the particular engagement or occasion is deemed vital to American national interest or that of her allies. (2) If the United States decides it is necessary to put combat troops in a given situation, it should do so wholeheartedly, and with the clear intention of winning. (3) If the United States does decide to commit forces to combat overseas, it should have clearly defined political and military objectives. (4) The relationship between

American objectives and the forces committed—size, composition, disposition—must be continually reassessed and adjusted if necessary. (5) Before the United States commits forces abroad, there must be some reasonable assurance they will have the support of the American people and their elected representatives in Congress. (6) The commitment of U.S. forces to combat should be the last resort.

These six tests have been adopted by the military informally and formally as contained in various speeches of the Joint Chiefs (for example). They represent a serious and apparently permanent constraint on the use of power by the United States (see Weinberger 1985).

Bibliography

Albrecht-Carrié, Rene. 1973. *A Diplomatic History of Europe Since The Congress of Vienna*. New York: Harper and Row.

Allison, Graham T. 1971. *Essence of Decision: Explaining the Cuban Missile Crisis*. Boston: Little, Brown.

Altfeld, Michael. 1983. "Arms Races?—And Escalation? A Comment on Wallace." *International Studies Quarterly* 27:225–31.

———. 1984. "The Decision to Ally: A Theory and Test." *Western Political Quarterly* 37:523–44.

Andreano, Ralph, ed. 1967. *The Economic Impact of the American Civil War*. 2d ed. Cambridge, MA: Schenkman Publishing Company.

Arbetman, Marina. 1990. "The Political Economy of Exchange Rate Fluctuations." Ph.D. diss., Vanderbilt University.

Balke, Nathan S., and Robert J. Gordon. 1989. "The Estimation of Prewar Gross National Product: Methodology and New Evidence." *Journal of Political Economy* 97:38–92.

Banks, Arthur S. 1971. *Cross-Polity Time-Series Data*. Cambridge, MA: MIT Press.

Barraclough, Geoffrey. 1961. "Europe and the Wider World in the Nineteenth and Twentieth Centuries." In A.O. Sarkissian, ed., *Studies in Diplomatic History and Historiography*. London: Longman.

Bauer, L., and H. Matis. 1989. *Geburt der Neuzeit*. Munich: Deutscher Taschenbuch Verlag.

Beard, Charles A., and Mary R. Beard. 1930. *The Rise of American Civilization*. New York: Macmillan.

Berkowitz, Bruce. 1985. "Proliferation, Deterrence, and the Likelihood of Nuclear War." *Journal of Conflict Resolution* 29:1.

Berry, Thomas, Sr. 1968. *Estimated Annual Variations in Gross National Product, 1789–1909*. Richmond, VA: Bostwick Press.

———. 1978. *Revised Annual Estimates of American Gross National Product*, "Preliminary Annual Estimates of Four Major Components of Demand, 1789–1889." Bostwick Paper No. 3. Richmond, VA: Bostwick Press.

Betts, Richard K. 1987. *Nuclear Blackmail and Nuclear Balance*. Washington, DC: Brookings Institution.

Bienen, Henry, and Nicholas van de Walle. 1991. *Of Time and Power*. Stanford, CA: Stanford University Press.

347

Binder, Leonard. 1958. "The Middle East as a Subordinate International System." *World Politics* 10:408–29.

Blainey, Geoffrey. 1973. *The Causes of War.* New York: Free Press.

———. 1988. *The Causes of War.* 3d ed. New York: Macmillan.

Boulding, Kenneth. 1962. *Conflict and Defense: A General Theory.* New York: Harper and Brothers.

Bowman, Larry W. 1968. "The Subordinate State System of Southern Africa." *International Studies Quarterly* 12:231–61.

Brams, Steven J. 1975. *Game Theory and Politics.* New York: Free Press.

———. 1985. *Superpower Games: Applying Game Theory to Superpower Conflict.* New Haven, CT: Yale University Press.

Brecher, Michael. 1963. "International Relations and Asian Studies: The Subordinate State System of Southern Asia." *World Politics* 15:213–35.

Bremer, Stuart A. 1980. "National Capabilities and War Proneness." In J. David Singer, ed., *The Correlates of War II: Testing Some Realpolitik Models.* New York: Free Press.

———. 1992. "Dangerous Dyads: Conditions Affecting the Likelihood of Interstate War, 1816–1965." *Journal of Conflict Resolution* 36:309–41.

———. 1993. "Democracy and Militarized Interstate Conflict, 1816–1965." *International Interactions* 18:231–49.

Brodie, Bernard, ed. 1946. *The Absolute Weapon.* New York: Harcourt, Brace and Company.

Bueno de Mesquita, Bruce. 1975. "Measuring Systemic Polarity." *Journal of Conflict Resolution* 19:187–216.

———. 1978. "Systemic Polarization and the Occurrence and Duration of War." *Journal of Conflict Resolution* 22:241–67.

———. 1980. "Theories of International Conflict: An Analysis and an Appraisal." In Ted R. Gurr, ed., *Handbook of Political Conflict.* New York: Free Press.

———. 1981a. *The War Trap.* New Haven, CT: Yale University Press.

———. 1981b. "Risk, Power Distributions, and the Likelihood of War." *International Studies Quarterly* 25:541–68.

———. 1985. "The War Trap Revisited." *American Political Science Review* 79:1113–30.

———. 1989. "The Contribution of Expected Utility Theory to the Study of International Conflict." In Manus Midlarsky, ed., *Handbook of War Studies.* Boston: Unwin Hyman.

———. 1990a. "Pride of Place: The Origins of German Hegemony." *World Politics* 43:28–52.

———. 1990b. "Big Wars, Little Wars: Avoiding Selection Bias." *International Interactions* 16:159–69.

———, and David Lalman. 1986. "Reason and War." *American Political Science Review* 80:1113–50.

———. 1988. "Empirical Support for Systemic and Dyadic Explanations of International Conflict." *World Politics* 41:1–20.

———, and David Lalman. 1992. *War and Reason: Domestic and International Imperatives.* New Haven, CT: Yale University Press.

————. 1994. "Power Relationships, Democratic Constraints, and War." In Frank W. Wayman and Paul F. Diehl, eds., *Reconstructing Realpolitik*. Ann Arbor, MI: University of Michigan Press.

————, and William H. Riker. 1982. "An Assessment of the Merits of Selective Nuclear Proliferation." *Journal of Conflict Resolution* 26:283–306.

Bueno de Mesquita, Bruce, and Randolph M. Siverson. 1993. "War and the Fate of Political Leaders." Paper presented at the annual meeting of the American Political Science Association.

Bundy, McGeorge. 1983. "The Bishops and the Bomb." *New York Review of Books*, June 16:3–8.

————, George F. Kennan, Robert McNamara, and Gerard Smith. 1982. "Nuclear Weapons and the Atlantic Alliance." *Foreign Affairs* 60:753–68.

Chan, Steve. 1987. "Growth with Equity: A Test of Olson's Theory for the Asia-Pacific Rim Countries." *Journal of Peace Research* 24:135–49.

Choucri, Nazli, and Robert C. North. 1975. *Nations in Conflict: National Growth and International Violence*. San Francisco: W. H. Freeman and Company.

————. 1989. "Lateral Pressure in International Relations: Concept and Theory." In Manus Midlarsky, ed., *Handbook of War Studies*. Boston: Unwin Hyman.

Claude, Inis L., Jr. 1962. *Power and International Relations*. New York: Random House.

Cochran, Thomas C. 1961. "Did the Civil War Retard Industrialization?" *Mississippi Valley Historical Review* 48:197–210.

Conway's All The World's Fighting Ships. Various years. Edited by Robert Gardiner. New York: Mayflower Books.

Conybeare, John. 1992. "A Portfolio Diversification Model of Alliances." *Journal of Conflict Resolution* 36:53–85.

Crislip, Mark N. 1994. "Reputation, Power Transition and War." Paper presented at the annual meeting of the Midwest Political Science Association, Chicago, IL.

Dacey, Raymond. 1976. "A Historiography of Negro Slavery, 1918–1976." In Benjamin Taylor and Thurman White, eds., *Issues and Ideas in America*. Norman, OK: University of Oklahoma Press.

Dehio, Ludwig. 1962. *The Precarious Balance: Four Centuries of the European Power Struggle*. New York: Vintage Books.

Delper, Helen, ed. 1974. *Encyclopedia of Latin America*. New York: McGraw Hill Book Company.

Deutsch, Karl. 1978. *The Analysis of International Relations*. 2d ed. Englewood Cliffs, NJ: Prentice-Hall.

————, and J. David Singer. 1964. "Multipolar Power Systems and International Stability." *World Politics* 16:390–406.

Diehl, Paul. 1983a. *Arms Races and the Outbreak of War, 1816–1980*. Ph.D. diss., University of Michigan, Ann Arbor.

————. 1983b. "Arms Races and Escalation: A Closer Look." *Journal of Peace Research* 20:205–12.

————. 1985. "Contiguity and Military Escalation in Major Power Rivalries." *Journal of Politics* 47:1203–11.

————, and Jean Kingston. 1987. "Messenger or Message?: Military Buildups and the Initiation of Conflict." *Journal of Politics* 49:801–13.

Doran, Charles F. 1971. *The Politics of Assimilation: Hegemony and Its Aftermath*. Baltimore: Johns Hopkins University Press.

————. 1983a. "Power Cycle Theory and the Contemporary State System." In William R. Thompson, ed., *Contending Approaches to World System Analysis*. Beverly Hills, CA: Sage Publications.

————. 1983b. "War and Power Dynamics: Economic Underpinnings." *International Studies Quarterly* 27:419–41.

————. 1989. "Systemic Disequilibrium, Foreign Policy Role, and the Power Cycle: Challenges for Research Design." *Journal of Conflict Resolution* 33:371–401.

————. 1991. *Systems in Crisis*. New York: Cambridge University Press.

————, and Wes Parsons. 1980. "War and the Cycle of Relative Power." *American Political Science Review* 74:947–65.

Dougherty, James F., and Robert L. Pfaltzgraff, Jr. 1981. *Contending Theories of International Relations*. New York: Harper and Row.

Doyle, Michael W. 1983. "Kant, Liberal Legacies, and Foreign Affairs." *Philosophy and Public Affairs* 12:205–35.

Duncan, Julian Smith. 1932. *Public and Private Operation of Railways in Brazil*. New York: Columbia University Press.

East, Maurice A. 1972. "Status Discrepancy and Violence in the International System." In J. Rosenau, V. Davis, and M. East, eds., *The Analysis of International Politics*. New York: Free Press.

Easterlin, Richard. 1960. "Interregional Differences in Per Capita Income, Population, and Total Income, 1840–1950." In *Trends in the American Economy in the Nineteenth Century,* Studies in Income and Wealth, Vol. 24. Princeton, NJ: National Bureau of Economic Research.

————. 1961. "Regional Income Trends, 1840–1950." In Seymour Harris, ed., *American Economic History*. New York: McGraw Hill.

Einstein, Albert. 1960. *Einstein on Peace*. New York: Simon and Schuster.

Ellsberg, Daniel. 1959. "The Theory and Practice of Blackmail." Lecture at the Lowell Institute, Boston, MA, March 10. Reprinted in Oran R. Young, ed., *Bargaining: Formal Theories of Negotiation*. Urbana, IL: University of Illinois Press, 1975.

————. 1961. "The Crude Analysis of Strategic Choice." *American Economic Review* 51:472–78.

Engerman, Stanley L. 1966. "The Economic Impact of the Civil War." *Explorations in Entrepreneurial History,* 2d Ser., 3:176–99.

————. 1971. "Some Economic Factors in Southern Backwardness in the Nineteenth Century." In John F. Kain and John R. Meyer, eds., *Essays in Regional Economics*. Cambridge, MA: Harvard University Press.

Ferris, Wayne. 1973. *The Power Capabilities of Nations-States.* Lexington, MA: Lexington Books.

Fogel, Robert, and Stanley Engerman. 1974. *Vol. 1, Time on the Cross. Vol. 2, The Economics of American Negro Slavery.* Boston: Little, Brown.

Freedman, Lawrence. 1981. *The Evolution of Nuclear Strategy.* New York: St. Martin's Press.

Fudenberg, Drew, and Jean Tirole. 1991. *Game Theory.* Cambridge, MA: MIT Press.

Gaddis, John Lewis. 1987. *The Long Peace: Inquiries into the History of the Cold War.* New York: Oxford University Press.

Gamson, William, and Andre Modigliani. 1971. *Untangling the Cold War: A Strategy for Testing Rival Theories.* Boston: Little, Brown.

Garnham, David A. 1976a. "Power Parity and Lethal International Violence, 1969–1973." *Journal of Conflict Resolution* 20: 379–94.

———. 1976b. "Dyadic International War 1816–1965: The Role of Power Parity and Geographic Proximity." *The Western Political Quarterly* 29:231–42.

Geller, Daniel S. 1985. *Domestic Factors in Foreign Policy: A Cross-National Statistical Analysis.* Cambridge, MA: Schenkman Publishing Company.

———. 1992a. "Capability Concentration, Power Transition, and War." *International Interactions* 17:269–84.

———. 1992b. "Power Transition and Conflict Initiation." *Conflict Management and Peace Science* 12:1–16.

———. 1993. "Power Differentials and War in Rival Dyads." *International Studies Quarterly* 37:173–94.

George, Alexander, and Richard Smoke. 1974. *Deterrence in American Foreign Policy.* New York: Columbia University Press.

Gilchrist, David, and David W. Lewis, eds. 1965. *Economic Change in the Civil War Era.* Grenville, DE: Eleutherian Mills-Hagley Foundation.

Gilpin, Robert. 1981. *War and Change in World Politics.* New York:Cambridge University Press.

———. 1988. "The Theory of Hegemonic War." *Journal of Interdisciplinary History* 18:591–613.

Gochman, Charles. 1975. "Status, Conflict, and War: The Major Powers, 1820–1970." Ph.D. diss., University of Michigan, Ann Arbor.

———. 1980. "Status, Capabilities, and Major Power Conflict." In J. D. Singer, ed., *Correlates of War, Vol. 2.* New York: Free Press.

———. 1990. "Capability Driven Disputes." In Charles Gochman and Alan Sobrosky, eds., *Prisoners of War?,* 141–59. Lexington, MA: Lexington Books.

———. 1991. "Interstate Metrics." *International Interactions* 17:93–112.

———, and Russell J. Leng. 1983. "Realpolitik and the Road to War: An Analysis of Attributes and Behavior." *International Studies Quarterly* 27:97–120.

Gochman, Charles, and Zeev Maoz. 1984. "Militarized Interstate Disputes, 1816–1976." *Journal of Conflict Resolution* 28:585–616.

Goertz, Gary. 1992. "Contextual Theories and Indicators in World Politics." *International Interactions* 17:285–303.

Goldin, Claudia G., and Frank K. Lewis. 1975. "The Economic Cost of the American Civil War: Estimates and Implications." *Journal of Economic History* 35:299–326.

———. 1978. "The Post-Bellum Recovery of the South and the Cost of the Civil War: Comment." *Journal of Economic History* 38:487–92.

Gray, Colin. 1979. "Nuclear Strategy: The Case for a Theory of Victory." *International Security* 4:64–87.

Griffith, William E. 1975. *The World and the Great Power Triangles.* Cambridge, MA: MIT Press.

Gulick, Edward Vose. 1955. *Europe's Classical Balance of Power.* New York: W. W. Norton and Company.

Haas, Ernst. [1953] 1961. "The Balance of Power: Prescription, Concept, or Propaganda?" In James Rosenau, ed., *International Politics and Foreign Policy.* New York: Free Press.

Haas, Michael. 1970. "International Subsystems: Stability and Polarity." *American Political Science Review* 64:98–123.

Hacker, Louis M. 1940. *The Triumph of American Capitalism.* New York: Columbia University Press.

Hardin, Russell, John Mearsheimer, Gerald Dworkin, and Robert Goodin. 1985. *Nuclear Deterrence: Ethics and States.* Chicago: University of Chicago Press.

Hart, B. H. Liddell. 1967. *Strategy.* London: Faber and Faber.

Hartmann, Frederick H. 1978. *The Relations of Nations.* 5th ed. New York: Macmillan Publishing Co.

Harvey, Frank, and Patrick James. 1992. "Nuclear Deterrence Theory: The Record of Aggregate Testing." *Conflict Management and Peace Science* 12:17–46.

Heath, Dwight B. 1972. *Historical Dictionary of Bolivia.* Metuchen, NJ: Scarecrow Press.

Hellman, Donald C. 1969. "The Emergence of an East Asian International Subsystem." *International Studies Quarterly* 13:421–34.

Hobson, John. [1902] 1965. *Imperialism: A Study.* Ann Arbor, MI: University of Michigan Press.

Holsti, Kalevi J. 1991. *Peace and War.* New York: Cambridge University Press.

Holsti, Ole. 1967. "Cognitive Dynamics and Images of the Enemy." In D. Finlay, ed., *Enemies in Politics.* Chicago: Rand McNally.

Hopkins, Raymond F., and Richard W. Mansbach. 1973. *Structure and Process in International Politics.* New York: Harper and Row.

Horn, Michael. 1987. *Arms Races and the International System.* Ph.D. diss., University of Rochester, Rochester, NY.

Houweling, Henk, and Jan Siccama. 1988a. "Power Transitions as a Cause of War." *Journal of Conflict Resolution* 31:87–102.

———. 1988b. *Studies of War.* Boston: Martinus Nijhoff Publishers.

———. 1991. "Power Transitions and Critical Points as Predictors of Great Power War." *Journal of Conflict Resolution* 35:642–58.

———. 1993. "A Neo-Functionalist Explanation of World Wars: A Critique and an Alternative." *International Interactions* 18:387–407.

Howard, Michael. 1983. *The Causes of War.* Cambridge, MA: Harvard University Press.

Hume, David. [1752] 1990. "Of the Balance of Power." In John Vasquez, ed., *Classics of International Relations.* Englewood Cliffs, NJ: Prentice-Hall.

Hummell, H. J. 1972. *Probleme der Mehrebenenanalyse.* Stuttgart: Teubner.

Huntington, Samuel, ed. 1982. *The Strategic Imperative.* Cambridge, MA: Harvard University Press.

———. 1988. "The U.S.—Decline or Renewal?" *Foreign Affairs* 67:76–96.

Hussein, Seifeldin, and Jacek Kugler. 1990. "Conditional Anarchy: The Importance of the Status Quo in World Politics." Paper presented at the annual meeting of the Peace Science Society (International).

Huth, Paul, D., Scott Bennett, and Christopher Gelpi. 1992. "System Uncertainty, Risk Propensity, and International Conflict among the Great Powers." *Journal of Conflict Resolution* 36:478–517.

Huth, Paul, and Bruce Russett. 1984. "What Makes Deterrence Work? Cases from 1900 to 1980." *World Politics* 36:496–526.

———. 1990. "Testing Deterrence Theory: Rigor Makes a Difference." *World Politics* 42:466–501.

Huxley, Anthony, ed. 1962. *Standard Encyclopedia of the World's Mountains.* London: Weidenfeld and Nicolson.

Hwang, Young-Bae. 1993. *The Search for Alliance Stability.* Ph.D. diss., Vanderbilt University.

Intriligator, Michael, and Dagobert Brito. 1981. "Nuclear Proliferation and the Probability of War." *Public Choice* 37:247–60.

———. 1984. "Can Arms Races Lead to the Outbreak of War?" *Journal of Conflict Resolution* 28:63–84.

———. 1987. "The Stability of Mutual Deterrence." In J. Kugler and F. Zagare, eds., *Exploring the Stability of Deterrence.* Boulder, CO: Lynne-Rienner.

Iusi-Scarborough, Grace, and Bruce Bueno de Mesquita. 1988. "Threat and Alignment Behavior." *International Interactions* 14:85–93.

Jackman, Robert W. 1993. *Power without Force: The Political Capacity of Nation States.* Ann Arbor: University of Michigan Press.

James, Patrick. 1993. "Neo-Realism as a Research Enterprise: Toward Elaborated Structrual Realism." *International Political Science Review* 14:123–48.

Jervis, Robert. 1972. "Bargaining and Bargaining Tactics." In J. R. Pennock, and J. W. Chapman, eds. *Coercion.* Nomos 14: Yearbook of the American Society for Political and Legal Philosophy. Chicago: Aldine-Atherton.

———. 1979. "Deterrence Theory Revisited." *World Politics* 31:289–324.

———. 1984. *The Illogic of American Nuclear Strategy.* Ithaca, NY: Cornell University Press.

Kadera, Kelly M. 1992. "The Paths to Preponderance and Transition." Paper presented at the annual meeting of the International Studies Association.

Kagan, D. 1969. *The Outbreak of the Peloponnesian War.* Ithaca, NY: Cornell University Press.

———. 1987. "World War I, World War II, World War III." *Commentary* 83(March):21–40.

Kahn, Herman. 1960. *On Thermonuclear War.* Princeton, NJ: Princeton University Press.

———. 1962. *Thinking about the Unthinkable.* New York: Horizon Books.

———. 1965. *On Escalation: Metaphors and Scenarios.* Rev. ed. Baltimore, MD: Penguin Books.

Kahneman, Daniel, and Amos Tversky. 1979. "Prospect Theory: An Analysis of Decision under Risk." *Econometrica* 47:263–91.

Kaplan, A. 1964. *The Conduct of Inquiry: Methodology for Behavioral Science.* San Francisco: Chandler.

Kaplan, Morton. 1957. *System and Process in International Politics.* New York: John Wiley.

Kegley, Charles W., ed. 1991. *The Long Postwar Peace.* New York: Harper Collins.

———, and Eugene Wittkopf. 1989. *The Nuclear Reader: Strategy, Weapons, War.* 2d ed. New York: St. Martin's.

Kennedy, Paul. 1988. *The Rise and Fall of the Great Powers.* New York: Random House.

Keohane, Robert. 1980. "The Theory of Hegemonic Stability and Change in International Economic Regimes." In Ole Holsti, Randolph Siverson, and Alexander George, eds., *Change in the International System.* Boulder, CO: Westview Press.

———. 1982. "Theory of World Politics: Structural Realism and Beyond." In A. Finifter, ed., *Political Science: The State of the Discipline.* Washington, DC: American Political Science Association.

———. 1984. *After Hegemony: Cooperation and Discord in the World Political Economy.* Princeton, NJ: Princeton University Press.

———, ed. 1986. *Neorealism and Its Critics.* New York: Columbia University Press.

Kilgour, D. Marc, and Frank C. Zagare. 1991. "Credibility, Uncertainty, and Deterrence." *American Journal of Political Science* 35:305–34.

Kim, Woosang. 1989. "Power, Alliance, and Major Wars, 1816–1975." *Journal of Conflict Resolution* 33:255–73.

———. 1991. "Alliance Transitions and Great Power War." *American Journal of Political Science* 35:833–50.

———. 1992. "Power Transitions and Great Power War from Westphalia to Waterloo." *World Politics* 45:153–72.

———, and James Morrow. 1992. "When Do Power Shifts Lead to War?" *American Journal of Political Science* 36:896–922.

Kindleberger, Charles. 1973. *The World in Depression, 1929–1939.* Los Angeles: University of California Press.

Kissinger, Henry. 1957. *Nuclear Weapons and Foreign Policy.* New York: Harper.

————. 1979. *White House Years.* Boston: Little, Brown and Company.

Kosiak, Steven. 1993. "Analysis of the Fiscal Year 1994 Defense Budget Request." CRS, The Library of Congress.

Krasner, Stephen D., ed. 1983. *International Regimes.* Ithaca, NY: Cornell.

Kruskal, William. 1968. "Tests of Significance." In *International Encyclopedia of Statistics.* New York: Macmillan.

Kugler, Jacek. 1973. "The Consequences of War: Fluctuations in National Capabilities Following Major Wars, 1880–1970." Ph.D. diss., University of Michigan.

————. 1984. "Terror Without Deterrence." *Journal of Conflict Resolution* 28:470–506.

————. 1990. "The War Phenomenon: A Working Distinction." *International Interactions* 16:201–13.

————. 1991. "The Study of War and Peace: Quo Vadis?" In William Crotty, ed., *Political Science Looking to the Future,* Vol. 2. Evanston, IL: Northwestern University Press.

————, and Marina Arbetman. 1987. "The Phoenix Factor Revisited." Paper presented at the annual meeting of the International Studies Association.

————. 1989a. "Choosing among Measures of Power: A Review of the Empirical Record." In Richard Stoll, and Michael Ward, eds., *Power in World Politics.* Boulder, CO: Lynne Rienner.

————. 1989b. "Exploring the Phoenix Factor with the Collective Goods Perspective." *Journal of Conflict Resolution* 33:84–112.

Kugler, Jacek, and William Domke. 1986. "Comparing the Strength of Nations." *Comparative Political Studies* 19:39–69.

Kugler, Jacek, and A. F. K.Organski. 1989a. "The End of Hegemony?" *International Interactions* 15:113–28.

————. 1989b. "The Power Transition: A Retrospective and Prospective Evaluation." In M. Midlarsky, ed., *Handbook of War Studies.* Boston: Unwin Hyman.

Kugler, Jacek, and Suzanne Werner. 1993. "Conditional Anarchy: The Constraining Power of the Status Quo." Paper delivered at the annual meeting of the Midwest Political Science Association.

Kugler, Jacek, and Frank C. Zagare, eds. 1987. *Exploring the Stability of Deterrence.* Boulder, CO: Lynne Rienner.

————. 1990. "The Long-Term Stability of Deterrence." *International Interactions* 15:255–78.

Lalman, David. 1988. "Conflict Resolution and Peace." *American Journal of Political Science* 32:590–615.

Lambelet, J. C. 1975. "Do Arms Races Lead to War?" *Journal of Peace Research* 12:123–28.

Lebovic, James H. 1986. "The Middle East: The Region as a System." *International Interactions* 12:267–89.

Lebow, Richard Ned. 1981. *Between Peace and War: The Nature of International Crisis.* Baltimore, MD: The Johns Hopkins University Press.

Lemke, Douglas. 1991. "Predicting Peace: Power Transitions in South America." Paper presented at the annual meeting of the International Studies Association, Vancouver, British Columbia.

———. 1993. *Multiple Hierarchies in World Politics.* Ph.D. diss., Vanderbilt University.

Leng, Russell. 1983. "When Will They Ever Learn? Coercive Bargaining in Recurrent Crises." *Journal of Conflict Resolution* 27:379–419.

Lenin, Vladimir I. [1917] 1975. *Imperialism, the Highest Stage of Capitalism.* Peking: Foreign Languages Press.

Levy, Jack S. 1981. "Alliance Formation and War Behavior: An Analysis of the Great Powers." *Journal of Conflict Resolution* 25:581–613.

———. 1983a. *War in the Modern Great Power System, 1495–1975.* Lexington, KY: University of Kentucky Press.

———. 1983b. "Misperception and the Causes of War." *World Politics* 36:76–99.

———. 1985. "Theories of General War." *World Politics* 37:344–74.

———. 1987. "Declining Power and the Preventive Motivation for War." *World Politics* 40:82–107.

———. 1989. "The Causes of War: A Review of Theories and Evidence." In P. Tetlock, et al., eds., *Behavior, Society, and Nuclear War.* New York: Oxford University Press.

———. 1992a. "An Introduction to Prospect Theory." *Political Psychology* 13:171–86.

———. 1992b. "Prospect Theory and International Relations: Theoretical Applications and Analytical Problems." *Political Psychology* 13:283–310.

Liska, George. 1962. *Nations in Alliance: The Limits of Interdependence.* Baltimore: Johns Hopkins Press.

Lowenthal, Mark M. 1993. "The Clinton Foreign Policy: Emerging Themes." CRS, The Library of Congress.

Luce, R. Duncan, and Howard Raiffa. 1957. *Games and Decisions.* New York: John Wiley and Sons.

Luraghi, Raimondo. 1972. "The Civil War and the Modernization of American Society: Social Structure and Industrial Revolution in the Old South Before and during the War." *Civil War History* 18:230–50.

McNamara, Robert. 1969. *Defense Program and the 1969 Defense Budget.* Washington, DC: U.S. Department of Defense.

Maddison, Angus. 1973. *Class Structure and Economic Growth: India and Pakistan since the Moghuls.* New York: W. W. Norton and Company.

———. 1982. *Phases of Capitalist Development.* Oxford: Oxford University Press.

———. 1983. "A Comparison of Levels of GDP Per Capita in Developed and Developing Countries." *Journal of Economic History* 43:27–41.

———. 1989. *The World Economy in the 20th Century.* Paris: OECD.

————. 1991. *Dynamic Forces in Capitalist Development: A Long-Run Comparative View.* New York: Oxford University Press.

Mansfield, Edward. 1992. "The Concentration of Capabilities and the Onset of Wars." *Journal of Conflict Resolution* 36:3–24.

————. 1994. *Power, Trade, and War.* Princeton, NJ: Princeton Universtiy Press.

Maoz, Zeev. 1982. *Paths to Conflict: International Dispute Initiation, 1816–1976.* Boulder, CO: Westview Press.

————. 1983. "Resolve, Capabilities, and the Outcome of Interstate Disputes, 1816–1976." *Journal of Conflict Resolution* 27:195–229.

————, and Bruce Russett. 1993. "Normative and Structural Causes of Democratic Peace, 1946–1986." *American Political Science Review* 87:624–38.

Mearsheimer, John J. 1983. *Conventional Deterrence.* Ithaca, NY: Cornell University Press.

————. 1990. "Back to the Future: Instability in Europe after the Cold War." *International Security* 15:5–56.

Merritt, Richard L., and Dina A. Zinnes. 1989. "Alternative Indices of National Power." In R. Stoll, and M. Ward, eds., *Power in World Politics.* Boulder, CO: Lynne Rienner.

Midlarsky, Manus I. 1975. *On War: Political Violence in the International System.* New York: Free Press.

————. 1990a. *The Onset of World War.* Boston: Unwin Hyman.

————. 1990b. "Systemic Wars and Dyadic Wars: No Single Theory." *International Interactions* 16:171–81.

Milstein, Jeffrey S. 1972. "American and Soviet Influence, Balance of Power, and Arab-Israeli Violence." In B. Russett, ed., *Peace, War, and Numbers.* Beverly Hills, CA: Sage Publications.

Modelski, George. 1983. "Long Cycles of World Leadership." In William R. Thompson, ed., *Contending Approaches to World System Analysis.* Beverly Hills, CA: Sage Publications.

————. 1987. *Long Cycles in World Politics.* London: Macmillan.

————, and William R. Thompson. 1988. *Seapower in Global Politics.* Seattle, WA: University of Washington Press.

————. 1989. "Long Cycles and Global War." In Manus Midlarsky, ed., *Handbook of War Studies.* Boston: Unwin Hyman.

————. 1992. "K-Waves in International Relations: Structural Change in the Global Economy and World Politics." Paper presented at the annual meeting of the International Studies Association.

————. 1994. *Innovation, Growth and War: The Co-Evolution of Global Politics and Economics.* Columbia, SC: University of South Carolina Press.

Morgan, T. Clifton, and Peter M. Dawson. 1990. "Bargaining Tough: Commitment Strategy in International Conflict." Paper presented at the annual meeting of the American Political Science Association.

Morgenstern, Oskar. 1959. *The Question of National Defense.* New York: Random House.

Morgenthau, Hans. Various years. *Politics among Nations: The Struggle for Power and Peace*. New York: Alfred A. Knopf.

Morrison, Denton E., and Ramon E. Henkel. 1969. "Significance Tests Reconsidered." *The American Sociologist* 4:131–40.

Morrow, James D. 1989. "A Twist of Truth." *Journal of Conflict Resolution* 33:500–529.

———. 1991. "Alliances and Asymmetry." *American Journal of Political Science* 35:904–33.

———. 1993. "Arms versus Allies: Tradeoffs in the Search for Security." *International Organization* 47:207–33.

Most, Benjamin A., and Harvey Starr. 1984. "International Relations Theory, Foreign Policy Substitutability, and 'Nice' Laws." *World Politics* 36:383–406.

Moul, William B. 1985. "Balances of Power and European Great Power War, 1815–1939: A Suggestion and Some Evidence." *Canadian Journal of Political Science* 43:481–528.

———. 1988. "Balances of Power and the Escalation to War of Serious Disputes among the European Great Powers, 1815–1939: Some Evidence." *American Journal of Political Science* 32:241–75.

———. 1989. "Measuring the 'Balances of Power': A Look at Some Numbers." *Review of International Studies* 15:101–21.

Mueller, John. 1988. "The Essential Irrelevance of Nuclear Weapons: Stability in the Postwar World." *International Security* 13:55–79.

———. 1989. *Retreat from Doomsday: The Obsolescence of Major War*. New York: Basic Books.

Mulhall, M. G. 1881. *Between the Amazon and the Andes*. London: Edward Stanford.

Niou, Emerson, Peter Ordeshook, and Gregory Rose. 1989. *The Balance of Power*. New York: Cambridge University Press.

Nye, Joseph. 1990. *Bound to Lead*. New York: Basic Books.

Olson, Mancur. 1982. *The Rise and Decline of Nations*. New Haven, CT: Yale University Press.

Organski, A. F. K. 1958. *World Politics*. New York: Alfred A. Knopf.

———. 1968. *World Politics*, 2d ed. New York: Alfred A. Knopf.

———. 1990. *The $36 Billion Bargain*. New York: Columbia University Press.

———. 1992. "The Second American Century." Paper presented at the Conference on Parity and War, 2–3 October, Claremont Graduate School.

———. 1993. "Europe and the Rest of the World." In P. Anderson, et al., eds., *The New History of Europe*. Turin, Italy: Giulio Einaudi Editore.

———. Forthcoming. "The Second American Century." Manuscript.

———, and Marina Arbetman. 1993. "The Second American Century." In W. Zimmerman, and H. Jacobson, eds., *Behavior, Culture, and Conflict in World Politics*. Ann Arbor: University of Michigan Press.

Organski, A. F. K., and Jacek Kugler. 1977. "The Costs of Major Wars: The Phoenix Factor." *American Political Science Review* 71:1347–66.

———. 1980. *The War Ledger*. Chicago: University of Chicago Press.

———. N.d. "National Capabilities, Transitions, and Overtakings." Claremont Graduate School, Claremont, CA. Mimeo.

Organski, A. F. K., Jacek Kugler, Timothy Johnson, and Youssef Cohen. 1984. *Births, Deaths, and Taxes: The Demographic and Political Transitions.* Chicago: University of Chicago Press.

Organski, Katherine, and A. F. K. Organski. 1961. *Population and World Power.* New York: Alfred A. Knopf.

Ostrom, Charles, and Robert Marra. 1986. "U.S. Defense Spending and the Soviet Estimate." *American Political Science Review* 80:819–42.

Phillips, Ulrich. [1918] 1966. *American Negro Slavery.* New York: D. Appleton and Company. Reprint, Baton Rouge, LA: Louisiana State University Press.

Ponder, Daniel. 1992. "State Political Capacity and Federal Allocation of Monetary Resources: A Theory of Distribution." Paper presented at the annual meeting of the American Political Science Association.

Powell, Robert. 1987. "Crisis Bargaining, Escalation, and MAD." *American Political Science Review* 81:717–35.

———. 1990. *Nuclear Deterrence Theory: The Search for Credibility.* New York: Cambridge University Press.

Quester, George. 1966. *Deterrence Before Hiroshima.* New York: Wiley.

Randall, J. G., and D. Donald. 1969. *The Civil War and Reconstruction.* Lexington, MA: D.C. Heath and Company.

Ransom, Roger, and Richard Sutch. 1977. *One Kind of Freedom.* New York: Cambridge University Press.

Rapoport, Anatol. 1964. *Strategy and Conscience.* New York: Harper and Row.

Rasler, Karen, and William R. Thompson. 1992a. "Concentration, Polarity, and Transitional Warfare." Paper presented at the annual meeting of the International Studies Association.

———. 1992b. "Assessing the Costs of War: A Preliminary Cut." In G. Ausenda, ed., *Effects of War on Society.* Republic of San Marino: AIEP Editore.

———. 1994. *Transition and Global Struggle.* Lexington, KY: University of Kentucky Press.

Ray, G. Whitfield. 1914. *Through Five Republics on Horseback.* Cleveland, OH: Evangelical Publishing House.

Ray, James Lee, and Ayse Vural. 1986. "Power Disparities and Paradoxical Conflict Outcomes." *International Interactions* 12:315–42.

Reynolds, Paul D. 1969. *A Primer in Theory Construction.* Indianapolis: Bobbs-Merrill.

Richardson, Lewis Fry. 1960a. *Statistics of Deadly Quarrels.* Pittsburgh: Boxwood Press.

———. 1960b. *Arms and Insecurity.* Pittsburgh: Boxwood Press.

Rock, Stephen R. 1989. *Why Peace Breaks Out: Great Power Rapprochement in Historical Perspective.* Chapel Hill, NC: University of North Carolina Press.

Rood, J. Q. T., and J. Siccama. 1989. *Verzwakking van de Sterkste: Oorzaken en Gevolben van Amerikaans Machtsverval,* (Weakening of the strongest: Causes and consequences of American decline in power). The Hague: Netherlands Institute of International Relations.

Roosevelt, Theodore. 1914. *Through the Brazilian Wilderness.* New York: Scribners and Sons.

Rosecrance, Richard, and Arthur A. Stein. 1993. *The Domestic Bases of Grand Strategy.* Ithaca, NY: Cornell University Press.

Rosen, Steven. 1977. "A Stable System of Mutual Nuclear Deterrence in the Arab-Israeli Conflict." *American Political Science Review* 71:367–83.

Rouyer, Alwin. 1987. "Political Capacity and the Decline of Fertility in India." *American Political Science Review* 81:453–70.

Russett, Bruce. 1967. *International Regions and the International System.* Chicago: Rand McNally.

———. 1985. "The Mysterious Case of Vanishing Hegemony." *International Organization* 39:207–31.

———, and Harvey Starr. 1981. *World Politics: The Menu for Choice.* San Francisco: W. H. Freeman.

Scheiber, Harry N. 1965. "Economic Change in the Civil War Era: An Analysis of Recent Studies." *Civil War History* 11:396–411.

Schelling, Thomas C. 1960. *The Strategy of Conflict.* Cambridge, MA: Harvard University Press.

———. 1966. *Arms and Influence.* New Haven, CT: Yale University Press.

Schweller, Randall L. 1992. "Domestic Structure and Preventive War: Are Democracies More Pacific?" *World Politics* 44:235–69.

Sellers, James L. 1927. "The Economic Incidence of the Civil War in the South." *Mississippi Valley Historical Review* 14:179–91.

Selten, Reinhard. 1975. "A Re-examination of the Perfectness Concept for Equilibrium Points in Extensive Games." *International Journal of Game Theory* 4:25–55.

Shepsle, Kenneth, and Barry Weingast. 1981. "Structure-Induced Equilibrium and Legislative Choice." *Public Choice* 37:503–19.

Siegfried, Andre. 1933. *Impressions of South America.* New York: Harcourt, Brace and Company.

Sigler, John H. 1969. "News Flows in the North African International Subsystem." *International Studies Quarterly* 13:381–97.

Singer, J. David. 1958. "Threat-Perception and the Armament-Tension Dilemma." *Journal of Conflict Resolution* 2:90–105.

———. 1972. "The Correlates of War Project: Interim Report and Rationale." *World Politics* 24:243–70.

———. 1980. "Accounting for International War: The State of the Discipline." *Journal of Peace Research* 18:1–18.

———, Stuart Bremer, and John Stuckey. 1972. "Capability Distribution, Uncertainty, and Major Power War, 1820–1965." In Bruce Russett, ed., *Peace, War, and Numbers.* Beverly Hills, CA: Sage Publications.

Singer, J. David, and Melvin Small. 1966a. "Formal Alliances, 1815–1939: A Quantitative Description." *Journal of Peace Research* 1:257–82.

————. 1966b. "The Composition and Status Ordering of the International System: 1815–1940." *World Politics* 18:236–82.

————. 1968. "Alliance Aggregation and the Onset of War." In J. David Singer, ed., *Quantitative International Politics*. New York: Free Press.

————. 1972. *The Wages of War*. New York: Wiley.

Sivard, Ruth Leger. 1989. *World Military and Social Expenditures*. Washington, DC: World Priorities.

Siverson, Randolph M., and Paul Diehl. 1989. "Arms Races, the Conflict Spiral, and the Onset of War." In M. Midlarsky, ed., *Handbook of War Studies*. Boston: Unwin Hyman.

Siverson, Randolph M., and Ross Miller. 1994. "The Escalation of Disputes to War." *International Interactions* 19:77–97.

Siverson, Randolph M., and Harvey Starr. 1991. *The Diffusion of War*. Ann Arbor: University of Michigan Press.

Siverson, Randolph M., and Michael P. Sullivan. 1983. "The Distribution of Power and the Onset of War." *Journal of Conflict Resolution* 27:473–94.

Siverson, Randolph M., and Michael R. Tennefoss. 1984. "Power, Alliance, and the Escalation of International Conflict, 1815–1965." *American Political Science Review* 78:1057–69.

Skocpol, Theda. 1981. *States and Social Revolutions*. New York: Cambridge University Press.

Small, Melvin, and J. David Singer. 1973. "The Diplomatic Importance of States, 1816–1970: An Extension and Refinement of the Indicator." *World Politics* 25:577–99.

————. 1982. *Resort to Arms: International and Civil Wars, 1816–1980*. Beverly Hills, CA: Sage Publications.

Smith, Adam. [1776] 1981. *An Inquiry into the Nature and Causes of the Wealth of Nations*. Indianapolis: Liberty Classics.

Smoker, Paul. 1969. "Fear in the Arms Race: A Mathematical Study." In J. Rosenau, ed., *International Politics and Foreign Policy*. New York: Free Press.

Snyder, Glenn H. 1961. *Deterrence and Defense*. Princeton, NJ: Princeton University Press.

————. 1972. "Crisis Bargaining." In Charles Hermann, ed., *International Crises: Insights from Behavioral Research*. New York: Free Press.

————, and Paul Diesing. 1977. *Conflict among Nations: Bargaining, Decision Making and System Structure in International Crises*. Princeton, NJ: Princeton University Press.

Starr, Harvey. 1978. "'Opportunity' and 'Willingness' as Ordering Concepts in the Study of War." *International Interactions* 4:363–87.

————, and Benjamin Most. 1976. "The Substance and Study of Borders in International Relations Research." *International Studies Quarterly* 20:581–620.

Stoessinger, John. 1974. *Why Nations Go to War*. New York: St. Martin's Press.

Strange, Susan. 1982. "Cave! Hic Dragones: A Critique of Regime Analysis." *International Organization* 36:479–96.

Summer, Robert, and Alan Heston. 1978. *International Comparisons of Real Product and Purchasing Power*. Baltimore, MD: Johns Hopkins University Press.

Taylor, A. J. P. 1954. *The Struggle for Mastery in Europe, 1848–1918*. Oxford: Oxford University Press.

Thomas, Emory M. 1971. *The Confederacy as a Revolutionary Experience*. Columbia, SC: University of South Carolina Press.

Thompson, William R. 1973. "The Regional Subsystem: A Conceptual Explication and a Propositional Inventory." *International Studies Quarterly* 17:89–118.

———. 1983a. "Succession Crises in the Global Political System: A Test of the Transition Model." In Albert Bergesen, ed., *Crises in the World-System*. Beverly Hills, CA: Sage Publications.

———. 1983b. "Cycles, Capabilities, and War: An Ecumenical View." In William R. Thompson, ed., *Contending Approaches to World System Analysis*. Beverly Hills, CA: Sage Publications.

———. 1988. *On Global War: Historical-Structural Approaches to World Politics*. Columbia, SC: University of South Carolina Press.

———. 1990. "The Size of War, Structural and Geopolitical Contexts, and Theory Building/Testing." *International Interactions* 16:183–99.

———. 1992. "Dehio, Long Cycles, and the Geohistorical Context of Structural Transition." *World Politics* 44:127–52.

Thucydides. 1951. *History of the Peloponnesian War*. New York: Random House.

Toynbee, Arnold J. 1954. *A Study of History*. Vol. 9. New York: Oxford University Press.

U.S. Arms Control and Disarmament Agency. 1991. "Military Expenditures, Armed Forces, GNP, Central Government Expenditures, and Population, 1979–1989." Library of Congress Numbers: JX1974.A1U52 78-645925.

van Damme, Eric. 1991. "Refinements of Nash Equilibria." No. 9107, Center for Economic Research, Tilberg University, Netherlands.

Vasquez, John A. 1986. "Capability, Types of War, Peace." *Western Political Quarterly* 39:313–27.

———. 1987. "The Steps to War: Toward a Scientific Explanation of Correlates of War Findings." *World Politics* 40:108–45.

———. 1991. "The Deterrence Myth: Nuclear Weapons and the Prevention of Nuclear War." In Charles W. Kegley, ed., *The Long Postwar Peace: Contending Explanations and Projections*. New York: Harper Collins.

———. 1993. *The War Puzzle*. New York: Cambridge University Press.

Väyrynen, Raimo. 1983. "Economic Cycles, Power Transitions, Political Management, and Wars between Major Powers." *International Studies Quarterly* 27:389–418.

von Hippel, Frank. 1983. "The Effects of Nuclear War." In D. Hafemesiste and D. Schoeer, eds., *Physics, Technology, and the Nuclear Arms Race*. New York: American Institute of Physics.

Wagner, R. Harrison. 1986. "The Theory of Games and the Balance of Power." *World Politics* 38:546–76.

Wallace, Michael D. 1972. "Status, Formal Organization, and Arms Levels as Factors Leading to the Onset of War, 1820–1964." In B. Russett, ed., *Peace, War, and Numbers.* Beverly Hills, CA: Sage Publications.

———. 1975. "Clusters of Nations in the Global System, 1865–1964." *International Studies Quarterly* 19:67–110.

———. 1979. "Arms Races and Escalation: Some New Evidence." *Journal of Conflict Resolution* 23:3–16.

———. 1980. "Some Persisting Findings: A Reply to Professor Weede." *Journal of Conflict Resolution* 24:289–92.

———. 1982. "Armaments and Escalation, Two Competing Hypotheses." *International Studies Quarterly* 26:37–56.

———. 1983. "Arms Races and Escalation: A Reply to Altfeld." *International Studies Quarterly* 27:233–35.

Wallensteen, Peter. 1981. "Incompatibility, Confrontation, and War: Four Models and Three Historical Systems, 1816–1976." *Journal of Peace Research* 18:57–90.

Wallerstein, Immanuel. 1980. *The Modern World System.* Vol. 2. New York: Academic Press.

———. 1984. *The Politics of the World Economy.* Cambridge: Cambridge University Press.

Walt, Stephen M. 1987. *The Origins of Alliances.* Ithaca, NY: Cornell University Press.

Waltz, Kenneth N. 1959. *Man, the State, and War: A Theoretical Analysis.* New York: Columbia University Press.

———. 1964. "The Stability of a Bipolar World." *Daedalus* 93:881–909.

———. 1979. *Theory of International Politics.* Reading, MA: Addison-Wesley.

———. 1981. "The Spread of Nuclear Weapons: More May be Better." *Adelphi Paper #171.* London: The International Institute for Strategic Studies.

———. 1993. "The Emerging Structure of International Politics." *International Security* 18:44–79.

Ward, Michael D. 1984. "Differential Paths to Parity: A Study of Contemporary Arms Races." *American Political Science Review* 78:297–317.

Wayman, Frank. 1975. *Military Involvement in Politics: A Causal Model.* Beverly Hills, CA: Sage Professional Papers in International Studies.

———. 1982. "Power Transitions, Rivalries, and War, 1816–1970." Paper presented at the Institute for the Study of Conflict Theory and International Security Meeting, Urbana-Champaign, IL.

———. 1989. "Power Shifts and War." Paper presented at the annual meeting of the International Studies Association.

Weede, Erich. 1976. "Overwhelming Preponderance as a Pacifying Condition among Contiguous Asian Dyads, 1950–1969." *Journal of Conflict Resolution* 20:395–411.

———. 1980. "Arms Races and Escalation: Some Persisting Doubts." *Journal of Conflict Resolution* 24:285–88.

Weinberger, Caspar W. 1985. "The Uses of Military Power." In *Defense '85*. Arlington, VA: American Armed Forces Information Services.

Werner, Suzanne, and Douglas Lemke. 1994. "Power and Position: An Expected Utility Model of Peace and War." Paper presented at the annual meeting of the American Political Science Association, New York, NY.

Wheeler, H. 1980. "Postwar Industrial Growth." In J. D. Singer, ed., *The Correlates of War*. Vol. 2. New York: Free Press.

Wohlforth, W. C. 1987. "The Perception of Power: Russia in the Pre-1914 Balance." *World Politics* 39:353–81.

Wohlstetter, Albert. 1959. "The Delicate Balance of Terror." *Foreign Affairs* 37:211–34.

———. 1968. "Theory and Opposed-System Design." *Journal of Conflict Resolution* 12:302–31.

Wolfers, Arnold. 1951. "The Pole of Power and the Pole of Indifference." *World Politics* 4:39–63.

Wolfson, M., A. Puri, and M. Martelli. 1992. "The Nonlinear Dynamics of International Conflict." *Journal of Conflict Resolution* 36:119–49.

Worldmark Encyclopedia of the Nations. 1988. New York: Wiley.

Wright, Carroll D. 1897. *The Industrial Evolution of the United States*. Meadville, PA: Flood and Vincent.

Wright, Quincy. 1935. *The Causes of War and the Conditions of Peace*. London: Longmans, Green and Company.

———. 1965. *A Study of War. 2d ed. with a commentary on war since 1942*. Chicago: University of Chicago Press.

Wright, Winthrop R. 1974. *British-Owned Railways in Argentina: Their Effect on Economic Nationalism*. Austin: University of Texas Press.

Young, Oran R. 1968. *The Politics of Force: Bargaining during International Crises*. Princeton, NJ: Princeton University Press.

———. 1975. *Bargaining: Formal Theories of Negotiation*. Urbana: University of Illinois Press.

Zagare, Frank C. 1987. *The Dynamics of Deterrence*. Chicago: University of Chicago Press.

———. 1990. "Rationality and Deterrence." *World Politics* 42:238–60.

———, and D. Marc Kilgour. 1993. "Asymmetric Deterrence." *International Studies Quarterly* 37:1–27.

Zinnes, Dina A. 1967. "An Analytical Study of the Balance of Power Theories." *Journal of Peace Research* 4:270–88.

Contributors

Marina Arbetman is Assistant Professor of Political Science at Tulane University.

Bruce Bueno de Mesquita is Senior Fellow at the Hoover Institution, Stanford University.

Vesna Danilovic is Assistant Professor of Political Science at Texas A & M University.

Daniel S. Geller is Professor of Political Science at the University of Mississippi.

Henk W. Houweling is Associate Professor of International Relations at the University of Amsterdam, The Netherlands.

Kelly M. Kadera is Assistant Professor of Political Science at the University of Iowa.

Woosang Kim is Associate Professor of Political Science at SookMyung Women's University in Seoul, Korea.

Jacek Kugler is the Elisabeth Helm Rosecrans Professor of International Relations and Political Economy at the Center for Politics and Economics, Claremont Graduate School.

Douglas Lemke is Assistant Professor of Political Science at Florida State University.

Ross A. Miller is Assistant Professor of Political Science at Santa Clara University.

James D. Morrow is Senior Research Fellow at the Hoover Institution, Stanford University.

A. F. K. Organski is Professor of Political Science at the University of Michigan.

Jan G. Siccama is Director of Research at the Netherlands Institute of International Relations "Clingendael," The Hague.

Randolph M. Siverson is Professor of Political Science at the University of California–Davis.

Ronald Tammen is Professor of Political Science at the National War College.

William R. Thompson is Professor of Political Science and Director of the Center for the Study of International Relations at Indiana University, Bloomington.

John A. Vasquez is Professor of Political Science at Vanderbilt University.

Frank Whelon Wayman is Professor of Political Science at the University of Michigan–Dearborn.

Suzanne Werner is Assistant Professor of Political Science at Emory University.

Frank C. Zagare is Professor of Political Science at the University at Buffalo, State University of New York.

Index